The Dynamic Bacterial Genome

This book provides an in-depth analysis of the mechanisms and biological consequences of genome rearrangements in bacteria. Genome rearrangements take place as a result of the actions of discrete genetic elements such as conjugative transposons, plasmids, phage, and nonconjugative transposons. Bacteria also contain systems to mediate genetic rearrangements such as the general recombination pathway and specialized endogenous recombination mechanisms. The biological effect of these rearrangements is far-reaching and impacts on bacterial virulence, antibiotic resistance, and the ability of the bacteria to avoid the attentions of the host immune system (e.g., antigenic variation). These rearrangements also provide the raw material on which natural selection can act.

Each chapter examines the mechanisms involved in genome rearrangements and the direct biological consequences of these events. Because genome rearrangements are so important in evolution, at least one of the chapters views the phenomenon from an evolutionary angle. This book provides the reader with a holistic view of genome rearrangements (i.e., studies on both the biological consequences of genome rearrangement and the mechanisms underlying these processes are presented).

The book is written by leading research workers in the field and is aimed at final-year undergraduates, postgraduate and postdoctoral workers, and established biologists.

Peter Mullany is a Reader in Molecular Microbiology at University College London. His main research interests are the molecular biology and molecular ecology of conjugative transposons. He has been at the forefront of this work for the past 10 years and has pioneered the genetic analysis of the important human pathogen *Clostridium difficile*.

Over the past decade, the rapid development of an array of techniques in the fields of cellular and molecular biology have transformed whole areas of research across the biological sciences. Microbiology has perhaps been influenced most of all. Our understanding of microbial diversity and evolutionary biology, and of how pathogenic bacteria and viruses interact with their animal and plant hosts at the molecular level, for example, have been revolutionized. Perhaps the most exciting recent advance in microbiology has been the development of the interface discipline of cellular microbiology, a fusion of classic microbiology, microbial molecular biology, and eukaryotic cellular and molecular biology. Cellular microbiology is revealing how pathogenic bacteria interact with host cells in what is turning out to be a complex evolutionary battle of competing gene products. Molecular and cellular biology are no longer discrete subject areas but vital tools and an integrated part of current microbiological research. As part of this revolution in molecular biology, the genomes of a growing number of pathogenic and model bacteria have been fully sequenced, with immense implications for our future understanding of microorganisms at the molecular level.

Advances in Molecular and Cellular Microbiology is a series edited by researchers active in these exciting and rapidly expanding fields. Each volume focuses on a particular aspect of cellular or molecular microbiology and provides an overview of the area, as well as examines current research. This series will enable graduate students and researchers to keep up with the rapidly diversifying literature in current microbiological research.

CELLULAR MICROBIOLOGY

MCM

ADVANCES IN MOLECULAR AND

Series Editors

Professor Brian Henderson
University College London

Professor Michael Wilson
University College London

Professor Sir Anthony Coates
St. George's Hospital Medical School, London

Professor Michael Curtis
St. Bartholemew's and Royal London Hospital, London

Published Titles

1. *Bacterial Adhesion to Host Tissues.* Edited by Michael Wilson 0521801079
2. *Bacterial Evasion of Host Immune Responses.* Edited by Brian Henderson and Petra Oyston 0521801737
3. *Dormancy and Low-Growth States in Microbial Disease.* Edited by Anthony R. M. Coates 0521809401
4. *Susceptibility to Infectious Diseases.* Edited by Richard Bellamy 0521815258
5. *Bacterial Invasion of Host Cells.* Edited by Richard J. Lamont 0521809541
6. *Mammalian Host Defense Peptides.* Edited by Deirdre Devine and Robert Hancock 0521822203
7. *Bacterial Protein Toxins.* Edited by Alistair J. Lax 052182091X

Forthcoming Titles in the Series

The Influence of Bacterial Communities on Host Biology. Edited by Margaret McFall-Ngai, Brian Henderson, and Edward Ruby 0521834651

The Yeast Cell Cycle. Edited by Jeremy Hyams 0521835569

Salmonella Infections. Edited by Pietro Mastroeni and Duncan Maskell 0521835046

Phagocytosis of Bacteria and Bacterial Pathogenicity. Edited by Joel Ernst and Olle Stendahl 0521845696

Quorum Sensing and Bacterial Cell-to-Cell Communication. Edited by Donald R. Demuth and Richard J. Lamont 0521846382

Advances in Molecular and Cellular Microbiology, Volume 8

The Dynamic Bacterial Genome

EDITED BY
PETER MULLANY
University College London

CAMBRIDGE
UNIVERSITY PRESS

CAMBRIDGE UNIVERSITY PRESS
Cambridge, New York, Melbourne, Madrid, Cape Town, Singapore,
São Paulo, Delhi, Dubai, Tokyo

Cambridge University Press
32 Avenue of the Americas, New York, NY 10013-2473, USA

www.cambridge.org
Information on this title: www.cambridge.org/9780521129619

First published 2005
This digitally printed version 2009

A catalog record for this publication is available from the British Library

Library of Congress Cataloging in Publication data

The dynamic bacterial genome / edited by Peter Mullany.
    p.   cm. – (Advances in molecular and cellular microbiology ; 8)
Includes bibliographical references and index.
ISBN 0-521-82157-6 (hardcover)
1. Bacterial genetics.   2. Genetic recombination.   3. Bacterial transformation.
[DNLM: 1. Genome, Bacterial.   2. Recombination, Genetic. QH 448 D997 2005]
I. Mullany, Peter, 1959–   II. Title.   III. Series.
QH434.D98   2005
572.8′293
[2   2004024708

ISBN 978-0-521-82157-5 Hardback
ISBN 978-0-521-12961-9 Paperback

Additional resources for this publication at www.cambridge.org/9780521129619

Contents

PART 3 Biological Consequences of the Mobile Genome

CONTENTS

Contributors

James R. Brown
Bioinformatics Division
Genetics Research
GlaxoSmithKline
1250 South Collegeville Road
UP1345
Collegeville, PA 19426
USA

Eric Déziel
INRS-Institut
Armand-Frappier
531 Boulevard des Prairies
Laval
Québec
Canada H7V 1B7

Irena Draskovic
Public Health Research Institute
225 Warren Street
Newark, NJ 07103
USA

David Dubnau
Public Health Research Institute
225 Warren Street
Newark, NJ 07103
USA

Laura S. Frost
Department of Biological Sciences
University of Alberta
Edmonton
Alberta
Canada T6G 2E9

Ian Grainge
Blanch Lane
South Mimms
Pottters Bar
Hertfordshire EN6 3LD
UK

Michael J. Gubbins
Department of Biological Sciences
University of Alberta
Edmonton
Alberta
Canada T6G 2E9

Jörg Hacker
Institut für molekulare
    Infektionsbiologie
Universität Würzburg
Röntgenring 11
D-97070 Würzburg
Germany

Bianca Hochhut
Institut für molekulare
  Infektionsbiologie
Universität Würzburg
Röntgenring 11
D-97070 Würzburg
Germany

Diarmaid Hughes
Department of Cell and Molecular
  Biology
Box 596
Biomedical Center
Uppsala University
S-751 24 Uppsala
Sweden

Makkuni Jayaram
Section of Molecular Genetics and
  Microbiology
University of Texas at Austin
USA

Peter Mullany
Division of Microbial Diseases
  Eastman Dental Institute for Oral
  Health Care Sciences
University College London (UCL)
University of London
256 Gray's Inn Road
London, WC1X 8LD
UK

Tobias Norström
Department of Cell and Molecular
  Biology
Box 596
Biomedical Center
Uppsala University
S-751 24 Uppsala
Sweden

Marie-Agnès Petit
U571
Faculté de médecine Necker-Enfants
  Malades
156 rue de Vaugirard
75015 Paris
France

Adam P. Roberts
Division of Microbial Diseases
Eastman Dental Institute for Oral
  Health Care Sciences
University College London (UCL)
University of London
256 Gray's Inn Road
London, WC1X 8LD
UK

Sally J. Rowland
University of Glasgow
56 Dumbarton Road
Glasgow G11 6NU
Scotland
UK

W. Marshall Stark
University of Glasgow
56 Dumbarton Road
Glasgow G11 6NU
Scotland
UK

Richard Villemur
INRS-Institut
Armand-Frappier
531 Boulevard des Prairies
Laval
Québec
Canada H7V 1B7

William R. Will
Department of Biological Sciences
University of Alberta
Edmonton
Alberta
Canada T6G 2E9

Steven Zimmerly
Department of Biological Sciences
University of Calgary
2500 University Drive NW
Calgary
Canada AB T2N 1N4

CONTRIBUTORS

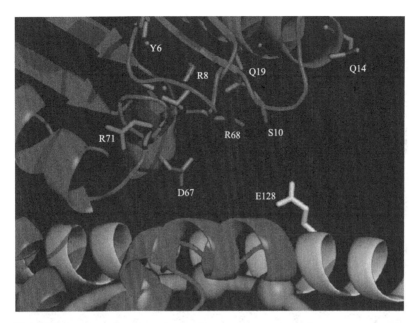

Figure 2.4.

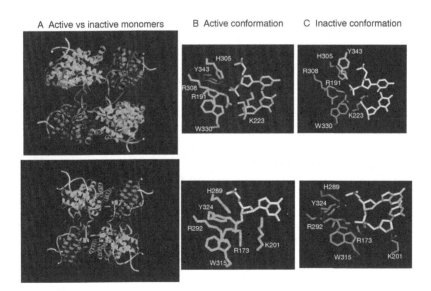

Figure 2.5.

A colour version of these plates is available for download from www.cambridge.org/9780521129619

c

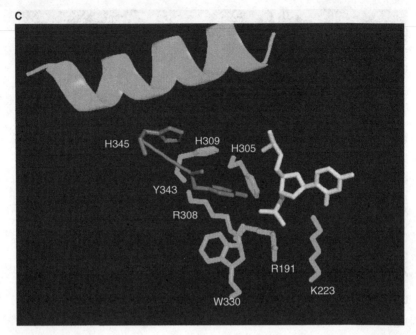

Figure 2.6c.

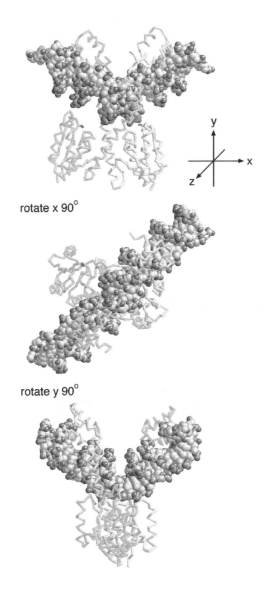

rotate x 90°

rotate y 90°

Figure 3.3.

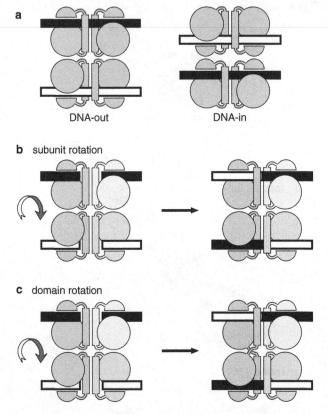

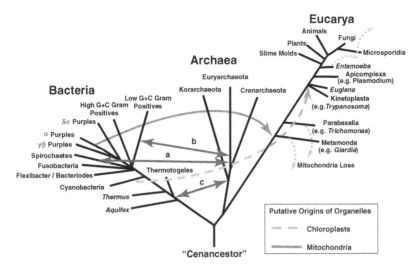

Figure 11.1.

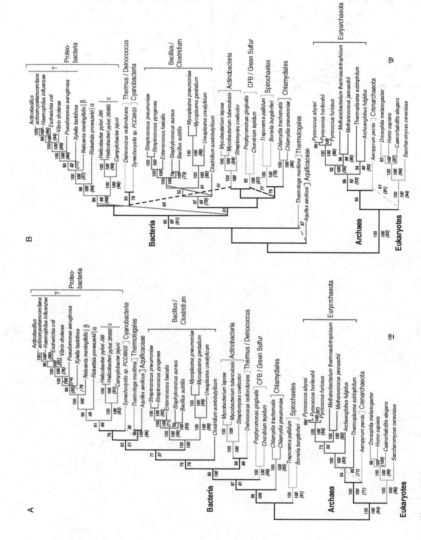

Figure 11.2.

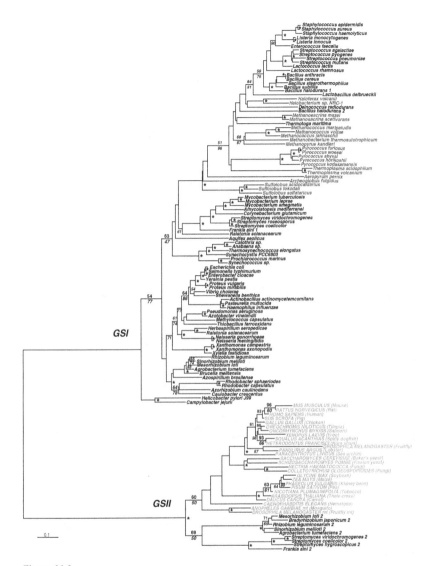

Figure 11.3.

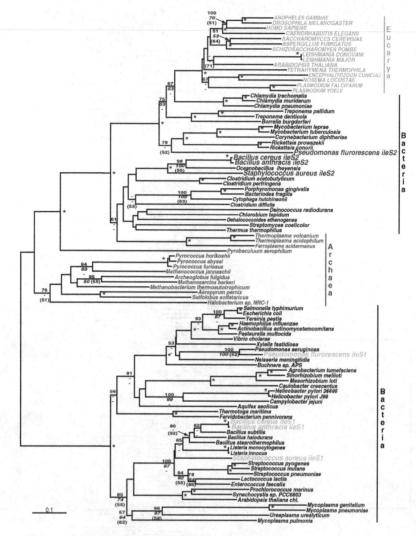

Figure 11.4.

Part 1 Basic Mechanisms of Genome Rearrangement in Bacteria

CHAPTER 1

Mechanisms of homologous recombination in bacteria

Marie-Agnès Petit

Homologous recombination promotes the pairing between identical – or nearly identical – DNA sequences and the subsequent exchange of genetic material between them. It is an important and widely conserved function in living organisms, from bacteria to humans, that serves to repair double-stranded breaks or single-stranded gaps in the DNA, arising as a consequence of ionizing radiations, ultraviolet (UV) light, or chemical treatments creating replication-blocking adducts (Kuzminov, 1999). More recently, homologous recombination functions were also found in bacteria to rescue replication forks that have stalled for various reasons, such as a missing factor (e.g., the helicase), or a particular difficulty upstream of the fork, such as supercoiling or intense traffic of proteins (Michel et al., 2001).

Besides its molecular role, homologous recombination has played a major role in genome dynamics, by changing gene copy numbers through deletions, duplications, and amplifications: Intrachromosomal recombination between ribosomal operons or between mobile elements scattered into the genome leads to deletion or tandem duplications of large regions within the genome, up to several hundred kilobases (Roth et al., 1996). The duplications are unstable. Mostly they recombine back to the parental organization, and, therefore, remain undetected, except when appropriate selection, by gene dosage mostly, is exerted (Petes and Hill, 1988). In contrast, such duplications are ideal substrate for the diversification of genes: One gene is kept intact whereas the other is mutagenized, which leads to the birth of gene families. Once the duplicated segment has sufficiently diverged, it becomes more stable because of the lack of perfect homology to recombine the duplicated segment. Tandem duplications are also the starting point for further gene amplification, the repetition up to 20- to 100-fold of the tandem array (Kodama et al., 2002).

Finally, homologous recombination is also critical in terms of evolution, by allowing the generation of new allele combinations, and, as a consequence, the possibility to evolve and adapt to new environments, a hallmark of living organisms. In eukaryotes this happens mainly through meiosis, whereas in bacteria and archea the so-called "horizontal transfer" of genes is taking place on a larger scale (Ochman, Lawrence, and Groisman, 2000). Homologous recombination is one of the mechanisms through which such gene transfers occur, in particular during generalized transduction, conjugation, and natural transformation.

Much of what is known at the molecular level about homologous recombination in bacteria is from the in-depth work realized over the last 50 years on *Escherichia coli* (*E. coli*). For more recent reviews on this topic, the reader is referred to Kuzminov (1999) and Cox (2001). This chapter begins with a brief description of the knowledge based on the *E. coli* paradigm, but its main focus is on how other eubacteria resemble or differ from the paradigm. Because this book is on genomic rearrangements, plasmid recombination, which is a field in itself, is excluded.

HOMOLOGOUS RECOMBINATION: THE DNA ACTORS

Toward a definition of homologous recombination

The more processes of homologous recombination are known at the molecular level, the more difficult they are to be defined precisely. Concerning the DNA partners, in the original definition, homologous recombination concerned only events between pairs of chromosome homologs, and, therefore, was restricted to diploid cells. It then appeared that recombination could also concern two sequences at different loci (either in the same or in different chromosomes), the so-called "ectopic recombination." Finally, and especially in bacteria, homologous recombination was found to be a major way to integrate incoming DNA into a genome.

At the molecular level, homologous pairing and strand exchange may occur between two DNA molecules without any consequence at the genetic level, the so-called "non–cross-over" products (see the section "The DNA intermediates" in this chapter). The process is silent phenotypically, but essential molecularly, as it leads to DNA repair. During such a process, however, some point mutations of one molecule engaged into the homologous pairing may be transferred to the other molecule, and lead to gene conversion. This is not only frequent in fungi but also takes place in bacteria (Abdulkarim and Hughes, 1996). Finally, the enzymes that process homologous recombination intermediates are able to pair sequences that are not identical, but partially diverged or homeologous (see the following section "The DNA products"). To summarize, homologous recombination does not necessarily recombine DNA and does not necessarily involve chromosome homologs, or identical sequences. In bacteria, one may define homologous recombination as all *recA*-dependent events, but even this simple assessment is not always true (see "RecA-independent homologous recombination" in this chapter).

The DNA products

Homologous recombination has its primary consequence at the DNA level; therefore, the main products of the process are first described. Most bacterial genomes are circular, and recombination products leading to linear chromosomes are lethal, so the focus is on circular products.

When the bacterial chromosome recombines with incoming DNA, as is the case during horizontal transfer, two main products are expected: If the entering DNA molecule contains two different stretches of homology with the bacterial chromosome (Fig. 1.1A), recombination can proceed by "double cross-over" (DCO). In this case, the intervening part of the chromosome is exchanged with the incoming DNA, and, therefore, lost. The DCO product is stable because no repeated sequences flank the incoming DNA. If the entering molecule contains a single stretch of homologous DNA and is circular – for instance, a nonreplicative plasmid – recombination occurs by single crossing over (sometimes called "Campbell-type" recombination, Fig. 1.1B). It produces a recombined chromosome in which the incoming

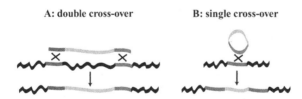

Figure 1.1. Two examples of integration of incoming DNA (flat lines, grey color) into the bacterial chromosome (black wavy lines). **(A)** The incoming DNA is linear and contains two regions of homology (dark grey) with the chromosome, each recombines (between the regions shown as a cross), and the resulting recombinant has integrated the foreign DNA by double cross-over. **(B)** The incoming DNA is circular and contains one region of homology (dark grey) with the chromosome. Upon recombination by single cross-over (shown as a cross), the foreign DNA is flanked by two copies of the homologous region, oriented in parallel.

DNA is flanked by directly repeated sequences, sometimes called "pop-in" recombinant. No DNA has been lost, but the resulting recombinant is unstable because it can "pop out" by homologous recombination. In cases where the introduced DNA confers a selective advantage, such recombinants are maintained. This process is widely used by geneticists to interrupt genes in bacteria such as *Bacillus subtilis, Lactococcus lactis, Deinococcus radiodurans*, and so on.

Chromosomal DNA can also generate intramolecular recombinants. This happens due to recombination between members of a gene family dispersed in the genome, typically between *rrn* operons or between mobile elements. If the two identical copies are inverted with respect to one another, the product is an inversion of the intervening sequence. If the two copies are oriented the same way, the process is called unequal crossing over because recombination probably takes place in an "unequal way" between sister chromatids behind a replication fork (Fig. 1.2). It leads to one chromosome containing a duplicated stretch flanked by the sequences that served to initiate the cross-over, and the other chromosome deleted for this same stretch of DNA, most likely unable to give a progeny. The chromosome containing the duplicated region is called a merodiploid or partial diploid; it is highly unstable and tends to recombine back to its original configuration. However, the frequency of production of these merodiploid is quite high: At any given time for one particular duplication, around 10^{-4} of an *E. coli* K12 population

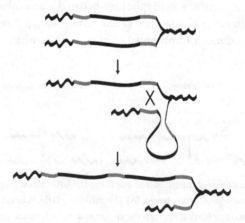

Figure 1.2. Intrachromosomal recombination. If two identical copies of a gene or a mobile element (grey and flat lines) are present on a chromosome (black wavy lines), they can recombine together (shown as cross) behind a replication fork to produce one chromosome with a deletion of the intervening region, and the other with a duplicated region.

MARIE-AGNÈS PETIT

is in the merodiploid state (Anderson and Roth, 1977; Petes and Hill, 1988). Such rearrangements offer a chance for the selection of new chromosomal variants and evolution.

The DNA intermediates

It is now generally admitted that a free DNA end, either a double-stranded extremity or a single-stranded gap, is the prerequisite to initiate homologous recombination. Two models based on this assumption and adapted from the review of Kuzminov (1999) are presented in Fig. 1.3. One starts with a double-stranded extremity (Fig. 1.3A), and the other starts with the single-stranded gap (Fig. 1.3B). In both cases, the initial event and the final event are the

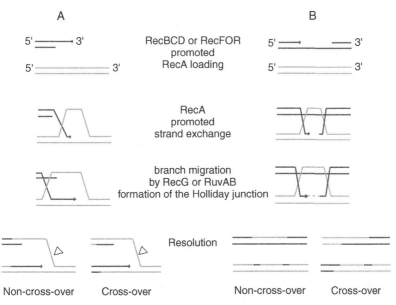

Figure 1.3. Two models showing possible intermediates of the recombination process, as adapted from Kuzminov (1999). Recombination starts from a double-stranded break (column A) or a single-stranded gap (column B) blue molecule, which is processed by RecBCD (left) or RecFOR (right) to load RecA and promote strand pairing with an intact grey molecule (step 2). Each strand of the DNA duplex is drawn, and 5′ and 3′ extremities are indicated on the first lanes. The black arrow shows the 3′ extremity of an invading molecule. Due to the action of the RecG and/or RuvAB helicase, Holliday junctions are created (step 3) and resolved by RuvC or another nuclease into crossed-over or non–crossed-over products (step 4). Dotted lane indicates DNA synthesis, and the white triangle stands for a putative nuclease that would cleave the D loop.

most documented ones, whereas the steps in between are speculative – in particular, the short patch of DNA synthesis by unknown polymerase and the cleavages by unknown endonuclease. Initial to all homologous recombination is the invasion of a single-stranded DNA into a duplex molecule so it pairs and forms the so-called "heteroduplex." Next – and less well characterized – this intermediate is converted into a Holliday junction (HJ), in which the second strand of the invading molecule has also paired with its recipient. This four-stranded structure is able to move around (branch migration) and then to be processed by a specific nuclease into either of two products, depending on the orientation of the cleavage. One cleavage will result in product molecules having received a small patch of the donor DNA, the heteroduplex region; it is called *gene conversion*. The other cleavage orientation will result in the exchange of the flanking sequences, or "crossing over," between the two recombining molecules.

The third model (Fig. 1.4) accounts for the repair of blocked replication fork (Seigneur et al., 1998). Recombination genes are not essential in bacteria (bacteria deleted for *rec* genes are viable), but they can become essential for the viability of replication mutants. In such mutants, replication fork progression is hindered, and a process called *replication fork reversal* is supposed to take place, in which the DNA intermediate is structurally identical to the HJ (Fig. 1.4B). This intermediate is processed by recombination enzymes and rescues the replication fork by removing the replication block either by recombination (Fig. 1.4E) or by trimming the extremity of the new strands (Fig. 1.4D) to allow restart. If the recombination enzyme RecBCD (see Chapter 2) does not process this intermediate, it is subjected to a cleavage by RuvABC (see Chapter 2), which may be lethal (Fig. 1.4C). The little revolution brought about by this model is that recombination enzymes reveal themselves as being closely interconnected with the replication process and more generally involved in the normal life cycle of a bacterium, rather than being specialized in some aspects of lesion repairs, or even more specialized processes, such as conjugation or transformation.

GENETICS AND BIOCHEMISTRY OF HOMOLOGOUS RECOMBINATION IN *E. COLI*

A wealth of genes and proteins play a role in homologous recombination processes in *E. coli*. They are briefly listed, and refer to the recombination models described in Fig. 1.3. Three distinct and successive steps are involved in homologous recombination: presynapsis, during which enzymes prepare the DNA substrate for RecA; synapsis, where RecA bound to DNA promotes

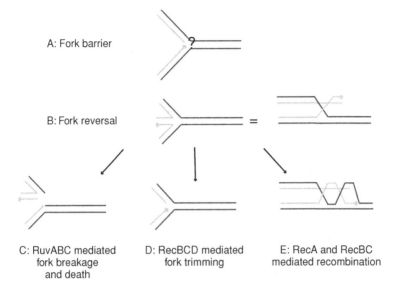

A: Fork barrier

B: Fork reversal

C: RuvABC mediated
fork breakage
and death

D: RecBCD mediated
fork trimming

E: RecA and RecBC
mediated recombination

Figure 1.4. A model for the repair of a blocked replication fork by recombination enzymes, adapted from Seigneur et al. (1998). A replication fork is drawn with newly replicated strands in a light grey color, the arrow showing its 3′ extremity. The question mark stands for the block, which leads to a fork reversal, where the two replicated strands have paired (step B). This intermediate can be drawn such that the newly paired region faces its unreplicated homolog, ahead of the reversed fork (step B, right part). Three possible fates for this intermediate have been proposed: breakage if RuvABC arrives on the DNA before RecBCD (step C), fork trimming if RecBCD acts on the exposed double-strand end (step D), and recombination if RecBCD and RecA act in concert (step E). For simplicity the reversed fork is shown with blunt extremities, but a normal fork would tend to produce a 3′ protruding end because of the advance of the leading strand relative to the lagging strand. Such an extremity may be processed by exonucleases to produce a blunt end or be used directly by RecA in step E.

the active search for a homologous molecule; and postsynapsis, where intermediates are processed into products.

Presynapsis

Two sets of proteins are needed in *E. coli* for the processing of DNA ends and the efficient loading of RecA:

- The RecBCD heterotrimer is a dual enzyme that is very well characterized biochemically (Kowalczykowski et al., 1994; Myers and Stahl, 1994) and structurally (Singleton et al., 2004). It is a potent exonuclease (ExoV) strictly dependent on blunt or nearly blunt ends for its activity, as well as a highly processive helicase. The RecB subunit encodes the nuclease activity, which degrades preferentially in the 3′ to 5′ direction, and the RecD subunit modulates RecB: Upon interaction with a specific DNA sequence, the Chi site, RecD converts the polarity of DNA degradation by RecB, such that a 3′ single-stranded extremity is created, and this promotes the loading of RecA (Anderson and Kowalczykowski, 1997). In addition, both RecB and RecD act as DNA helicases, of opposite polarity and different speed (Dillingham et al., 2003; Taylor and Smith, 2003). RecC is inert enzymatically, it serves as a structural component allowing the physical separation of the two DNA strands, and it is proposed to contain the Chi recognition site (Singleton et al., 2004). How exactly the RecBCD complex recognizes the Chi site is not yet understood. *recB* and *recC* mutants are sensitive to UV and gamma irradiation, and deficient for homologous recombination when the DNA substrate has a double-stranded blunt end, such as during generalized transduction and conjugation (Fig. 1.3A). Interestingly, *recB* or *recC* mutations profoundly affect cell growth, with up to 80% of dead cells in a liquid culture (Capaldo, Ramsey, and Barbour, 1974), and tend to yield suppressor mutations. Such is not the case for *recA* mutants and points to an additional role of the RecBCD complex in the cell, besides recombination, probably its exonucleolytic role, removing useless – or potentially dangerous – linear DNA. The strict dependence of the RecBCD complex for double-stranded ends makes this complex the first actor to act on linear DNA (Fig. 1.3, left side), and one of the key enzymes for the rescue of arrested replication forks (Figs. 1.4D and 1.4E) and for preventing RuvABC mediated fork breakage (Fig. 1.4C).

- The second group of proteins comprises RecF, RecO, and RecR. Although less well characterized at the molecular level, RecF, RecO, and RecR play a key role in preparing substrates for RecA on gapped DNA (Fig. 1.3, right side), to which RecBCD has no access (Morimatsu and Kowalczykowski, 2003). Their role would consist of competing away the single-stranded binding protein SSB to favor RecA loading. In addition to this role, RecFOR may also process double-strand breaks in cells in which the RecBCD complex is absent (Amundsen and Smith, 2003). For this purpose, at least two additional functions are recruited, the RecJ 5′ to 3′ nuclease, and the RecQ helicase (or, in its absence, UvrD or Helicase IV). These accessory functions would serve to trim the blunt extremity into a 3′ single strand for RecA. More recently, it was shown that RecFOR and RecJ could collaborate with RecBCD when the RecB component has lost its nuclease activity (Ivancic-Bace et al., 2003). *recF*, *recO*, and *recR* mutants are less sensitive than *recB* or *recC* mutants to radiation, but double mutants *recB recF* are nearly as sensitive as the *recA* mutant strain. *recF*, *recO*, or *recR* mutants do not exhibit any particular growth defects. This may reflect the low frequency of gap formation, the exclusive RecFOR substrate, in *E. coli*.

Synapsis

The RecA protein forms a stable filament on single-stranded DNA (ss DNA), which extends in the 5′ to 3′ direction (Roca and Cox, 1997). It promotes ss DNA pairing with a homologous double-stranded DNA, which is the key step of homologous recombination. Once the two DNA molecules have been placed in register, strand exchange can start. Homologous recombination is completely abolished in *recA* mutants, except in two particular cases mentioned in the section "RecA-independent homologous recombination." *recA* mutants are highly sensitive to UV and gamma radiation, and also affected for their growth, with a reduced doubling time and 50% of dead cells in liquid culture (Capaldo, Ramsey, and Barbour, 1974).

The minimal length of DNA on which RecA binds *in vitro* is 8nt, and the smaller stable duplex made by RecA is 15 base pair (bp) long (Hsieh et al., 1992). *In vivo*, the minimal length of homology on which RecA can act is probably as small as 23 to 27 bp, but data vary according to the system used (Lloyd and Low, 1996). RecA also promotes strand exchange between sequences that are not identical *in vitro*, and *in vivo*, an editing process mediated by MutS and MutL aborts such intermediates by a mechanism that remains to be elucidated. As a consequence, genetic exchanges between closely related species with nearly identical DNA sequence (called homeologous sequence, diverged up to 15% or even more) are highly increased when the recipient strain is mutated for the *mutS* or *mutL* function (Rayssiguier, Thaler, and Radman, 1989; Vulic et al., 1997). MutS and MutL were also found to edit intrachromosomal rearrangements between two slightly diverged *rhs* sequences (Petit et al., 1991). A more recent study on a wide spectrum of natural *E. coli* isolates has revealed that such mutant strains are present in 3% to 5% of isolates (Denamur et al., 2002), and they should favor horizontal gene transfer between related species.

In addition to its role at the heart of recombination, RecA of *E. coli* is also responsible for inducing the SOS response (see "The SOS response"), and for promoting the first step of replication fork reversal in some cases (McGlynn and Lloyd, 2000; Robu, Inman, and Cox, 2001; Seigneur, Ehrlich, and Michel, 2000).

Postsynapsis

Downstream of the RecA-promoted strand exchange, two alternative enzymes process the intermediates, RuvABC and RecG (Sharples, Ingleston, and Lloyd, 1999). RuvA is a DNA-binding protein specific for HJs, and RuvB is a DNA helicase that catalyzes branch migration when bound to RuvA. This helicase promotes the branch migration of HJ and delivers them to the specific nuclease RuvC, which resolves the recombination intermediate by cleaving symmetrically across the junction. Depending on which strands are cleaved, different products are expected, as drawn on Figs. 1.3A and 1.3B. RecG is also a DNA helicase, which favors branch migration of HJ and three-stranded branched structures (Lloyd and Sharples, 1993). Whether a nuclease is also involved to cleave the junctions processed by RecG is not known at present. What prompted the conclusion that *ruvABC* and *recG* encode redundant functions was the genetic observation that single mutants were only partially affected for recombination and partially sensitive to UV radiation, whereas the double mutant was as deficient and as sensitive as a *recA* strain, and affected for viability (Lloyd, 1991).

The outcome of recombination events appears to differ markedly with each situation, the RuvABC complex favoring the cross-over products when recombination is initiated from a double-stranded break, and the non–cross-over products when recombination is initiated from a gap (Cromie and Leach, 2000). Concerning RecG, one study suggests that it favors the cross-overs when recombination is initiated from a gap (Michel et al., 2000).

During the replication fork repair process (Fig. 1.4), a toxic role of RuvABC has been revealed: It recognizes the putative regressed fork intermediate, which has an HJ structure, and cleaves it, which leads to a linear chromosome. This aberrant role is countered by RecBCD, which either degrades the tail or initiates recombination with RecA (Seigneur et al., 1998).

Involvement of DNA replication

Some recombination intermediates (e.g., the one drawn in Fig. 1.3, last step, left side) are converted into replication forks due to the action of a group of seven proteins called collectively the PriA-dependant primosome, composed of PriA, PriB, PriC, DnaT, DnaC, DnaB, and DnaG (Marians, 1999,

2000). These proteins allow the loading of the replicative helicase, DnaB, and the DNA primase, DnaG, which are both absolutely required for starting coordinated replication of two DNA strands. The importance of this replication step during homologous recombination has been underlined by the observation that *priA* mutants are defective for conjugation and generalized transduction (Kogoma et al., 1996). In addition, *priA* mutants exhibit extremely slow growth and readily produce suppressor mutations, a phenotype more severe than *recA*, and even *recB* or *recC* mutants, which suggests that PriA plays an additional role elsewhere in the cell. A simple possibility is the replication restart at forks that lost the DnaB helicase. Once the DnaB and DnaG proteins are loaded, the DNA polymerase III holoenzyme takes over the replication step (Xu and Marians, 2003), and nothing is known at present concerning where and how replication stops.

Distinct from this main involvement of DNA replication during "ends-out" recombination, some short patches of DNA synthesis may also be needed for recombination starting from a gap, as shown on the right part of Fig. 1.3. The enzymes involved in this short DNA synthesis remain to be identified.

The SOS response

When *E. coli* cells are stressed by exposure to UV, gamma radiations, or chemical agents cross-linking the DNA, the SOS response, or SOS regulon, is induced. It consists of set of 31 (Fernandez De Henestrosa et al., 2000) to 50 genes (Courcelle et al., 2001), half of unknown functions, whose transcription is under the control of the LexA repressor. This repressor efficiently autocleaves into its inactive form when it contacts the RecA nucleofilament bound to DNA. RecA acts as a co-protease of LexA and this form of RecA is referred to as "RecA star." In this elegant way, once the cell senses DNA damage, by way of the RecA bound to a single-stranded portion of the DNA, the SOS genes are derepressed, among which four are directly involved in homologous recombination: RecA, present normally at 8000 molecules per cell is induced 10-fold, RuvA and RuvB are induced 2- to 3-fold, and RecN is induced 20-fold. Among the unknown SOS functions, one may encode a factor inhibiting the nuclease activity of RecBCD, and, therefore, favor its recombination activity (Rinken and Wackernagel, 1992). Once the recombination process is over and RecA has left the DNA, the LexA protein, which is itself induced by SOS, is no longer cleaved, and repression resumes. The DinI protein may also contribute to the closing off of the response by inhibiting the co-protease activity of RecA (Voloshin et al., 2001).

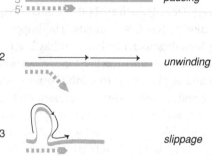

1 3' ——→——————→ *pausing*
   5' ▪▪▪▪▪▪▪▪▷

2 ——————→——————→ *unwinding*
   ▪▪▪▪▪▪,,

3 ⟆ *slippage*
   ▪▪▪▪▪▪▷

Figure 1.5. A model for recombination by replication slippage between tandem repeats, adapted from Viguera, Canceill, and Ehrlich (2001). Slippage is proposed to occur while a DNA polymerase replicates single-stranded DNA (shown as a thick grey line), as present on the lagging strand of a replication fork. If the DNA polymerase pauses in a region of tandem repeats (indicated with black arrows over the single-stranded DNA), the newly replicated strand (dotted thick grey line) may unwind (step 2) and pair erroneously with the downstream repetition (step 3). Then replication resumes, and the heteroduplex molecule leads after one more round of replication to a chromosome with a deletion of one repetition.

RecA-independent homologous recombination

Two cases of RecA-independent homologous recombination are described in *E. coli*. The first concerns recombination between short (around 1 kb long) tandem repeats, which recombine approximately as efficiently via a RecA-dependent process and a RecA-free process (Bierne et al., 1997; Saveson and Lovett, 1997). This RecA-free process is believed to result from replication slippage (Fig. 1.5) of the DNA polymerase when the tandem region is replicated: After a replication pause, the tip of the newly synthetized strand may be unwound and reanneal erroneously with the downstream copy of the tandem (Canceill and Ehrlich, 1996; d'Alencon et al., 1994; Viguera, Canceill, and Ehrlich, 2001). The close proximity between the recombining sequences is probably a strict requirement for such a recombination, and it may explain why such tandem repeats are rare in bacterial genomes.

The second case in which *E. coli* recombines its homologous DNA without using RecA concerns the activation of two cryptic genes – encoded by the Rac prophage, *recE* and *recT* – by the *sbcA* mutation that turns on the transcription of the operon. These functions are mentioned at the end of the next paragraph.

HOMOLOGOUS RECOMBINATION IN BACTERIA OTHER THAN *E. COLI* AND IN BACTERIOPHAGES

Bacteria are famous for their ability to conquer all kinds of habitats, as well as to live in highly complex ecosystems. As recombination appears more intrinsically linked with the everyday life of bacteria, it is of a high interest to survey how far the functions uncovered for *E. coli* are relevant or adapted to each species. For instance, a pathogen such as *Helicobacter pylori* has a huge propensity to mutate and rearrange its genome, and this may correlate to its pathogenicity (Loughlin et al., 2003). Also, the invasive *Salmonella typhimurium* pathogen appears to rely strongly on recombination functions to overcome the stresses endured on entry into macrophages (Schapiro, Libby, and Fang, 2003). The increasing number of fully sequenced genomes, as well as the powerful bioinformatic tools available today will allow a general overview over 50 eubacterial genomes (Table 1.1). The updated version of COG (cluster of orthologous groups, http://www.ncbi.nlm.nih.gov/COG/) has been used for this purpose (Tatusov, Koonin, and Lipman, 1997; Tatusov et al., 2001). Bacteriophages, especially when they are temperate or remain as remnants in bacterial genomes, are able to contribute to homologous recombination in bacteria and are briefly mentioned at the end of this section. Archea, although sharing ecological niches with bacteria and exchanging DNA with them through horizontal transfer, were not included in this study because they encode eukaryotic-like proteins with respect to homologous recombination, a field beyond the scope of this chapter.

The RecBCD and AddAB classes of exonucleases/helicases

S. typhimurium recBCD genes are essential for its infectivity (Buchmeier et al., 1993). This function is apparently needed to resist the stress due to nitric oxide (NO) encountered in macrophages, and a nice set of genetic evidence supports the model of recombination depicted in Fig. 1.4, where the consequence of NO would be to provoke replication fork arrests and a RecG-dependent reversal. Unless protected by RecBCD, this reversed fork is subjected to RuvC-dependent cleavage and subsequent death (Schapiro, Libby, and Fang, 2003).

A functional analog of RecBCD, called AddAB, has been characterized genetically and biochemically in *Bacillus subtilis*, a gram-positive bacterium isolated from soil, which is naturally competent (Chedin and Kowalczykowski, 2002). AddA and RecB share 21% identical amino acids, but AddB and RecC

Table 1.1. *Distribution of recombination functions among 50 bacteria*

Group	High GC Gram+			Low GC Gram+										"Ancient" bacteria						Chlam & Spiro				δ / ε		α						β		γ								
	Actinobact.			bacilli / clostridia						parasites																																	
Name	Cgl	Mtu	Mle	Cac	Lla	Spy	Sau	Lin	Bsu	Uur	Mpu	Mpn	Mge	Tma	Fnu	Aae	Sym	Nos	Dra	Ctr	Cpn	Tpa	Bbu	Hpy	Cje	Atu	Sme	Bme	Mlo	Ccr	Rpr	Rso	Nme	Ec	Ype	Sty	Vch	Pae	Hi	Xfa	Bu
O.P. or O.E.(a)			*			*				*	*	*	*									*	*								*								*		*
RecB																																									
AddA																																									
RecR																																									
RecO																																									
RecF																																									
RecJ																																									
RecQ																																									
RecN																																									
RecA																																									
RuvA																																									
RuvB																																									
RuvC																																									
RecU																																									
RecG																																									
LexA																																									
PriA																																									
PriB																																									
PriC																																									
DnaD																																									
DnaB																																									

(a) O.P. obligate parasit; O.E. obligate endosymbiont

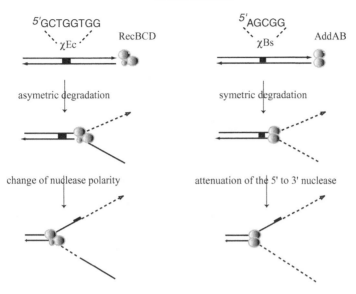

Figure 1.6. A summary of the biochemical properties of RecBCD of *E. coli* and AddAB of *B. subtilis*, adapted from Chedin and Kowalczykowski (2002). The two strands of a DNA molecule are drawn, with an arrow showing the 3′ extremities. The black rectangle stands for the Chi site, and dotted lines for the DNA degraded by nucleases. The main difference between the two sets of enzymes is visible in step 2: Degradation is asymmetric for RecBCD, and symmetric for AddAB. See text for more details.

are unrelated. Even if a *B. subtilis* ORF aligns with *E. coli recD*, it seems unrelated to the activity of the AddAB enzyme (Chedin and Kowalczykowski, 2002). Both AddA and AddB contain nuclease motives (Quiberoni et al., 2001), and it is supposed that they degrade strands of opposite polarity. Upon encounter of a specific sequence, different from the *E. coli* one but also called a Chi site, AddA, but not AddB, stops degrading DNA so that a 3′ single-stranded extremity is exposed (Chedin, Ehrlich, and Kowalczykowski, 2000). Whether this extremity facilitates RecA loading remains to be shown. A summary of the compared activities of RecBCD and AddAB is presented in Fig. 1.6. It is remarkable how similarly both enzymes work to achieve their role of nuclease/recombinase, despite their low sequence similarity.

The *addA* or *addB* mutants are affected for growth, with 80% of the cells in a liquid culture unable to form colonies, but do not yield suppressor mutations as *recBC* mutants. Interestingly, *add* mutants are less sensitive to UV radiations compared with *E. coli recBC* mutants (M.-A. Petit, personal

Table 1.2. *Sequence of the Chi site in various bacteria*

Organism	Sequence
E. coli	5'-GCTGGTGG-3'
H. influenzae	5'-GNTGGTGG-3'
B. subtilis	5'-AGCGG-3'
L. lactis	5'-GCGCGTG-3'

observation, 2003), and recombination during transformation is almost un-affected (Fernandez, Ayora, and Alonso, 2000). AddAB is not induced upon competence induction in *B. subtilis* (Ogura et al., 2002). Finally, when ex-pressed in *E. coli*, AddAB partially complements the defects of a *recBC* mutant (Kooistra, Haijema, and Venema, 1993). Similar phenotypes have been re-ported for *rexAB* mutants (orthologs of *addAB*) in *Lactococcus lactis* (el Karoui, Ehrlich, and Gruss, 1998) and *Streptococcus pneumoniae* (Halpern et al., 2004).

Critical to the action of these nucleases/helicases is the recognition of a DNA sequence, the Chi site, which converts the nuclease into a recombinase. Interestingly, in all bacteria where it was looked for, the Chi sequence differs (Table 1.2). This has been proposed as a way of self-recognition for a bacterial species: Incoming DNA would be readily recombined into the genome if it contains Chi sites, and would be degraded otherwise (el Karoui et al., 1999).

Among fully sequenced bacterial genomes, three categories can be made, with respect to the presence of such a nuclease/recombinase: Those contain-ing the RecBCD orthologs, those containing the AddA ortholog, and those free of any of these two functions (Table 1.3). Interestingly, no clear division among bacterial branches appears: Proteobacteria gamma and most beta contain RecBCD, whereas alpha and epsilon contain AddA. Among gram-positive bacteria, most contain the AddA enzyme, but actinobacteria contain the RecBCD version (or no version at all). The two sequenced spirochetes con-tain one enzyme type each. Whether bacteria free of RecBCD or the AddA counterpart encode a functional analog remains to be investigated. *Deinococ-cus radiodurans*, one of the most recombination-proficient bacteria, belongs to this group. During this search, it was observed that *recC*, *recB*, and *recD* genes are often clustered in a single operon, as well as *addB* and *A*. Some species in which *addA* is not preceded by an *addB* ortholog (AddB-orphan species) instead encode at this position an ORF containing a *recB* exonuclease

Table 1.3. *Distribution of RecBCD and AddA among 50 eubacteria*

	Group I: RecBCD	Group II A: AddA and AddB	Group II B: AddA and COG2887	Group III: no RecBC or AddA ortholog
Genes order	recC recB recD	addB addA	cog addA	
Species	Bbu* Buc Ctr Cpn Eco EcZ Ecs Hin Mtu MtC Nme NmA Pae Pmu Vch Sty Xfa Ype	Bha Bsu Cac Lla Lin Spy Spn Sau	Atu Bme Ccr Cje Fnu Hpy jHp Mlo Rso Rpr Rco Sme Tpa	Aae Cgl Dra Mle Mpu Mpn Mge Nos Syn Tma Uur
Number of species per category	18	8	13	11

* Bacterial species abbreviations: **Aae**, *Aquifex aeolicus*; **Atu**, *Agrobacterium tumefaciens*; **Bha**, *Bacillus halodurans*; **Bsu**, *Bacillus subtilis*; **Bbu**, *Borrelia burgdorferi*; **Bme**, *Brucella melitensis*; **Buc**, *Buchneria*; **Cac**, *Clostridium acetobutylicum*; **Ccr**, *Caulobacter crescentus*; **Cgl**, *Corynebacterium glutamicum*; **Cje**, *Campylobacter jejuni*; **Cpn**, *Chlamydia pneumoniae*; **Ctr**, *Chlamydia trachomatis*; **Dra**, *Deinococcus radiodurans*; **Eco**, **EcZ** and **Ecs**, *Escherichia coli*; **Fnu**, *Fusobacterium nucleatum*; **Hin**, *Haemophilus influenzae*; **Hpy** and **jHp**, *Helicobacter pylori*; **Lin**, *Listeria inocua*; **Lla**, *Lactococcus lactis*; **Mge**, *Mycoplasma genitalium*; **Mle**, *Mycobacterium leprae*; **Mlo**, *Mesorhizobium loti*; **Mpn**, *Mycoplasma pneumoniae*; **Mpu**, *Mycoplasma pulmonis*; **Mtu** and **MtC**, *Mycobacterium tuberculosis*; **Nme** and **NmA**, *Neisseria meningitides*; **Nos**, *Nostoc*; **Pae**, *Pseudomonas aeruginosa*; **Pmu**, *Pasteurella multocida*; **Rso**, *Ralstonia solanacearum*; **Rpr**, *Rickettsia prowazekii*; **Rco**, *Rickettsia conorii*; **Sau**, *Staphylococcus aureus*; **Spy**, *Streptococcus pyogenes*; **Spn**, *Streptococcus pneumoniae*; **Sty**, *Salmonella typhimurium*; **Sme**, *Sinorhizobium meliloti*; **Syn**, *Synechocystis*; **Tma**, *Thermotoga maritime*; **Tpa**, *Treponema palladium*; **Uur**, *Ureaplasma urealyticum*; **Vch**, *Vibrio cholerae*; **Xfa**, *Xylella fastidiosa*; **Ype**, *Yersinia pestis*.

motif (COG number 2887). Such an ORF, if not always adjacent to *addA*, is present in all AddB-orphan strains. Whether this ORF encodes an *addB*-like function is an open question. In addition, a phylogenetic tree of AddA (as presented in the page of AddA orthologs, COG number 1074) shows that all AddA-orphan species form a group separate from the species containing both AddA and AddB. For these reasons, I propose to group such bacterial species into a subcategory IIB.

The RecFOR complex

Bona fide orthologs of RecF, RecO, and RecR have been described in *B. subtilis*. Mutants in each of these genes behave similarly and define an epistatic group, as in *E. coli*. Interestingly, a fourth gene, *recL*, belongs to this epistatic group, but its genetic locus remains to be found (Alonso, Luder, and Tailor, 1991). A striking difference between *E. coli* and *B. subtilis* concerns the relative contribution of *recBCD/addAB* and *recFOR* functions to the resistance to UV radiation or chemical agents: *B. subtilis recFOR* mutants are very sensitive to UV radiation, as much as *recBCD* mutants of *E. coli*, and conversely, as mentioned previously, *addAB* mutants are as modestly sensitive as *E. coli recFOR* mutants. However, mutants missing both the RecBCD/Add and RecFOR functions are affected to a comparable extent in both hosts, and as affected as *recA* mutants, which suggests the absence of a third presynaptic activity in both bacteria.

Interestingly, the RecFOR functions are more widely spread among fully sequenced bacterial genomes than the RecBCD/Add function. Only 5 among 50 genomes do not contain any ortholog of at least one of the three functions. Among them, four are obligate intracellular parasites or obligate endosymbionts (Table 1.1). In some cases, one or even two of the three genes are missing, which could reflect either a loss of selective advantage for the RecFOR function and/or some undetected additional function for each unit of the complex. The RecR unit is present in all 45 species, which might favor the second possibility. RecO is missing in 4 of these 45 species (in particular in the delta/epsilon subdivision of Proteobacteria) and RecF is missing in all RecO minus species, and also in 4 more (in particular, in the beta subdivision of Proteobacteria). The loss of RecF, RecO, or RecR may be correlated with the loss of gene conversion at rDNA loci, which is observed in obligate endosymbionts (Dale et al., 2003). Indeed, this loss of homologous recombination function may constitute a first step toward chromosomal degeneration.

The RecA protein

In all bacteria in which it has been investigated, RecA is the central protein of homologous recombination and DNA repair. The *recA* gene from one species often complements the defects of another species *recA* mutant, even as distantly related as *B. subtilis* and *E. coli* (de Vos, de Vries, and Venema, 1983). One remarkable exception is the *recA* gene from *H. pylori*, which is inactive in *E. coli*, and appears to require posttranslation modification for activity (Schmitt et al., 1995). The *Pseudomonas aeruginosa* RecA, when expressed in *E. coli*, is threefold more active than *E. coli* RecA (Baitin Zaitsev, and Lanzov, 2003). The regulation of *recA* gene expression varies almost with each species in which it has been studied. In naturally competent species such as *B. subtilis* and *S. pneumoniae*, the *recA* gene is induced upon competence induction (Lovett, Love, and Yasbin, 1989). The *H. influenzae recA* gene is not induced during competence, but its expression is controlled by both LexA and cyclic AMP receptor protein (Zulty and Barcak, 1993). Finally, the *recA* gene expression of *D. radiodurans* is controlled positively by IrrE (Earl et al., 2002). Biochemical studies have revealed that *D. radiodurans* RecA behaves very differently from the *E. coli* enzyme, as it has double-stranded DNA as its preferred substrate. However, later stages of strand exchange appear comparable (Kim and Cox, 2002). The RecA protein is remarkably conserved among bacteria, but it is not universal, as among the COG 50 genomes list, it lacks in the *Buchneria* species.

The editing of recombination between diverged sequences by the MutLS proteins differs markedly between bacterial species. Although very efficient in *E. coli* during conjugation, the HexAB proteins of *Streptococcus pneumoniae* (MutLS orthologs) have a moderate effect during natural transformation, due essentially to a limitation in amounts of proteins available (Humbert et al., 1995). The MutLS proteins of *B. subtilis* are almost inoperative during transformation (Majewski and Cohan, 1998). This aspect is further developed in Chapter 3.

The RuvABC complex and the RecG helicase

In *H. pylori*, a species exhibiting high levels of genetic diversity, RuvC appears critical for continued survival *in vivo* in the mouse stomach (Loughlin et al., 2003). Here, the role of homologous recombination may differ from the one played in *S. typhimurium*, where the important function for infectivity is *recBC*, and where indeed addition of the *ruvC* mutation in a *recBC* mutant

Table 1.4. *Components of PriA-dependent primosomes*

	E. coli	B. subtilis
Initiation	PriA	PriA
Intermediate proteins	PriB & PriC & DnaT	DnaD
Helicase loader	DnaC	DnaB & DnaI
Replicative helicase	DnaB	DnaC

suppresses the attenuation phenotype (Schapiro, Libby, and Fang, 2003). Although *Salmonella* may use recombination to repair a stalled replication fork, *H. pylori* may use it to recombine genes. Interestingly, although RuvA and RuvB are almost universal among eubacteria, the RuvC element, required to cleave the HJ, is absent in all low G+C gram-positive bacteria. Interestingly, the RecU protein of *B. subtilis* was reported to cleave HJ (Ayora et al., 2004), an ortholog present in five of the six species representing low GC gram-positive bacteria in Table 1.1, so that it may indeed constitute a functional analog of RuvC in this branch.

The *mmsA* gene of *S. pneumoniae* encodes an ortholog of RecG, which has also similar biochemical characteristics (Hedayati, Steffen, and Bryant, 2002). This function is required for efficient natural transformation (Martin et al., 1996). RecG is found in 43 of the 50 sequenced genomes, and absent mostly in obligate parasites. This underlines the important role of this helicase in bacteria, which is probably many-fold and still far from fully understood at present.

Primosomal proteins

The primosomal proteins of *B. subtilis* have been analyzed in detail. A PriA ortholog is present and active, and the *priA* mutant exhibits the same low viability and tendency to accumulate suppressor mutations as its *E. coli* counterpart (Polard et al., 2002). However, the next components of the primosomal "cascade," namely DnaD, DnaI, and DnaB, are unrelated to the *E. coli* ones (Bruand et al., 2001). They differ in number, structure, and sequence from the PriB, PriC, DnaT, and DnaC proteins, but correspond to functional analogs (Table 1.4): They are involved in *B. subtilis* as mediators between PriA and the replicative helicase (Marsin et al., 2001; Velten et al., 2003). A limited domain, centered on an ATP-binding site, is common to DnaC of *E.*

coli and DnaI of *B. subtilis,* which has led sometimes to their grouping (as done in COG).

The analysis of the 50 fully sequenced bacterial genomes suggests that PriA is ubiquitous in bacteria: All but 4 obligate parasits encode a PriA ortholog. Interestingly, none of the mediator primosomal proteins, acting after PriA to load the replicative helicase, are present in more than a few species, as if each branch had formed its own set of proteins (Table 1.1).

SOS response

A LexA ortholog, the negative regulator of SOS response, is found in many bacterial species. However, in the case of *D. radiodurans,* this protein is not the repressor of the SOS response, and RecA expression in particular does not depend on LexA (Narumi et al., 2001), but on a positive regulator. The precise set of genes induced by LexA has been determined in *B. subtilis* (Dubnau and Lovett, 2002). It comprises 21 genes, and among recombination proteins, RecA and RuvAB are induced, as in *E. coli,* but not the RecN protein.

Some other unknown recombination proteins

A study of DNA repair mutants affected in homologous recombination in *B. subtilis* has been systematically undertaken over the years by the group of Dr. Juan Alonso (Fernandez, Ayora, and Alonso, 2000). Apart from the AddAB and RecFLOR functions already mentioned, some other genes, which define new epistatic groups, have been described. The *ruvA, recU,* and *recD* genes form the group epsilon, which is likely to take over the postsynaptic step of recombination (Ayora et al., 2004). The most interesting epistatic groups are group gamma (*recH* and *recP* genes) and zeta (*recS* gene, a *recQ* ortholog). Some genes locations (*recH, P, D*), roles, and functions are not yet defined, but they may reveal new ways to process DNA for homologous recombination.

The phage versions of recombination genes

Forterre has proposed that plasmid or virus DNA informational proteins could sometimes displace cellular analogs (Forterre, 1999). This possibility, combined with the huge reservoir of genes probably at hand with viral genomes, renews the interest for the study of phage recombination genes, which can only be briefly mentioned here. The best studied functional analog of RecA is RecT, encoded by the *E. coli rac* prophage. The RecT enzyme promotes not only *in vitro* single-stranded annealing (Hall and Kolodner, 1994), but also single-stranded invasion into duplex DNA (Noirot and

Table 1.5. *Phage recombination genes*

Phage name	Lambda	rac	DLP12	T4	T7	SPP1
Exonuclease	Redα	RecE				G34.1P
Nuclease/helicase				Gp46&47		
RecFOR analog	Orf					
Recombinase& ATPase				UvsX		
Recombinase	Redβ	RecT				G35P
Resolvase	Rap		Rus	Gp49	Gp3	

Kolodner, 1998), and the intermediates of the strand exchange reaction look strikingly similar to those generated by RecA (Noirot et al., 2003). However, unlike RecA, it does not have an ATP hydrolysis activity. Together with RecE, its associated exonuclease of the 5′ to 3′ polarity (Muyrers et al., 2000), it promotes *in vivo* double-stranded break repair (Takahashi et al., 1993), and the recombination between short homologous sequences (Zhang et al., 1998). The *rac* prophage belongs to the same family as the temperate bacteriophage lambda, which encodes orthologs of RecT/RecE as Redβ and Redα, respectively. Lambda also encodes a RecBCD inhibitor, Gam, so cells harboring the three Lambda functions are now widely used to introduce PCR fragments flanked by short, 50 bp long, homologies into the *E. coli* chromosome (Datsenko and Wanner, 2000; Yu et al., 2000). Lambda also possesses a functional analog of RecFOR encoded by the *orf* gene (Sawitzke and Stahl, 1997). Finally, homologous recombination is intrinsically connected with the replication cycle of some lytic bacteriophages like T4 of *E. coli* (Bleuit et al., 2001) and SPP1 of *B. subtilis* (Ayora et al., 2002). The genome of bacteriophage T4 encodes a RecA analog, UvsX, which hydrolyses ATP, like RecA, and gp46 and 47 RecBCD analogs (Miller et al., 2003), and SPP1 encodes a G35P ATP-independent recombinase and a G34.1P nuclease (Ayora et al., 2002; Martinez, Alonso, and Ayora, personal communication).

Finally some phages or prophages encode nucleases able to cleave HJs: gp49 for T4, gp3 for T7, Rap for lambda, and RusA for the DLP12 prophage (Sharples, 2001). Most of them are cleaving various branched DNA substrates, except for RusA, whose activity is restricted to HJ (Bolt and Lloyd, 2002). A summary of phage recombination proteins is presented in Table 1.5. One remarkable property of all these phage functions is that they do not exhibit sequence similarity with bacterial analogs. This underlines the rich diversity of bacteriophage "solutions" leading to the same set of functions, presynapsis, synapsis, and postsynapsis of homologous recombination, and the difficulty to predict them by sequence analysis.

CONCLUSION

Although bacteria have some common key functions such as RecA or PriA, they also each possess some level of diversity in other components. The main differences observed concern the RecBCD versus AddA helicase/nuclease, which are distributed among bacterial groups with no apparent logic. This activity seems absent from a substantial number of species, 11 among the 50 analyzed here, and in particular, in *D. radiodurans*. The second best studied set of proteins active at the presynapsis stage, RecFOR, appears more conserved, and also more active in bacteria other than *E. coli*. More generally, this study has shown a number of cases of functional analogs, and, therefore, convergent evolution (RecBCD and AddAB, RecA and RecT, PriB/PriC/DnaT/DnaC and DnaD/DnaI/DnaB, the various RuvC analogs in phages), and the list may grow in the future. Nevertheless, a big leap exists between these bacterial enzymes and the one at play in archea and eukaryots. Some attempts at proposing functional analogs have been made (Cromie, Connelly, and Leach, 2001). More recently, Rad52, Rad55, and Rad57 proteins were proposed to act in a way parallel to the RecFOR enzymes (Morimatsu and Kowalczykowski, 2003). Whether these proposals hold true will certainly be seen in the near future.

ACKNOWLEDGMENTS

I want to thank Juan Alonso, Meriem El Karoui, Jann Martinsohn, Ivan Matic, Bénédicte Michel, Xavier Veaute, and Marin Vulic for critical reading of this manuscript, and Emmanuelle Le Chatelier for her precious help with Table 1.1.

REFERENCES

Abdulkarim, F., and Hughes, D. (1996). Homologous recombination between the tuf genes of *Salmonella typhimurium*. *J Mol Biol*, **260**, 506–522.

Alonso, J.C., Luder, G., and Tailor, R.H. (1991). Characterization of *Bacillus subtilis* recombinational pathways. *J Bacteriol*, **173**, 3977–3980.

Amundsen, S.K., and Smith, G.R. (2003). Interchangeable parts of the *Escherichia coli* recombination machinery. *Cell*, **112**, 741–744.

Anderson, D.G., and Kowalczykowski, S.C. (1997). The translocating RecBCD enzyme stimulates recombination by directing RecA protein onto ssDNA in a chi-regulated manner. *Cell*, **90**, 77–86.

Anderson, R.P., and Roth, J.R. (1977). Tandem genetic duplications in phage and bacteria. *Annu Rev Microbiol*, **31**, 473–505.

Ayora, S., Missich, R., Mesa, P., Lurz, R., Yang, S., Egelman, E.H., and Alonso, J.C. (2002). Homologous-pairing activity of the Bacillus subtilis bacteriophage SPP1 replication protein G35P. *J Biol Chem*, **277**, 35969–35979.

Ayora, S., Carrasco, B., Doncel, E., Lurz, R., and Alonso, J.C. (2004). Bacillus subtilis RecU protein cleaves Holliday junctions and anneals single-stranded DNA. *Proc Natl Acad Sci U S A*, **101**, 452–457.

Baitin, D.M., Zaitsev, E.N., and Lanzov, V.A. (2003). Hyper-recombinogenic RecA protein from *Pseudomonas aeruginosa* with enhanced activity of its primary DNA binding site. *J Mol Biol*, **328**, 1–7.

Bierne, H., Seigneur, M., Ehrlich, S.D., and Michel, B. (1997). uvrD mutations enhance tandem repeat deletion in the *Escherichia coli* chromosome via SOS induction of the RecF recombination pathway. *Mol Microbiol*, **26**, 557–567.

Bleuit, J.S., Xu, H., Ma, Y., Wang, T., Liu, J., and Morrical, S.W. (2001). Mediator proteins orchestrate enzyme-ssDNA assembly during T4 recombination-dependent DNA replication and repair. *Proc Natl Acad Sci U S A*, **98**, 8298–8305.

Bolt, E.L., and Lloyd, R.G. (2002). Substrate specificity of RusA resolvase reveals the DNA structures targeted by RuvAB and RecG in vivo. *Mol Cell*, **10**, 187–198.

Bruand, C., Farache, M., McGovern, S., Ehrlich, S.D., and Polard, P. (2001). DnaB, DnaD and DnaI proteins are components of the *Bacillus subtilis* replication restart primosome. *Mol Microbiol*, **42**, 245–255.

Buchmeier, N.A., Lipps, C.J., So, M.Y., and Heffron, F. (1993). Recombination-deficient mutants of *Salmonella typhimurium* are avirulent and sensitive to the oxidative burst of macrophages. *Mol Microbiol*, **7**, 933–936.

Canceill, D., and Ehrlich, S.D. (1996). Copy-choice recombination mediated by DNA polymerase III holoenzyme from *Escherichia coli*. *Proc Natl Acad Sci U S A*, **93**, 6647–6652.

Capaldo, F.N., Ramsey, G., and Barbour, S.D. (1974). Analysis of the growth of recombination-deficient strains of *Escherichia coli* K-12. *J Bacteriol*, **118**, 242–249.

Chedin, F., Ehrlich, S.D., and Kowalczykowski, S.C. (2000). The *Bacillus subtilis* AddAB helicase/nuclease is regulated by its cognate Chi sequence in vitro. *J Mol Biol*, **298**, 7–20.

Chedin, F., and Kowalczykowski, S.C. (2002). A novel family of regulated helicases/nucleases from gram-positive bacteria: Insights into the initiation of DNA recombination. *Mol Microbiol*, **43**, 823–834.

Courcelle, J., Khodursky, A., Peter, B., Brown, P.O., and Hanawalt, P.C. (2001). Comparative gene expression profiles following UV exposure in wild-type and SOS-deficient *Escherichia coli*. *Genetics*, **158**, 41–64.

Cox, M.M. (2001). Recombinational DNA repair of damaged replication forks in *Escherichia coli*: Questions. *Annu Rev Genet*, **35**, 53–82.

Cromie, G.A., Connelly, J.C., and Leach, D.R. (2001). Recombination at double-strand breaks and DNA ends: Conserved mechanisms from phage to humans. *Mol Cell*, **8**, 1163–1174.

Cromie, G.A., and Leach, D.R. (2000). Control of crossing over. *Mol Cell*, **6**, 815–826.

d'Alencon, E., Petranovic, M., Michel, B., Noirot, P., Aucouturier, A., Uzest, M., and Ehrlich, S.D. (1994). Copy-choice illegitimate DNA recombination revisited. *Embo J*, **13**, 2725–2734.

Dale, C., Wang, B., Moran, N., and Ochman, H. (2003). Loss of DNA recombinational repair enzymes in the initial stages of genome degeneration. *Mol Biol Evol*, **20**, 1188–1194.

Datsenko, K.A., and Wanner, B.L. (2000). One-step inactivation of chromosomal genes in *Escherichia coli* K-12 using PCR products. *Proc Natl Acad Sci U S A*, **97**, 6640–6645.

de Vos, W.M., de Vries, S.C., and Venema, G. (1983). Cloning and expression of the *Escherichia coli* recA gene in *Bacillus subtilis*. *Gene*, **25**, 301–308.

Denamur, E., Bonacorsi, S., Giraud, A., Duriez, P., Hilali, F., Amorin, C., Bingen, E., et al. (2002). High frequency of mutator strains among human uropathogenic *Escherichia coli* isolates. *J Bacteriol*, **184**, 605–609.

Dillingham, M.S., Spies, M., and Kowalczykowski, S.C. (2003). RecBCD enzyme is a bipolar DNA helicase. *Nature*, **423**, 893–897.

Dubnau, D., and Lovett, C.M., Jr. (2002). "Transformation and recombination." In Sonenshein, A.L., Hoch, J.A., and Losick, R. (eds.), *Bacillus subtilis and its closest relatives. From genes to cells.* Washington, DC: ASM Press, pp. 453–471.

Earl, A.M., Mohundro, M.M., Mian, I.S., and Battista, J.R. (2002). The IrrE protein of *Deinococcus radiodurans* R1 is a novel regulator of recA expression. *J Bacteriol*, **184**, 6216–6224.

el Karoui, M., Biaudet, V., Schbath, S., and Gruss, A. (1999). Characteristics of Chi distribution on different bacterial genomes. *Res Microbiol*, **150**, 579–587.

el Karoui, M., Ehrlich, D., and Gruss, A. (1998). Identification of the lactococcal exonuclease/recombinase and its modulation by the putative Chi sequence. *Proc Natl Acad Sci U S A*, **95**, 626–631.

Fernandez, S., Ayora, S., and Alonso, J.C. (2000). *Bacillus subtilis* homologous recombination: genes and products. *Res Microbiol*, **151**, 481–486.

Fernandez De Henestrosa, A.R., Ogi, T., Aoyagi, S., Chafin, D., Hayes, J.J., Ohmori, H., and Woodgate, R. (2000). Identification of additional genes belonging to the LexA regulon in *Escherichia coli*. *Mol Microbiol*, **35**, 1560–1572.

Forterre (1999). Displacement of cellular proteins by functional analogues from plasmids or viruses could explain puzzling phylogenies of many DNA informational proteins. *Mol. Microbiol.* **33**, 457–465.

Hall, S.D., and Kolodner, R.D. (1994). Homologous pairing and strand exchange promoted by the *Escherichia coli* RecT protein. *Proc Natl Acad Sci U S A*, **91**, 3205–3209.

Halpern, D., Gruss, A., Claverys, J.P., and El-Karoui, M. (2004). rexAB mutants in Streptococcus pneumoniae. *Microbiology*, **150**, 2409–2414.

Hedayati, M.A., Steffen, S.E., and Bryant, F.R. (2002). Effect of the *Streptococcus pneumoniae* MmsA protein on the RecA protein-promoted three-strand exchange reaction. Implications for the mechanism of transformational recombination. *J Biol Chem*, **277**, 24863–24869.

Hsieh, P., Camerini-Otero, C.S., and Camerini-Otero, R.D. (1992). The synapsis event in the homologous pairing of DNAs: RecA recognizes and pairs less than one helical repeat of DNA. *Proc Natl Acad Sci U S A*, **89**, 6492–6496.

Humbert, O., Prudhomme, M., Hakenbeck, R., Dowson, C.G., and Claverys, J.P. (1995). Homeologous recombination and mismatch repair during transformation in Streptococcus pneumoniae: Saturation of the Hex mismatch repair system. *Proc Natl Acad Sci U S A*, **92**, 9052–9056.

Ivancic-Bace, I., Peharec, P., Moslavac, S., Skrobot, N., Salaj-Smic, E., and Brcic-Kostic, K. (2003). RecFOR function is required for DNA repair and recombination in a RecA loading-deficient recB mutant of *Escherichia coli*. *Genetics*, **163**, 485–494.

Kim, J.I., and Cox, M.M. (2002). The RecA proteins of *Deinococcus radiodurans* and *Escherichia coli* promote DNA strand exchange via inverse pathways. *Proc Natl Acad Sci U S A*, **99**, 7917–7921.

Kodama, K., Kobayashi, T., Niki, H., Hiraga, S., Oshima, T., Mori, H., and Horiuchi, T. (2002). Amplification of hot DNA segments in *Escherichia coli*. *Mol Microbiol*, **45**, 1575–1588.

Kogoma, T., Cadwell, G.W., Barnard, K.G., and Asai, T. (1996). The DNA replication priming protein, PriA, is required for homologous recombination and double-strand break repair. *J Bacteriol*, **178**, 1258–1264.

Kooistra, J., Haijema, B.J., and Venema, G. (1993). The *Bacillus subtilis* addAB genes are fully functional in *Escherichia coli*. *Mol Microbiol*, **7**, 915–923.

Kowalczykowski, S.C., Dixon, D.A., Eggleston, A.K., Lauder, S.D., and Rehrauer, W.M. (1994). Biochemistry of homologous recombination in *Escherichia coli*. *Microbiol Rev*, **58**, 401–465.

Kuzminov, A. (1999). Recombinational repair of DNA damage in *Escherichia coli* and bacteriophage lambda. *Microbiol Mol Biol Rev*, **63**, 751–813.

Lloyd, R.G. (1991). Conjugational recombination in resolvase-deficient ruvC mutants of *Escherichia coli* K-12 depends on recG. *J Bacteriol*, **173**, 5414–5418.

Lloyd, R.G., and Low, K.B. (1996). "Homologous recombination." In Neidhart, F.C. (ed.), *Escherichia coli and Salmonella*, Vol. 2. Washington, DC: ASM Press, pp. 2236–2255.

Lloyd, R.G., and Sharples, G.J. (1993). Dissociation of synthetic Holliday junctions by *E. coli* RecG protein. *EMBO J*, **12**, 17–22.

Loughlin, M.F., Barnard, F.M., Jenkins, D., Sharples, G.J., and Jenks, P.J. (2003). *Helicobacter pylori* mutants defective in RuvC Holliday junction resolvase display reduced macrophage survival and spontaneous clearance from the murine gastric mucosa. *Infect Immun*, **71**, 2022–2031.

Lovett, C.M., Jr., Love, P.E., and Yasbin, R.E. (1989). Competence-specific induction of the *Bacillus subtilis* RecA protein analog: Evidence for dual regulation of a recombination protein. *J Bacteriol*, **171**, 2318–2322.

Majewski, J., and Cohan, F.M. (1998). The effect of mismatch repair and heteroduplex formation on sexual isolation in *Bacillus*. *Genetics*, **148**, 13–18.

Marians, K.J. (1999). PriA: At the crossroads of DNA replication and recombination. *Prog Nucleic Acid Res Mol Biol*, **63**, 39–67.

Marians, K.J. (2000). PriA-directed replication fork restart in *Escherichia coli*. *Trends Biochem Sci*, **25**, 185–189.

Marsin, S., McGovern, S., Ehrlich, S.D., Bruand, C., and Polard, P. (2001). Early steps of *Bacillus subtilis* primosome assembly. *J Biol Chem*, **276**, 45818–45825.

Martin, B., Sharples, G.J., Humbert, O., Lloyd, R.G., and Claverys, J.P. (1996). The mmsA locus of *Streptococcus pneumoniae* encodes a RecG-like protein involved in DNA repair and in three-strand recombination. *Mol Microbiol*, **19**, 1035–1045.

McGlynn, P., and Lloyd, R.G. (2000). Modulation of RNA polymerase by (p)ppGpp reveals a RecG-dependent mechanism for replication fork progression. *Cell*, **101**, 35–45.

Michel, B., Flores, M.J., Viguera, E., Grompone, G., Seigneur, M., and Bidnenko, V. (2001). Rescue of arrested replication forks by homologous recombination. *Proc Natl Acad Sci U S A*, **98**, 8181–8188.

Michel, B., Recchia, G.D., Penel-Colin, M., Ehrlich, S.D., and Sherratt, D.J. (2000). Resolution of Holliday junctions by RuvABC prevents dimer formation in rep mutants and UV-irradiated cells. *Mol Microbiol*, **37**, 180–191.

Miller, E.S., Kutter, E., Mosig, G., Arisaka, F., Kunisawa, T., and Ruger, W. (2003). Bacteriophage T4 genome. *Microbiol Mol Biol Rev*, **67**, 86–156.

Morimatsu, K., and Kowalczykowski, S.C. (2003). RecFOR proteins load RecA protein onto gapped DNA to accelerate DNA strand exchange: A universal step of recombinational repair. *Mol Cell*, **11**, 1337–1347.

Muyrers, J.P., Zhang, Y., Buchholz, F., and Stewart, A.F. (2000). RecE/RecT and Redalpha/Redbeta initiate double-stranded break repair by specifically interacting with their respective partners. *Genes Dev*, **14**, 1971–1982.

Myers, R.S., and Stahl, F.W. (1994). Chi and the RecBC D enzyme of *Escherichia coli*. *Annu Rev Genet*, **28**, 49–70.

Narumi, I., Satoh, K., Kikuchi, M., Funayama, T., Yanagisawa, T., Kobayashi, Y., Watanabe, H., and Yamamoto, K. (2001). The LexA protein from *Deinococcus radiodurans* is not involved in RecA induction following gamma irradiation. *J Bacteriol*, **183**, 6951–6956.

Noirot, P., Gupta, R.C., Radding, C.M., and Kolodner, R.D. (2003). Hallmarks of homology recognition by RecA-like recombinases are exhibited by the unrelated *Escherichia coli* RecT protein. *Embo J*, **22**, 324–334.

Noirot, P., and Kolodner, R.D. (1998). DNA strand invasion promoted by *Escherichia coli* RecT protein. *J Biol Chem*, **273**, 12274–12280.

Ochman, H., Lawrence, J.G., and Groisman, E.A. (2000). Lateral gene transfer and the nature of bacterial innovation. *Nature*, **405**, 299–304.

Ogura, M., Yamaguchi, H., Kobayashi, K., Ogasawara, N., Fujita, Y., and Tanaka, T. (2002). Whole-genome analysis of genes regulated by the *Bacillus subtilis* competence transcription factor ComK. *J Bacteriol*, **184**, 2344–2351.

Petes, T.D., and Hill, C.W. (1988). Recombination between repeated genes in microorganisms. *Annu Rev Genet*, **22**, 147–168.

Petit, M.A., Dimpfl, J., Radman, M., and Echols, H. (1991). Control of large chromosomal duplications in *Escherichia coli* by the mismatch repair system. *Genetics*, **129**, 327–332.

Polard, P., Marsin, S., McGovern, S., Velten, M., Wigley, D.B., Ehrlich, S.D., and Bruand, C. (2002). Restart of DNA replication in gram-positive bacteria: Functional characterisation of the *Bacillus subtilis* PriA initiator. *Nucleic Acids Res*, **30**, 1593–1605.

Quiberoni, A., Biswas, I., El Karoui, M., Rezaiki, L., Tailliez, P., and Gruss, A. (2001). In vivo evidence for two active nuclease motifs in the double-strand break repair enzyme RexAB of *Lactococcus lactis*. *J Bacteriol*, **183**, 4071–4078.

Rayssiguier, C., Thaler, D.S., and Radman, M. (1989). The barrier to recombination between *Escherichia coli* and *Salmonella typhimurium* is disrupted in mismatch-repair mutants. *Nature*, **342**, 396–401.

Rinken, R., and Wackernagel, W. (1992). Inhibition of the recBCD-dependent activation of Chi recombinational hot spots in SOS-induced cells of *Escherichia coli*. *J Bacteriol*, **174**, 1172–1178.

Robu, M.E., Inman, R.B., and Cox, M.M. (2001). RecA protein promotes the regression of stalled replication forks in vitro. *Proc Natl Acad Sci U S A*, **98**, 8211–8218.

Roca, A.I., and Cox, M.M. (1997). RecA protein: Structure, function, and role in recombinational DNA repair. *Prog Nucleic Acid Res Mol Biol*, **56**, 129–223.

Roth, J.R., Benson, N., Galitski, T., Haak, K., Lawrence, J., and Miesel, L. (1996). "Rearrangements of the bacterial chromosome: Formation and applications." In Neidhart, F.C. (ed.), *Escherichia coli and Salmonella*, Vol. 2. Washington, DC: ASM Press, pp. 2256–2276.

Saveson, C.J., and Lovett, S.T. (1997). Enhanced deletion formation by aberrant DNA replication in *Escherichia coli*. *Genetics*, **146**, 457–470.

Sawitzke, J.A., and Stahl, F.W. (1997). Roles for lambda Orf and *Escherichia coli* RecO, RecR and RecF in lambda recombination. *Genetics*, **147**, 357–369.

Schapiro, J.M., Libby, S.J., and Fang, F.C. (2003). Inhibition of bacterial DNA replication by zinc mobilization during nitrosative stress. *Proc Natl Acad Sci U S A*, **100**, 8496–8501.

Schmitt, W., Odenbreit, S., Heuermann, D., and Haas, R. (1995). Cloning of the *Helicobacter pylori* recA gene and functional characterization of its product. *Mol Gen Genet*, **248**, 563–572.

Seigneur, M., Bidnenko, V., Ehrlich, S.D., and Michel, B. (1998). RuvAB acts at arrested replication forks. *Cell*, **95**, 419–430.

Seigneur, M., Ehrlich, S.D., and Michel, B. (2000). RuvABC-dependent double-strand breaks in dnaBts mutants require recA. *Mol Microbiol*, **38**, 565–574.

Sharples, G.J. (2001). The X philes: Structure-specific endonucleases that resolve Holliday junctions. *Mol Microbiol*, **39**, 823–834.

Sharples, G.J., Ingleston, S.M., and Lloyd, R.G. (1999). Holliday junction processing in bacteria: Insights from the evolutionary conservation of RuvABC, RecG, and RusA. *J Bacteriol*, **181**, 5543–5550.

Singleton, M.R., Dillingham, M.S., Gaudier, M., Kowalczykowski, S.C., and Wigley, D.B. (2004). Crystal structure of RecBCD enzyme reveals a machine for processing DNA breaks. *Nature*, **432**, 187–193.

Takahashi, N.K., Kusano, K., Yokochi, T., Kitamura, Y., Yoshikura, H., and Kobayashi, I. (1993). Genetic analysis of double-strand break repair in *Escherichia coli*. *J Bacteriol*, **175**, 5176–5185.

Tatusov, R.L., Koonin, E.V., and Lipman, D.J. (1997). A genomic perspective on protein families. *Science*, **278**, 631–637.

Tatusov, R.L., Natale, D.A., Garkavtsev, I.V., Tatusova, T.A., Shankavaram, U.T., Rao, B.S., Kiryutin, B., Galperin, M.Y., Fedorova, N.D., and Koonin, E.V. (2001). The COG database: New developments in phylogenetic classification of proteins from complete genomes. *Nucleic Acids Res*, **29**, 22–28.

Taylor, A.F., and Smith, G.R. (2003). RecBCD enzyme is a DNA helicase with fast and slow motors of opposite polarity. *Nature*, **423**, 889–893.

Velten, M., McGovern, S., Marsin, S., Ehrlich, S.D., Noirot, P., and Polard, P. (2003). A two-protein strategy for the functional loading of a cellular replicative DNA helicase. *Mol Cell*, **11**, 1009–1020.

Viguera, E., Canceill, D., and Ehrlich, S.D. (2001). Replication slippage involves DNA polymerase pausing and dissociation. *EMBO J*, **20**, 2587–2595.

Voloshin, O.N., Ramirez, B.E., Bax, A., and Camerini-Otero, R.D. (2001). A model for the abrogation of the SOS response by an SOS protein: A negatively charged helix in DinI mimics DNA in its interaction with RecA. *Genes Dev*, **15**, 415–427.

Vulic, M., Dionisio, F., Taddei, F., and Radman, M. (1997). Molecular keys to speciation: DNA polymorphism and the control of genetic exchange in enterobacteria. *Proc Natl Acad Sci U S A*, **94**, 9763–9767.

Xu, L., and Marians, K.J. (2003). PriA mediates DNA replication pathway choice at recombination intermediates. *Mol Cell*, **11**, 817–826.

Yu, D., Ellis, H.M., Lee, E.C., Jenkins, N.A., Copeland, N.G., and Court, D.L. (2000). An efficient recombination system for chromosome engineering in *Escherichia coli*. *Proc Natl Acad Sci U S A*, **97**, 5978–5983.

Zhang, Y., Buchholz, F., Muyrers, J.P., and Stewart, A.F. (1998). A new logic for DNA engineering using recombination in *Escherichia coli*. *Nat Genet*, **20**, 123–128.

Zulty, J.J., and Barcak, G.J. (1993). Structural organization, nucleotide sequence, and regulation of the *Haemophilus influenzae* rec-1+ gene. *J Bacteriol*, **175**, 7269–7281.

CHAPTER 2

Introduction to site-specific recombination

Makkuni Jayaram and Ian Grainge

Recombination provides a means for creating genetic variety. Exchange of information within gene pools expands their diversity and enhances the choices available for natural selection to act on. Recombination can be broadly divided into two classes: the highly pervasive "homologous" recombination and the more specialized "site-specific recombination."

HOMOLOGOUS RECOMBINATION

Before we discuss site-specific recombination, a brief overview of general (or homologous) recombination is useful for an appreciation of the distinctions between the two systems. Homologous recombination is a nearly universal mechanism employed by living organisms to reshuffle their genetic information. Within a cell, recombination can occur between two homologous chromosomes, between two sister chromatids formed by DNA replication, and between extrachromosomal elements such as plasmids or viral genomes. In eukaryotes, the rate of recombination during mitosis is relatively low and is markedly increased during meiosis. In fact, genetic exchange between homologs and chiasma formation appear to be a prerequisite for the proper reductional segregation of chromosomes and the generation of haploid gametes. The consequences of the resulting genetic configuration and the corresponding fitness contribution to an individual will be manifested directly, and almost immediately, in a haploid organism. For a diploid organism, the expression of the novel genetic makeup must await the fusion between the male and female gametes to produce a zygote. Recombination is therefore one of the forces that drive Darwinian evolution.

According to a generally accepted view, the primary function of recombination in prokaryotes, and during the mitotic cell division in eukaryotes, is the

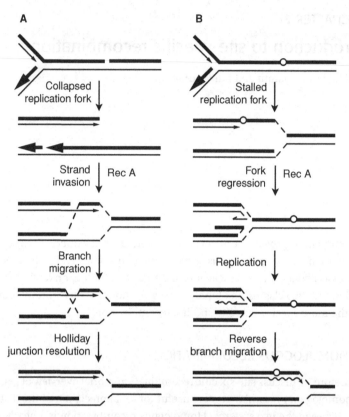

Figure 2.1. Reviving a collapsed replication fork by recombination. Two potential pathways for restarting replication following the arrest of an advancing fork from a strand nick **(A)** or a blocking lesion **(B)** are diagrammed. The broken duplex in **A** can invade an intact homolog and reestablish a replication fork by branch migration and Holliday junction resolution. The arrested replication fork in **B** can regress into a four-branch structure in which limited DNA synthesis is possible. By reverse branch migration, the fork can be translated to a point beyond the lesion. (Adapted from Lusetti and Cox, 2002.)

repair of stalled or collapsed replication forks (Lusetti and Cox, 2002; Fig. 2.1). Single-stranded nicks and other lesions in the template DNA can cause premature termination of an advancing replication fork. The double-stranded break formed from a preexisting nick may be repaired by strand invasion, branch migration, and Holliday junction resolution (Fig. 2.1A). A different type of lesion can produce a regressed fork, nicknamed a "chicken foot" for its shape, as shown in Fig. 2.1B. Normal replication may be restored by limited DNA synthesis, followed by reverse branch migration. The *Escherichia coli* (*E. coli*) RecA protein, as well as its functional relatives from other organisms, are central players in these events, and are assisted by a host of other repair/recombination proteins.

In summary, homologous recombination serves the dual purpose of maintaining genome integrity on the one hand and of promoting genetic diversity on the other. Throughout the centuries, plant and animal breeders have applied recombination to select for crops that are disease or drought resistant, farm animals that produce more milk or yield more meat, and beasts of burden with improved strength and stamina. So, too, the modern-day biotechnologists use recombination for genome engineering, creation of transgenic plants and animals, and the production of biofactories. These applications of recombination, with enormous potential benefits to mankind, do not demand a detailed appreciation of the intrinsic biochemical and conformational subtleties of the process. In this chapter, our principal concern will be the often ignored or overlooked mechanistic logic, geometric design, and topological attributes of the recombination reaction. We rely on a class of specialized recombination systems, the site-specific recombinases, to highlight these aspects.

SITE-SPECIFIC RECOMBINATION

Unlike homologous recombination that uses rather long stretches of homology between partner DNA molecules, site-specific recombination targets relatively short DNA sites with well-defined sequences. Whereas a large number of proteins with distinct biochemical activities cooperate to carry out homologous recombination, it is usually a single protein or a pair of proteins that carry out the catalytic steps of site-specific recombination. In some instances, the site-specific recombinase may be aided by one or two accessory proteins in synapsing the DNA partners and assembling the chemically competent recombination complex. Because of their overall simplicity, these systems have served as models for investigating the mechanisms of phosphoryl transfer reactions during recombination and related reactions.

CONSERVATIVE SITE-SPECIFIC RECOMBINASES

As the name implies, during conservative site-specific recombination, specific phosphodiester bonds are broken and exchanged between two DNA partners, at the same time conserving the energy of these scissile bonds. To accomplish this, the recombinase mediates the strand cleavage and joining steps through a transesterification mechanism. The reaction is completed without degradation or synthesis of DNA and without the requirement for exogenous input of high-energy co-factors such as ATP.

Only two classes of conservative site-specific recombinases have been discovered so far: the invertase/resolvase family (also called the serine family) and the lambda integrase family (also called the tyrosine family). A few members of the two families have been studied in biochemical and structural detail with respect to the arrangement of the recombination complex, the organization of the recombinase active site, the chemical mechanisms of strand exchange, and the conformational changes accompanying the reaction. Interestingly, these recombination systems are present almost exclusively in prokaryotes. Whereas no serine recombinase has been detected in eukaryotes, the eukaryotic tyrosine recombinases have been identified only in a small number of related yeasts belonging to a common genus (Broach and Volkert, 1991; Jayaram et al., 2004). Either conservative site-specific recombination emerged after the splitting of the evolutionary lineages leading to prokaryotes and eukaryotes, or it was somehow lost in the latter lineage. The resurgence of tyrosine recombination within a restricted yeast genus perhaps represents a freak evolutionary accident that imported the system through the horizontal transmission of a genetic element harboring it. Consistent with this notion, the yeast site-specific recombinases are encoded by extrachromosomal selfish DNA elements that apparently confer no advantage to their hosts.

THE SERINE AND TYROSINE FAMILIES: SUBSTRATE RECOGNITION AND REACTION PATHWAYS

According to the old nomenclature, the two families are distinguished by the type of reactions they carry out. The invertase/resolvase family mediates DNA inversion or DNA resolution (or excision) (Grindley, 2002; Johnson, 2002). The integrase family derives its name from the Int protein of phage lambda, the prototype member of the family, which integrates the lambda genome into the bacterial chromosome (Azaro and Landy, 2002). Under appropriate conditions, the same protein also catalyzes phage excision. The new names define the families by the type of chemistry they use during recombination: One relies on serine and the other on tyrosine as their active site nucleophiles during DNA breakage.

The general reaction pathways followed by the serine and tyrosine families are illustrated in Fig. 2.2. For both families, the chemical steps of the reaction take place at core DNA sites that are quite similar in their overall organization. Two identical (or nearly identical) sequences, comprising between one and two turns of DNA, border a short DNA segment (less than a turn; commonly referred to as the strand exchange region or spacer) in a head-to-head (inverted) fashion. Each repeated element provides the binding

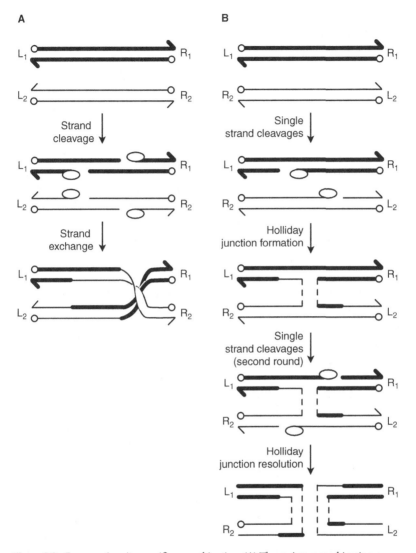

Figure 2.2. Conservative site-specific recombination. **(A)** The serine recombination reaction is initiated by double-stranded breaks in the two DNA partners arranged in a parallel fashion. Their left and right ends are marked by L and R, respectively. A 180-degree relative rotation of the cleaved recombination complex followed by strand joining completes recombination. Note that a half-twist is introduced in each DNA substrate, and a right-handed crossing is formed between them. **(B)** The tyrosine recombination reaction is initiated by the cleavage and exchange of one pair of strands to form a Holliday junction intermediate. Cleavage and exchange of the second pair of strands completes recombination. Note that the sites are arranged in an antiparallel manner, and there is no DNA crossing introduced between the recombination products. The knobs and split arrowheads represent the 5' and 3' ends, respectively, of DNA strands.

site for a monomer of the recombinase. The serine recombinases, in general, appear to bind their target sites as dimers. In contrast, the tyrosine recombinases exist in solution and bind DNA as monomers, and establish strong monomer–monomer interactions following binding. In a rare case, exemplified by the XerC/XerD system (Barre and Sherratt, 2002), the target site contains the binding sites for two distinct recombinase subunits, namely, the XerC and XerD proteins. The active recombinase is constituted by these two proteins acting in concert. The recombination reaction occurs in the context of two core sites and the four bound recombinase proteins brought together in a synaptic complex with a fixed geometry. In those recombination systems that use accessory proteins and DNA sites, the multiple DNA–protein interactions result in a complex synapse that is often characterized by a precisely defined DNA topology. The target phosphodiester bonds on the two DNA strands of a substrate (defining the left and right limits of strand exchange) are specified by the specific interaction of a recombinase subunit with its binding element.

The typical serine recombinase makes concerted double-stranded breaks in DNA, forming a 5′-phosphoserine bond and an adjacent 3′-hydroxyl group on each strand (Fig. 2.2A). In a long-held, although unproved, strand exchange model, the cleaved recombination complex rotates 180 degrees about its horizontal axis to bring the DNA strands in the strand joining configuration. This conformational transition may be described as an *isomerization step* – a term that has been more often used in the context of the Holliday junction dynamics during tyrosine recombination (see below). The 3′-hydroxyl groups now carry out nucleophilic attacks on the 5′-O-serine phosphodiester bonds to restore the DNA strands in the recombined state and release the recombinase from its covalent linkage to DNA. Note that the strand exchange step introduces two positive half-twists in the double helix, one each in each of the exchanged segments (equivalent to the relaxation of a negative supercoil), and one right-handed crossing between the double helical partners. If the recombining sites are in head-to-head orientation, as is the requirement for an invertase, the outcome is the inversion of DNA between the recombining sites. If the sites are in head-to-tail orientation, as is mandatory for a resolvase, the consequence is the deletion of DNA between the recombining sites.

In contrast to the serine recombinase, the tyrosine recombinase makes single-stranded DNA breaks on "like" strands (either top or bottom) in the recombining partners, using an active site tyrosine as the cleavage nucleophile (Fig. 2.2B). The cleavage reaction follows the classical type Ib topoisomerase mechanism (Wang, 2002), and the products are a 3′-O-phosphotyrosine bond and a 5′-hydroxyl group. The attack by the hydroxyl groups on the phosphotyrosine bonds in an intersubstrate orientation results in the formation of a Holliday intermediate. Reorganization of the recombination complex (or the isomerization of the Holliday junction), followed by the break/union of the second pair of strands, completes recombination. Note that, unlike in the serine recombination case, the two DNA sites are arranged in opposite orientations (antiparallel geometry), and the recombination products do not cross each other. We have more to say about recombination geometry and topology later.

NUCLEOTIDYL TRANSFER MECHANISM
DURING RECOMBINATION

Conceptually, it is convenient to think of the formation of each recombinant strand (and four such strands are formed during a complete recombination event) as being the sum of two nucleophilic transesterification reactions (Fig. 2.3). First, a nucleophile from the recombinase active site attacks a specific $3'$-$5'$ phosphodiester bond within a DNA strand to generate a protein–DNA phosphodiester linkage (either $5'$-O-phosphoserine or $3'$-O-phosphotyrosine) and a DNA hydroxyl group ($3'$-OH or $5'$-OH). Subsequently, the hydroxyl group attacks the phosphoprotein diester bond to reform the DNA phosphodiester linkage and free the recombinase protein. At each step, the scissile phosphate has to be oriented by active site residues that help maintain the positive charge on the phosphate atom. Concomitantly, the serine/tyrosine or the $3'$-/$5'$-hydroxyl nucleophile must also be activated and oriented. The resulting transition state will contain a pentacoordinated phosphate with a trigonal bipyramidal geometry (Baker and Mizuuchi, 2002; Mizuuchi, 1992b). Part of the driving force for the strand cleavage and joining steps can be provided by the stabilization of the corresponding leaving groups – the ionized forms of the DNA hydroxyl (during cleavage) and the recombinase nucleophile (during joining), respectively.

Consistent with a number of well-characterized phosphoryl transfer reactions in nucleic acids, the proposed inline nucleophilic substitutions (SN_2) are likely facilitated by acid–base catalysis. According to this scheme, the cleavage step and joining step should be followed by the inversion of phosphate configuration. As a result, the complete recombination event should be accompanied by retention (inversion followed by inversion). This result has been verified for tyrosine recombinases and a type Ib topoisomerase, although not for a serine recombinase, by using the R_P and S_P forms of chiral DNA substrates in which one of the nonbridging oxygen atoms at the

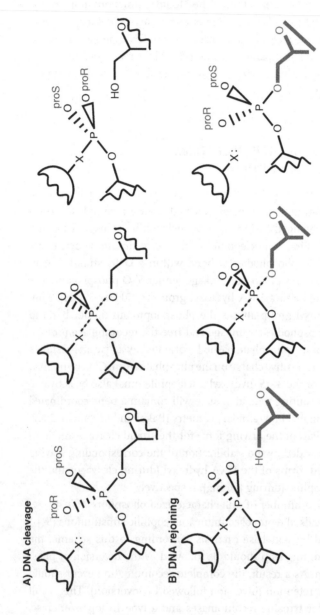

Figure 2.3. The inline nucleophilic substitution reactions at the scissile phosphodiester bond during conservative site-specific recombination. (A) The nucleophile X represents the active site serine or tyrosine. The proposed pentacoordinate transition state (*middle*) breaks down to give the cleaved intermediate (*right*). The leaving group is the 5′-OH (for the tyrosine family) or the 3′-OH (for the serine family). Note the inversion of phosphate configuration. (B) The mechanism is the same as in A, except that the nucleophile is the 5′-OH (for the tyrosine family) or the 3′-OH (for the serine family) from the DNA partner and the leaving group is X.

reactive phosphate is substituted with sulfur (Jayaram et al., 2002; Mizuuchi, 1992a; Mizuuchi and Adzuma, 1991; Stivers et al., 2000).

ACTIVE SITE CONFIGURATIONS OF SERINE AND TYROSINE RECOMBINASES

A combination of strategies – directed mutagenesis, biochemical analyses, and structural studies – has yielded a wealth of information on the organization of the recombination active sites. Among the serine recombinases, the most extensively studied members are the resolvase protein of the $\gamma\delta$ or Tn3 transposon, and the Hin and Gin invertases of *Salmonella* and phage Mu, respectively (Grindley, 2002; Johnson, 2002; Kanaar et al., 1990). The best characterized tyrosine recombinases are the Int protein of phage lambda, the Cre protein of phage P1, the XerC/XerD proteins of *E. coli*, and the Flp protein of yeast (Azaro and Landy, 2002; Barre and Sherratt, 2002; Jayaram et al., 2002; Van Duyne, 2002).

The first crystal structure to be solved among the site-specific recombinases was that of the catalytic domain of $\gamma\delta$ resolvase (Rice and Steitz, 1994b; Sanderson et al., 1990). This was followed by the solution of the co-crystal structure of resolvase complexed with *res* I DNA, the site where the chemistry of recombination occurs (Yang and Steitz, 1995). A Hin dimer bound to its target site, *hix*, has been modeled after the resolvase–*res* I complex (Feng et al., 1994; Johnson, 2002). The model incorporates information from the co-crystal structure of a 52 amino acid DNA-binding domain of a Hin monomer in association with a half-*hix* site. The cleavage nucleophile has been mapped to an invariant serine: Ser-10 in resolvase and Hin, and Ser-9 in Gin. Mutational analysis of Hin indicated that a cluster of amino acids around Ser-10 and Arg-67 are important for active site structure and/or function. In the resolvase–*res* I complex (Fig. 2.4), Arg-8, Asp-67, Arg-68 (corresponding to Arg-67 of Hin), and Arg-71 are all in the vicinity of Ser-10 without directly contacting it. Arg-71 is hydrogen bonded to the scissile phosphate. In the earlier resolvase structure, Arg-68 and Ser-10 were seen to coordinate a sulfate ion, which likely mimics the phosphate at the cleavage position. Glu-124 is projected from one resolvase subunit toward the active site of its neighbor and may serve to regulate recombination. Lack of Glu-124 relieves resolvase of its topological constraints, and recombination leads to both normal and aberrant products (Grindley, 2002). One disappointing feature of the crystal structure is that the two Ser-10 residues of the resolvase dimer are too far away from each other (more than 30 Å) and from their respective target phosphates ($\sim$11 Å and $\sim$17 Å). Obviously, the structure must undergo significant

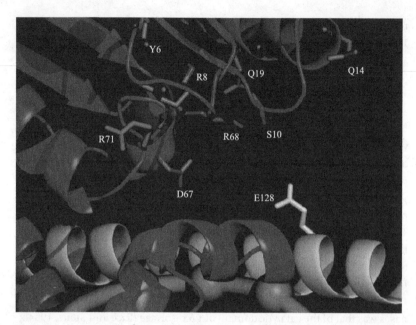

Figure 2.4. Arrangement of active site residues in resolvase bound to the *res* I site. The distribution of potential catalytic residues is displayed within the blue subunit of the resolvase dimer. Note that the E helix from the neighboring yellow subunit approaches the active site, projecting Glu-124 (not seen in this view) into it. The DNA backbone is shown in magenta with the phosphate groups as spherical knobs. Of the three neighboring phosphates shown here, the leftmost, partially hidden by the blue helix, is the scissile one. Note the large distance between Ser-10 and the scissile phosphate, indicating that this is not the functional state of the active site. For a stereo view of the resolvase active site, the reader is referred to Grindley (2002).

conformational adjustments before cleavage can be effected. The disposition of the serines also calls into question the accepted notion of concerted strand breaks depicted in Fig. 2.2A.

The combined information from solution studies and crystal structures offer an impressively detailed picture of the active sites of tyrosine recombinases (Jayaram et al., 2002; Rice, 2002; Van Duyne, 2002). In particular, the structures of the different Cre–DNA complexes (Gopaul et al., 1998; Guo et al., 1997, 1999) and the Flp–DNA complex (Chen et al., 2000) provide rich insights into the possible mechanistic features of recombination. As expected, comparison of the recombinase structures with those of vaccinia and human topoisomerases (type Ib; Cheng and Shuman, 1998; Redinbo et al., 1998; Stewart et al., 1998) readily explains the common chemistry employed

MAKKUNI JAYARAM AND IAN GRAINGE

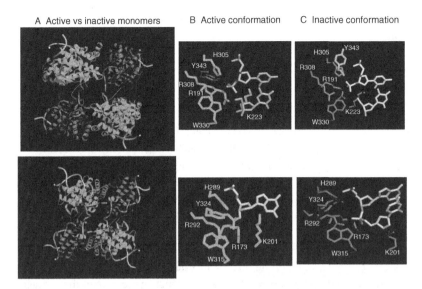

A Active vs inactive monomers   B Active conformation   C Inactive conformation

Figure 2.5. Active site configurations in Flp and Cre. The structural data for the recombination complexes formed by the Flp tetramer (**A**, *top row*) and the Cre tetramer (**A**, *bottom row*) are adapted from Chen et al. (2000) and Guo et al. (1997). The "active" monomers bound adjacent to the phosphodiester bonds that have been oriented for cleavage are shown in green; the "inactive" monomers are shown in magenta. The configurations of the active and inactive active sites are shown in panels **B** and **C**, respectively. The DNA and the scissile phosphate are represented in yellow. The primary distinction between the active and inactive states is in the orientation of the tyrosine nucleophile. In Flp, this tyrosine is provided in *trans* by the inactive monomer (see also Fig. 2.6).

by these enzymes for DNA cleavage and joining. From sequence alignments and functional characterization of site-directed mutants, it was apparent that two invariant arginines and a highly conserved histidine form an important catalytic triad (Fig. 2.5; Table 2.1), with an invariant tyrosine providing the cleavage nucleophile. These residues are Arg-191, His-305, Arg-308, and Tyr-343 in the Flp recombinase, and correspond to Arg-I, His/Lys-II, Arg-II, and Tyr in the more generalized nomenclature proposed by Van Duyne (2002). The histidine of the triad is replaced by a lysine in the two topoisomerases (Lys-220 and Lys-587 in the vaccinia and human enzymes, respectively). In the Cre and Flp structures, the triad residues are hydrogen bonded to the scissile phosphate. In the active conformation of the active site, the tyrosine nucleophile is ideally positioned for inline attack. In the inactive conformation, though, the tyrosine is displaced from its active configuration.

Table 2.1. *Conserved catalytic residues in Tyrosine recombinases and type Ib topoisomerases*

Recombinase/ topoisomerase	catalytic residues					
	Arg-I	Lys-β	His/Lys-II	Arg-II	His/Trp-III	Tyr
λ Int	Arg-212	Lys-235	His-308	Arg-311	His-333	Tyr-342
Cre	Arg-173	Lys-201	His-289	Arg-292	Trp-315	Tyr-324
Flp	Arg-191	Lys-223	His-305	Arg-308	Trp-330	Tyr-343
Vaccinia topoisomerase	Arg-130	Lys-167	Lys-220	Arg-223	His-265	Tyr-274
Human topoisomerase Ib	Arg-488	Lys-532	Lys-587	Arg-590	His-632	Tyr-723

Note: The nomenclature of the conserved residues as Arg-I, Lys-β, His/Lys-II, Arg-II, His/Trp-III is according to Van Duyne (2002). Tyr represents the cleavage nucleophile.

Recall that recombination proceeds in two steps by single-stranded exchanges, and only two of the active sites are functional at each step. This misorientation of tyrosine is sufficient to explain the inactive state of the active site, at least for Flp. When an exogenous small nucleophile such as hydrogen peroxide is supplied as a tyrosine mimic, strand breakage can occur at the scissile phosphate adjacent to the inactive Flp monomer (Jayaram et al., 2002; Lee et al., 1997).

The crystal structures revealed two additional catalytic residues, Lysβ (Lys-223 in Flp) and His/Trp-III (Trp-330 in Flp; histidine in the topoisomerases and in the large majority of the recombinases). Lysβ, so named because of its location between two adjacent β strands in Cre, contacts the scissile phosphate in the active Flp monomer, and forms a minor groove hydrogen bond in both Flp and Cre: with the adjacent base on the cleaved strand in Flp and on the noncleaved strand in Cre. In the inactive Flp and Cre monomers, though, the lysines are positioned quite differently. In Flp, the movement of this residue is modest, whereas in Cre it is pulled quite a distance away from the active site. Experiments with vaccinia topoisomerase using 5′-phosphorothiolate containing substrates suggest strongly that Lysβ functions as the general acid during cleavage by stabilizing the 5′-hydroxyl leaving group (Krogh and Shuman, 2000). Although there is no direct evidence, a similar function for Lysβ during recombination seems eminently plausible. The His/Trp-III residue forms a hydrogen bond to the scissile phosphate via the indole nitrogen in Cre and via the histidine side chain in the topoisomerases. In Flp, though, the tryptophan is closer to the 5'-hydroxyl leaving group in the active monomer and is farther displaced from DNA in the inactive monomer.

In summary, a set of conserved side chains in the tyrosine recombinase active site provides an array of hydrogen bonds to stabilize the pentacoordinate state of the scissile phosphate within the transition state. This feature has prompted the suggestion that the recombination mechanism does have an electrophilic component to it (Van Duyne, 2002). It is not unlikely that at least some of these features for phosphate orientation and transition state stabilization may occur within the serine recombinase active site as well. Assuming that Lysβ is the general acid during the cleavage step of tyrosine recombination, by the principle of reversibility, the same lysine may act as the general base during strand joining. What might be the general base and acid during strand cleavage and joining, respectively? The Cre structures suggest that His-II is positioned ideally to act as a proton acceptor during cleavage and a proton donor during joining. However, although mutations of His-305 in Flp show a large effect on strand joining, there is little or no effect on cleavage (Pan et al., 1993; Parsons et al., 1988).

SHARED AND NONSHARED ACTIVE SITES: DNA CLEAVAGE IN *CIS* VERSUS *TRANS*

Because recombination requires the breakage of four DNA strands and their exchange with a fixed directionality, it would be prudent to ensure the cleavage reaction is not initiated until a fully functional reaction complex has been assembled. At least, there should be some mechanism for restoring intact DNA strands in the event of aberrant or uncoordinated cleavage reactions, especially for those recombinases that make double-stranded DNA breaks. These safeguards for the serine and tyrosine recombinases come into play at the level of the reaction chemistry itself and at the level of the high-order DNA–protein complex within which this chemistry is triggered.

The utilization of transesterification rather than hydrolysis as the mechanism for cleavage is significant. In the latter case, the energy of the phosphodiester bond will be lost, and the products will be free phosphate and hydroxyl groups. To reform the bond, assistance from a DNA ligase and a high-energy co-factor such as ATP or NAD will be required. In contrast, the transesterification mechanism permits the ready reversal of the cleavage reaction and the resealing of the broken strand. Note that recombination, in effect, is the reversal of cleavage executed between the cleaved strands of two DNA partners.

A monomer of the Flp recombinase harbors an incomplete active site. A functional active site is a shared one, assembled at the interface of two monomers bound to the left and right arms of a recombination target site (Chen et al., 1992; Lee et al., 1999). Within a DNA substrate bound by two Flp molecules, one monomer orients the adjacent scissile phosphate for cleavage almost entirely by itself (activation in *cis*). The second monomer then provides the tyrosine nucleophile in *trans* in a precisely oriented fashion to complete the active site assembly (Fig. 2.6). This mode of active site assembly is accommodated through an oriented DNA bend induced by interactions between the Flp monomers and centered within the spacer region. The geometry of the bent complex is such that tyrosine at the second scissile phosphate (also supplied in *trans*) becomes misoriented. Thus, a dimer of Flp can assemble only one of the two possible active sites at a time. Two equivalent but oppositely oriented DNA bends can alternatively activate one or the other active site of a Flp dimer. This "half-of-the-sites" activity accounts for the two-step, single-stranded exchange mechanism for recombination.

In an active synaptic complex, composed of two DNA partners bent identically by the bound Flp dimers, one pair of strands can be cleaved and exchanged to form a Holliday junction. The reactive Flp dimers during this reaction are those associated with each DNA substrate. A limited readjustment of the DNA arms of the Holliday intermediate and the bound Flp subunits (isomerization) can induce the geometry required for its resolution by exchange of the second pair of strands. During this step, the active Flp dimers are those associated with the left arm of one substrate and the right arm of the second substrate.

Unlike Flp, Cre, lambda Int, and XerC/XerD normally cleave DNA in *cis* using active sites assembled entirely within one monomer (Blakely and Sherratt, 1996; Guo et al., 1997; Nunes-Duby et al., 1994). Nevertheless, the allosteric activation of a monomer by its neighbor and the manifestation of half-of-the-sites activity within the tetrameric protein assembly appear to be a common attribute of all these recombinases. A comparison of the crystal structures of Flp and Cre vividly underscores this aspect. For Flp, the helix M (housing the catalytic tyrosine and donated in *trans*) is well ordered in the active pair of monomers and places the tyrosines in their attacking position. In the inactive pair, the same helix is more disordered, and the tyrosines are displaced by almost 10 Å from their target phosphates (Chen et al., 2000). A similar displacement of the tyrosines is observed in the inactive pair of Cre

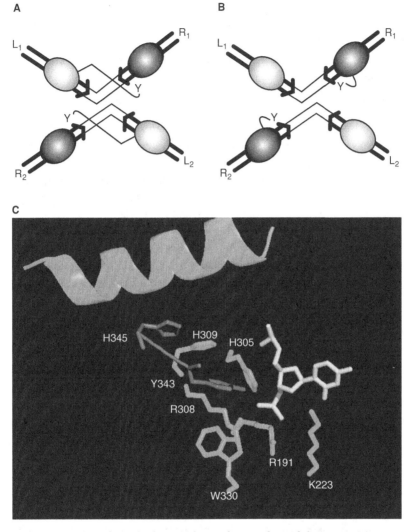

Figure 2.6. Bending of DNA by the recombinase dimer and strand cleavage in *cis* or *trans*. **(A)** In a schematic representation of the Flp synapse, two identically bent DNA substrates, each bound by a Flp dimer, are shown. The two "active" monomers are the darkly shaded ovals and are diagonally related to one another. Each receives the tyrosine nucleophile in *trans* from the partner monomer ("inactive") bound on the same DNA substrate. **(B)** The active interface in the Cre synapse is also formed between the two Cre monomers bound on the same substrate molecule. However, the active monomer provides the tyrosine nucleophile in *cis* for strand cleavage. **(C)** The trans delivery of Tyr-343 by Flp from one monomer (purple) to its partner (green) is facilitated by the juxtaposition of His-345 between helix E and His-309, which also stacks nicely above Tyr-343 to orient it correctly with respect to the scissile phosphate (represented in yellow).

monomers, even though the M helix in this case acts in the *cis* configuration (Guo et al., 1997). Interestingly, the M helix is better ordered in the inactive Cre monomers than in the active ones.

There is no evidence for a shared active site for the serine recombinases, and they cleave DNA in *cis* (Boocock et al., 1995). The presence of amino acids from one monomer, Glu-124 in particular, in close juxtaposition with the active site residues from a neighboring monomer in the crystal structure of resolvase initially suggested the prospect of an active site contributed by two resolvase monomers. However, as noted previously, Glu-124 is not a catalytic residue (Grindley, 2002). For the serine recombinases and a subset of the tyrosine recombinases, selection of a specific synapse topology is a key event in coordinating and regulating the subsequent chemical steps of recombination. We discuss this aspect in some detail later.

Regardless of whether the active site is shared and cleavage per se occurs in *cis* or *trans*, the activation of a quiescent active site by interprotomer interactions may be more or less a general feature among recombinases and related enzymes with a multisubunit composition. Mutations in XerC and XerD that likely destabilize helix M packing activate the mutated monomers within the tetramer for catalysis; these mutations also stimulate a topoisomerase-like DNA relaxing activity of a monomer (Arciszewska et al., 2000; Hallet et al., 1999). The length of the peptide linker between helix L and helix M appears to be important in the determination of *cis* or *trans* cleavage. A longer linker, present in Flp and related yeast site-specific recombinases, makes it possible for helix M (and the catalytic tyrosine) to reach the neighboring active site. As was proposed by Gopaul and Van Duyne (1999) and corroborated by the Flp crystal structure, a localized switch in the peptide connectivity can transform *cis* cleavage into *trans* with little change in the global configuration of the active site (Fig. 2.7). A simple analogy can be made to a jigsaw puzzle that gives the same final solution from two different sets of pieces.

EXCHANGING THE CLEAVED STRANDS

How are the cleaved strands correctly oriented across partner substrates for the joining reaction under the structural and dynamic constraints of the recombination complex? How extensive are the movement of the DNA and protein components during this process? What are the associated reconfigurations in DNA–protein and protein–protein contacts? The structures of Cre and Flp recombination complexes provide satisfactory answers to these questions for the tyrosine recombinases. The situation is less clear for the serine recombinases.

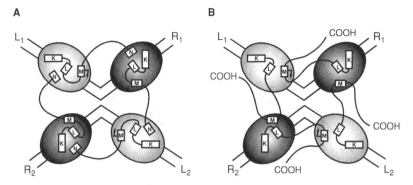

Figure 2.7. Switch between *cis* and *trans* cleavages. A rather simple change in the peptide connectivity, between the L and M helices, can account for the transition from *cis* cleavage **(A)** to *trans* cleavage **(B)**. The bent arrows indicate the active tyrosines poised for cleavage. For more details, see Van Duyne (2002) and Rice (2002).

The length of the exchange segment (spacer) during tyrosine recombination is in the 6 to 8 bp range. Although the exact spacer sequence is not critical under most circumstances, recombinant products are formed only between substrates with identical spacers. Either the exchange reaction is aborted and reversed when a mismatch is sensed, or an additional round of recombination of the mispaired products restores the parental configuration and normal base pairing. For a long time, it was believed that the Holliday intermediate branch migrates through the entire spacer region during the isomerization step. Thus, every base pair can be sampled sequentially, and the resolution step can be proscribed when this homology test fails. However, the associated rotation of the DNA arms would result in considerable disruption of protein–protein and perhaps DNA–protein contacts. Even if the broken contacts are replaced by equivalent new ones, and the initial and final forms of the Holliday junctions are isoenergetic, the activation barrier for this mode of isomerization would be considerable. Later experiments suggested that strand exchange takes place by segmental swapping of spacer nucleotides, approximately 3 from the left and right ends, following each cleavage event (Lee and Jayaram, 1995; Nunes-Duby et al., 1995; Voziyanov et al., 1999). This would eliminate the need for branch migration for a 6 bp spacer system, Cre, for example. For systems with 7 and 8 bp spacers, such as Int and Flp, branch migration will be limited to the central core of 1 and 2 bp, respectively.

Within the central cavity of the Cre crystal structure (Guo et al., 1997), each of the two cleaved strands is seen to have partially peeled away from

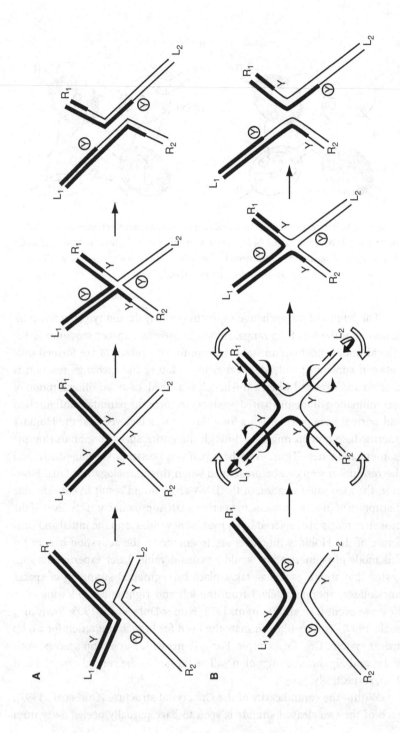

the parental complementary strand and positioned itself to pair with the complementary strand from the partner substrate. Results from modeling suggest that the displaced strand can form two Watson-Crick bp adjacent to the tyrosine-linked phosphate with no change in the quarternary structure of the remainder of the DNA–protein assembly. With modest additional movement of the DNA, a third bp can also be formed. The 5′-hydroxyl group of the exchanged spacer segment will now be correctly poised for strand joining. Following formation of the Holliday junction, it takes only limited rearrangement of the DNA arms, by a scissoring motion, and the associated protein subunits to establish the configuration for the second "cleavage-3 nt swap-joining" step to be carried out (Fig. 2.8A).

The Flp structure suggests that it can also accommodate a strand exchange mechanism that is more or less similar to that proposed for Cre (Chen et al., 2000; Rice, 2002). However, the isomerization of the Flp–Holliday junction complex requires more extensive DNA motions in addition to simple scissoring, including rotations that force the DNA arms into or below the plane of the junction and displace each protein monomer through ∼ ±15 degrees about the axis of the duplex to which it is bound (Fig. 2.8B). As a result, each monomer is translated by ∼1.5 Å along the duplex axis. The direction of rotation is clockwise for one pair of diagonally related Flp monomers, whose catalytic centers were responsible for the first exchange step (Fig. 2.8B); it is counterclockwise for the corresponding second pair whose catalytic centers

Figure 2.8. Isomerization of Holliday intermediate during tyrosine recombination. (A) The isomerization of the Holliday intermediate is schematically shown for the Cre reaction pathway. The cleavage-competent tyrosines are denoted by the circles drawn around them. The arms of the Holliday intermediate undergo a scissoring motion to activate the two tyrosine nucleophiles for the resolution step. (B) In the Flp reaction pathway, the conformational changes commensurate with isomerization are more elaborate. These are composed of the translation of each Flp monomer along the DNA axis through ∼1.5 Å (indicated by the thin arrows pointing toward or away from the center of the Holliday junction) and three types of rotations (shown by the distinct curved arrows) pivoting about the scissile phosphate. The scissoring movement of the DNA arms (∼7 degrees per arm) is coupled to their movement into or below the junction plane (∼15 degrees for each arm) and a rotation of ∼15 degrees around the DNA axis. The sense of rotation for the diagonally related pair of arms that move into the junction plane is clockwise (as one looks from the end of an arm toward the junction center). For the pair of arms that move below the plane, and whose scissile phosphates then become cleavage competent, the rotation is counterclockwise. A more incisive discussion of Holliday junction isomerization is given in Van Duyne (2002) and Rice (2002).

will be activated for the resolution step following isomerization. These are contrary to the direction of limited branch migration proposed by Voziyanov et al. (1999). This discrepancy may be related to the particular assumptions of the model, in which the pair of active tyrosines (and their target phospho-diesters) are placed closer to each other than the inactive pair. For activation, the inactive tyrosines have to be moved toward each other and toward the center of the Holliday junction. The crystal structure reveals the spacings to be the other way round, requiring tyrosine movement in the opposite direction. In any case, the potential disruption of protein contacts inherent in the branch migration model makes it somewhat less appealing than the spacer swapping model.

The question of how strands are exchanged during serine recombination has remained a matter of debate for quite some time now. Two rather disparate models have been proposed; each has its strong points and weaknesses. In the subunit exchange model (Fig. 2.9A), the DNA is placed outside the protein scaffold. Strand cleavage followed by right-handed rotation through 180 degrees brings the broken DNA ends in the strand joining configuration, and in the process, exchanges the resolvase subunits. An attractive feature of this model is that it can account for iterated rounds of recombination (rotations of 180, 360, 540 degrees, etc.) without dissociation of the synapse. Resolvase, as does the Gin or Hin invertase, carries out processive recombination at a low but readily detectable frequency. The resulting products have been instrumental in revealing the topology of recombination (see below). The unattractive feature of subunit exchange is that it requires the complete disruption of the dimer interfaces between the two resolvase subunits bound to each of the two *res* I sites. To overcome this difficulty, a domain swap model has been proposed by Martin Boocock (Grindley, 2002). The two cleaved half-sites and the attached amino-terminal catalytic domains are believed to be rotated as a rigid unit about a peptide loop as a hinge, leaving the carboxyl-terminal domains and much of the original dimer interactions as they were. The problem, though, is how to avoid the intertwining of the rotational hinges during processive recombination. The alternative to the subunit exchange model proposed by Rice and Steitz (1994a) places the *res* I DNA inside a fairly rigid protein scaffold formed by the resolvase tetramer (Fig. 2.9B). It has features reminiscent of the strand exchange mechanism revealed by the Cre and Flp structures. Its main strength is in avoiding the large-scale disruption of the protein contacts necessitated by the subunit exchange mechanism. However, the plausible motions of the cleaved DNA ends during joining pose contradictions with the experimentally observed topology of recombination (Rice and Steitz, 1994a; Stark et al., 1989). When the DNA crossing resulting from

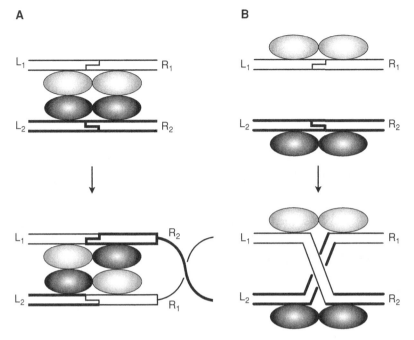

Figure 2.9. Strand exchange mechanisms for serine recombination. **(A)** In the subunit exchange model, there is a relative rotation of the broken half-sites at the right (and the associated resolvase monomers) with respect to those at the left. Note how, in the recombination products, the resolvase subunits have exchanged their partners. **(B)** In the DNA inside model, the outer protein scaffold is essentially stationary. The cleaved DNA strands move inside the scaffold to form recombinants.

strand exchange has the correct handedness (a node with a positive sign; see the "Topology and geometry of site-specific recombination" section), the two half twists introduced in the DNA have a negative sign (the correct sign is positive). When the half twists are made positive, the DNA crossing will become negative (and will have the wrong handedness). To circumvent this problem, it has been proposed that the center of each *res* I site within the synapse is underwound by half a turn. Relieving the torsional stress during exchange would produce the appropriate DNA rotations (Rice and Steitz, 1994a). This refinement does not help. Not only is there no evidence for DNA unwinding in the subsequently solved resolvase–*res* I crystal structure (Yang and Steitz, 1995), but also the revised model fails to accommodate even one 360-degree rotation without having to reset the synapse. The subunit exchange model and the DNA inside model (without subunit exchange) are thus tainted by energetic penalties for breaking large protein interfaces on the one hand and nagging discrepancies with experimental observations on the other. In short, the strand exchange mechanism for serine recombinases remains more or less an enigma. Because of its multiple unappetizing aspects, the DNA inside model is the least favored among the three models. It is important to clarify that the subunit exchange or domain exchange model will work equally well with the DNA on the inside or the outside. The distinction between the opposing models is whether DNA alone or DNA with the associated protein subunit (or protein domain) is exchanged.

TOPOLOGY AND GEOMETRY OF SITE-SPECIFIC RECOMBINATION

Analyses of the biological effects of DNA topology and, in particular, the topological consequences of site-specific recombination, were pioneered by Cozzarelli, White, and colleagues (Cozzarelli et al., 1984, 1990; Wasserman and Cozzarelli, 1986). More recently, the application of tangle calculus by Sumners, Cozzarelli, and colleagues (Sumners et al., 1995) has further enhanced the power of topology as a tool in deriving recombination mechanisms. The rationale is to first experimentally determine the topology of the DNA knots or links (catenanes) produced by a recombinase enzyme. The product topology is then subjected to mathematical analyses to deduce the nature of DNA crossings present within the recombination synapse, as well as those introduced by the recombinase enzyme in the process of strand exchange.

The serine recombinases require negatively supercoiled substrates, are selective with respect to the orientation of the recombination sites, and act only on sites located on the same DNA molecule. The complete resolvase recombination site is composed of three subsites: *res* I, *res* II, and *res* III. Two such sites, oriented head to tail, are the targets for resolvase recombination. Each of the three subsites is bound by a resolvase dimer; however, only the dimers bound at *res* I carry out strand cutting and exchange. Interactions of resolvase dimers present at the accessory sites (*res* II and *res* III) are responsible for arranging a functional synapse at the *res* I sites. The Hin or Gin invertase binds as a dimer to its target site *hix* or *gix*, respectively, and mediates recombination between two *hix* or two *gix* sites only when they are in head-to-head orientation. The reaction requires the *E. coli* Fis protein as an accessory factor. Interaction of a Fis dimer bound at its cognate site with the *hix* bound Hin dimers (or the *gix* bound Gin dimers) is responsible for organizing the active recombination complex. The topological analyses of serine recombination is aided by the fact that they do yield processive recombination (more than one round or reaction from the same

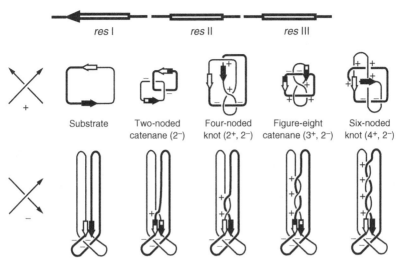

Figure 2.10. The resolvase synapse and the products of processive recombination. The organization of the res site is schematically diagrammed at the top. Whereas resolvase dimers bound at the res II and res III sites are important for organizing the recombination complex, the active recombinase is the dimer bound at res I. The topology of the resolvase recombination conforms to a three-noded synapse (3⁻), with a right-handed DNA swap during each exchange event. The res sites are shown by the unfilled and filled arrows in the substrate so an odd number of recombinations results in the "hybrid" arrows. To assign signs to the DNA crossings, the DNA axis in the substrate circle is assigned an arbitrary, fixed direction. The conventions for the + and − signs are shown at the left. The continuous and broken arrows correspond to the "overlying" and "underlying" axes segments, respectively. During the experimental procedures, supercoils are removed by topoisomerase I (or DNAse I nicking), leaving behind only the knot and catenane nodes in the recombination products. (Adapted from Grindley, 2002.)

synapse), so a series of sequential recombination products are available for characterization.

The product of the first round of recombination by resolvase is a two-noded 2 catenane (2^-): two linked DNA circles with two negative crossings (Fig. 2.10). The subsequent rounds of recombination yield a four-noded knot ($2^+, 2^-$), a figure eight catenane ($3^+, 2^-$), and a six-noded knot ($4^+, 2^-$), respectively. These product configurations are only compatible with a unique synapse in which three negative crossings are trapped by the resolvase dimers bound at res II and res III, the res I sites are arranged in a parallel orientation, and each DNA exchange step results in one positive crossing (Grindley, 2002). The sequential products for the Hin/Gin reaction are an unknotted inversion

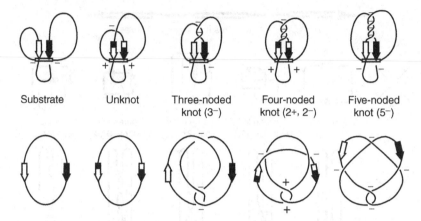

Substrate   Unknot   Three-noded knot (3⁻)   Four-noded knot (2⁺, 2⁻)   Five-noded knot (5⁻)

Figure 2.11. The Gin inversion synapse and the products of processive recombination. The synapse topology consists of two DNA crossings, and strand exchange follows a right-handed rotation as in the resolvase reaction. Because each recombination event inverts the DNA segment between the *gix* sites (changing the direction of the DNA axis), the synaptic nodes change in sign with each recombination event. For the same reason, the DNA crossings resulting from strand exchange have − signs as opposed to the + signs in the resolvase reaction. (Adapted from Johnson, 2002, and Kanaar et al., 1990.)

circle, a three-noded knot with all negative crossings (3^-), a four-noded knot with two positive and two negative crossings $(2^+, 2^-)$, and a five-noded knot with all negative crossings (5^-) (Fig. 2.11). The only synaptic configuration that accommodates the previous product topologies contains two trapped DNA crossings with a parallel orientation of the recombination sites (Johnson, 2002; Kanaar et al., 1990). Because each round of recombination causes relative inversion of the DNA segment between the *hix* or *gix* sites, the sign of the synaptic nodes alternates between − and +, as illustrated in Fig. 2.11. Each act of strand exchange introduces a right-handed crossing, as in the resolvase case, but with a negative sign. To generalize, the synapses arranged by the serine recombinases have a characteristic topological signature. These enzymes impose a parallel geometry on the recombination partners and carry out recombination with DNA rotation in the right-handed sense.

Topological analysis with tyrosine recombinases is much less straightforward than with their serine counterparts. They do not, in general, yield processive recombination. Furthermore, recombinases such as Cre and Flp have no topological restrictions with regard to substrate, are indifferent to site orientations, and are proficient at intermolecular reactions. Nevertheless, the known topology of the −3 resolvase synapse has been exploited to reveal the

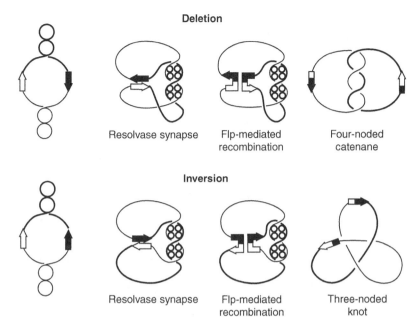

Deletion

Resolvase synapse   Flp-mediated   Four-noded
                            recombination   catenane

Inversion

Resolvase synapse   Flp-mediated   Three-noded
                            recombination   knot

Figure 2.12. Flp recombination from a preassembled resolvase synapse. In the *res–FRT* hybrid site, the *FRT* sites replacing *res* I are shown by the unfilled and filled arrows. Whereas the *FRT* site binds a Flp dimer, the *res* II and *res* III sites bind resolvase dimers. Flp-mediated deletion (*top*) or inversion (*bottom*) is performed after assembling the resolvase synapse. Further details can be found in Grainge et al. (2000).

geometry of the Flp and Cre target sites (*FRT* and *loxP*, respectively) in their respective synapses (Grainge et al., 2000, 2002; Kilbride et al., 1999). The underlying assumption is that the sites are essentially planar during recombination, one that was initially justified by the structure of the Cre synapse (Guo et al., 1997) and later supported by the structure of the Flp synapse (Chen et al., 2000). We describe here the experiments with Flp, but the results also hold true for Cre. The trick is to construct a hybrid *res–FRT* site by replacing the *res* I subsite with *FRT* but retaining *res* II and *res* III in their normal positions (Fig. 2.12). In a supercoiled plasmid containing two such sites with the normal orientations of *res* II–*res* III, the characteristic synapse with the three negative crossings can be assembled by the addition of resolvase. Subsequently, recombination is performed at the *FRT* sites using Flp, and the product topology is examined. When the *FRT* sites are in head-to-head orientation, the inversion reaction yields a three-noded knot with + signs. When the *FRT* sites are in head-to-tail orientation, the deletion reaction gives a four-noded catenane. The results are consistent with an antiparallel geometry for the *FRT* sites within the Flp synapse, with no DNA crossing being added during recombination (Fig. 2.12). The three external DNA crossings will synapse the head to head *FRT* sites in the antiparallel mode. Recombination preserves the three nodes in the inversion knot. For the head-to-tail *FRT* sites, a fourth node from the negatively supercoiled DNA must be trapped to orient them in the antiparallel fashion. Hence, recombination results in the four-noded catenane. The antiparallel geometry of the *FRT* and *loxP* sites deduced by the topology method agrees with the arrangements of these sites in the Flp and Cre crystal structures, respectively.

The fact that Flp and Cre perform recombination without introducing a DNA crossing offers a simple method of deriving the number of DNA crossings trapped in an "unknown" synapse, say, that formed by a recombinase or a transposase. In principle, the analysis is applicable to any high-order DNA–protein complex in which two DNA sites make a fixed number of plectonemic wraps with each other. As described with resolvase, the unknown synapse is first assembled in two matched substrates that differ only in the relative orientation of the recombination sites (head to head in one case and head to tail in the other), after which the deletion and inversion reactions are carried out. The prediction is that the number of crossings in the inversion knot and those in the deletion catenane will differ by one. The smaller of the two numbers then gives the DNA crossings within the unknown synapse. The additional crossing in one of the products reflects the need to keep the sites antiparallel for recombination. A more recent triumph of this "difference topology" method is the mapping of the DNA path within a complex that carries out a transposition reaction (Pathania et al., 2002).

The XerC/XerD recombinase, although a tyrosine recombinase, is reminiscent of resolvase in the manner in which it assembles the recombination synapse at plasmid target sites. In addition to the two recombinase proteins and the sites involved in strand exchange, two accessory proteins, PepA and ArgR, and their cognate sites are required for the reaction. The synapse organized by PepA–ArgR traps three negative supercoils, as does the resolvase synapse. The normal XerC/XerD reaction occurs between two head-to-tail sites in negatively supercoiled substrates. To be consistent with the crystal structure of Flp and Cre and the topological outcomes of the Flp and Cre reactions, we may assume sites at which DNA exchange occurs are also antiparallel for XerC/XerD. Hence, the functional synapse must contain one more crossing, or a total of four negative crossings. The predicted recombination product, a four-noded catenane (4^-), matches the experimental outcome (Barre and Sherratt, 2002; Colloms et al., 1997).

Tackling the topology of the reactions catalyzed by the lambda integrase has been a particularly daunting task. The problems go deeper than the lack of processive recombination and the random entrapment of supercoils during synapsis that then contribute to product topology. There is a wide spectrum of Int-mediated reactions, both intra- and intermolecular, and among the latter, inversion and deletion reactions. Furthermore, the nature of the recombination substrates, restrictions regarding their topological state, and the requirement for accessory proteins vary among these reactions. The integration of the lambda genome into the bacterial chromosome catalyzed by the Int protein requires a complex phage attachment site (attP) that has to be supercoiled and a much simpler bacterial attachment site (attB) that may or may not be supercoiled. The integration places two novel recombination sites attL and attR at the left and right borders of the integrant. The excision of the phage is mediated by Int in cooperation with a second phage-coded protein Xis. The attP DNA contains multiple protein-binding sites: the "arm" and "core" type binding sites for Int as well as binding sites for Xis and for the *E. coli* proteins IHF (integration host factor) and Fis (a protein we alluded to in the context of Hin/Gin inversion). For details of the organization of the recombination sites on which the Int protein acts and the types of recombination they partake in, the reader is referred to a more recent review by Azaro and Landy (2002).

Crisona et al. (1999) rose to the challenge posed by lambda Int. The sufficiently initiated reader is encouraged to study their paper describing a series of tour de force experiments and their interpretations. By using tangles to mathematically analyze the data, they derive a uniform model for Int recombination. They find that all Int reactions, even in the absence of supercoiling, are chiral. Each reaction yields only one of the two possible enantiomers of the product. For example, imagine that one examines the fraction, albeit small, of three-noded inversion knots formed from a relaxed or nicked circular substrate in a particular recombination reaction. Because the random occurrence of a right- or left-handed crossing is equally likely in such a substrate, one would expect to encounter knots with + crossings and − crossings with equal probability. Crisona et al. found the opposite: abundance of one knot type and the near exclusion of the other. They propose that this chirality results from a right-handed DNA crossing (equivalent to a negative supercoil or a + tangle) within or between recombination sites inside the synapse that favors the formation of a right-handed Holliday intermediate (Fig. 2.13). The differences between integrative and excisive recombinations that they observe can be accounted for by additional negative crossings trapped in the synaptic complex of the former. From a more limited set of reactions with

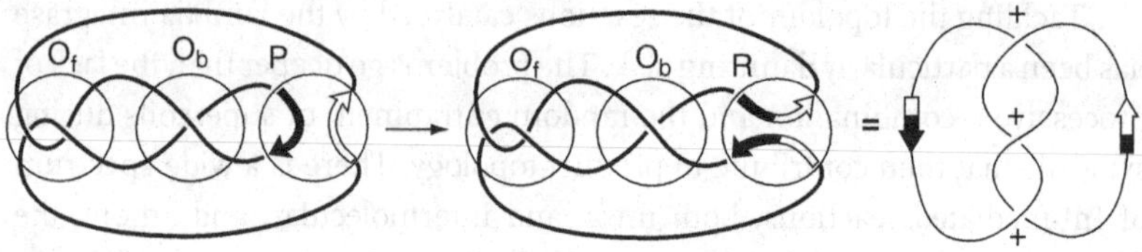

Figure 2.13. The tangle representation for Flp or Cre inversion reaction from a nicked circular substrate. The O_b and O_f tangles represent the recombinase-bound and free DNA, respectively. The sites are aligned in the antiparallel fashion in the parental tangle P (∞), which after recombination is converted to the R tangle (0). O_b is a +1 tangle corresponding to the right-handed crossing introduced by the bound protein. If two additional right-handed crossings are entrapped in the free DNA portion of the substrate, the recombinant product will be a three-noded knot with + crossings (3^+). If no crossings are trapped in the free DNA, the product will be an unknotted circle. For more details, see Crisona et al. (1997).

Cre and Flp, Crisona et al. generalized that the Int chirality applies to all tyrosine recombinases. However, the Cre and Flp crystal structures that together represent the uncleaved, cleaved, and Holliday intermediate states of the recombination complex are almost perfectly planar, and thus, do not reveal an obvious chirality.

EVOLUTION OF SITE-SPECIFIC RECOMBINATION

As we indicated at the beginning of this chapter, the origins of homologous recombination may perhaps be traced to the need for repairing DNA damages that pose impediments to genome replication. The beginnings of site-specific recombination, in contrast, must be rooted in the need to bring about precise genetic consequences through localized action on the genome. This fundamental distinction in physiological ends is reflected in the differences in enzymology that the two systems use and the reaction pathways they follow. By targeting two specific phosphodiester bonds within a defined short DNA stretch and using transesterification chemistry for first breaking them and then reforming them in their recombined configuration, a conservative site-specific recombinase performs a desired DNA rearrangement rapidly and decisively. It leaves no loose DNA ends to be processed and tied up by other enzymes. As such, the recombination active site must couple elementary chemical mechanisms for strand cutting and joining to the conformational transitions required for exchanging strands between two double helical partners. Given the rather limited number of chemically useful functional groups available to biological catalysts, the retention of key catalytic motifs among nucleases, topoisomerases, recombinases, and other enzymes that catalyze phosphoryl transfer reactions in nucleic acids is to be expected. The utilization of the Arg-I-His/Lys-II-Arg-II cluster as a key catalytic triad by the tyrosine recombinases and type Ib topoisomerases suggests mechanistic parallels to pancreatic ribonuclease and *Staphylococcus* nuclease reactions (Fersht, 1977; Grainge et al., 2000).

A rather subtle structural similarity between the Integrase/Tyrosine recombinases and the AraC family of transcriptional activators has come to light more recently (Gillette et al., 2000; Grishin, 2000). Several of the catalytic residues of the recombinases are located within two helix-turn-helix (HTH) motifs, suggesting that an ancient DNA-binding module might have duplicated and evolved to acquire enzymatic function. The tyrosine nucleophile for DNA breakage is harbored by the second helix of the carboxyl-terminal HTH in Cre. In the structures of other strand breaking and resealing enzymes (topoisomerase I, topoisomerase II, and archaeal topoisomerase VI), the catalytic tyrosine is present in an HTH region, although its position is not strictly conserved. These examples suggest parallel, independent evolution of homologous enzymatic domains with alternative placements of the catalytic residues.

The crystal structure of the *Nae* I restriction enzyme, an endonuclease that can be converted to a topoisomerase/recombinase by a leucine-to-lysine substitution, reveals a dimer with a bidomainal organization of each monomer subunit. The amino-terminal domain and the carboxyl-terminal domain contain potential DNA-binding motifs corresponding to the endonuclease and topoisomerase activities, respectively (Huai et al., 2000). The *Nae* I structure provides support for the evolutionary divergence of DNA processing enzymes from a limited number of ancestral proteins or for the convergence of a finite set of catalytic motifs in them.

CRYPTIC RIBONUCLEASE AND TOPOISOMERASE ACTIVITIES WITHIN SITE-SPECIFIC RECOMBINASES

Is it possible that an elementary nuclease active site might have been the progenitor for the more refined topoisomerase active site and the even more sophisticated recombinase active site? If so, might the latter enzymes still harbor evolutionary vestiges of their humble beginnings? Potential answers to these questions are suggested by the rather unexpected RNA and DNA cleavage activities that have been unmasked in Flp, Cre, the vaccinia topoisomerase, and the human topoisomerase Ib (Sau et al., 2001; Sekiguchi and Shuman, 1997; Wittschieben et al., 1998; Xu et al., 1998).

With the help of mixed DNA–RNA substrates, two site-specific RNA cleavage activities of the Flp protein have been revealed (Xu et al., 1998). Flp–RNase I closely follows the normal recombination mechanism. It requires the active site tyrosine, targets the same phosphodiester that partakes in DNA recombination, and appears to proceed by a phosphotyrosyl intermediate. In contrast, Flp–RNase II mimics the pancreatic RNase mechanism. It is independent of the active site tyrosine and targets the phosphodiester immediately to the 3′ side of the "recombination phosphate." The type I RNase activity (but not the type II activity) has also been demonstrated in the Cre recombinase, and a related activity has been observed in vaccinia topoisomerase (Sau et al., 2001; Sekiguchi and Shuman, 1997). Furthermore, there is an intrinsic, although normally cryptic, topoisomerase activity associated with Flp. In appropriately designed DNA substrates, and under special reaction conditions, Flp can carry out relaxation of negative supercoils (Xu et al., 1998). Site-specific or even sequence-independent topoisomerase activities have also been demonstrated for Cre, XerC/XerD, and lambda Int (Abremski et al., 1986; Azaro and Landy, 2002; Cornet et al., 1997; Landy, 1989). Finally, when the catalytic tyrosine of vaccinia topoisomerase is mutated to a glutamic acid, histidine, or cysteine, the variant protein shows site-specific endonuclease activity (Wittschieben et al., 1998).

Based on the observations listed previously, one might make a case with conviction that an early step in the evolution of the recombinase must have been the formation of a nuclease active site containing a group of catalytic residues, at least a subset of which has been conserved. The emergence of a precisely positioned protein nucleophile, the active site tyrosine, could then have set the stage for the maturation of this proactive site into a "topoisomerase" and further into a "recombinase" active site. We cannot, however, discount a related but alternative evolutionary argument. During the optimization of an active site intended to perform a certain chemical task, the design for a related task may be inadvertently incorporated. According to this view, the cryptic nuclease activity housed within the recombinase or topoisomerase active site simply reflects the fact that breaking DNA strands is inherent to the problem of recombination.

The RNA cleavage reactions of recombinases and topoisomerases have also fueled speculations on the possible relevance of these activities *in vivo* (Burgin, 1997; Sekiguchi and Shuman, 1997). The suggested roles include the triggering of repair pathways for the removal of ribonucleotides from DNA or involvement in RNA splicing reactions. Regardless of whether these enzymes once carried out reactions with which they are not presently identified, their active sites exemplify how novel and diverse activities can be derived from conserved chemical themes (Sherratt and Wigley, 1998).

DNA CLEAVAGE AND TRANSFER BY PARTIALLY CONSERVATIVE MECHANISMS

For a more balanced mechanistic perspective on conservative site-specific recombination, a brief comparison with strand transfer reactions that are only partly conservative is helpful (Fig. 2.14). DNA transposases, retroviral integrases, and the RAG1/RAG2 recombinase (that triggers rearrangements to create functional immunoglobin genes, antigen-binding receptors, and T-cell receptors) first break specific phosphodiester bonds hydrolytically (the nonconservative step; Craig, 2002a). The resulting 3'-hydroxyl groups then attack their target phosphodiester bonds in the conservative step to form joint molecules, which may be further processed in a number of ways. During replicative transposition, by phage Mu, for example, the 3'-hydroxyls formed at the two ends of the transposon are transferred to two phosphodiesters in the recipient DNA that are staggered by a few base pairs (Fig. 2.14A). Extensive DNA synthesis primed by the 3'-hydroxyls formed adjacent to the donor–recipient junction followed by ligation duplicates the transposon and the short target sequence. In contrast, transposition of certain mobile DNA elements follows the cut-and-paste mechanism (Craig, 2002b; Haniford, 2002; Reznikoff, 2002). For example, Tn7 is excised from the donor site by double-stranded cuts, generating 3'-hydroxyls at the transposon ends (the result of transposase cleavage) and 3 nt long 5' flaps (the result of cleavage by a transposon-encoded restriction enzyme-like nuclease; Fig. 2.14B). After strand joining, the short gaps resulting from the stagger of the target phosphodiesters are filled, the 5' overhanging nucleotides are removed, and free ends are ligated. A variation of this mechanism is employed during retroviral integration (Craigie, 2002). The integrase clips off two terminal nucleotides from the double-stranded DNA formed by reverse transcription to produce 3'-hydroxyls at either end flanked by 5' overhangs on the opposite strand. Subsequent events, strand transfer, gap filling, and removal of overhangs are analogous to those of Tn7 transposition. In transposons Tn5 and Tn10, 3'-hydroxyls formed at the transposon termini attack the opposing phosphodiester bonds on the uncleaved strand to release the transposon with hairpin termini (Fig. 2.14C). The hairpins are opened hydrolytically to create flush ends prior to strand transfer. During the initiation of V(D)J joining by the RAG1/RAG2 recombinase, exposure of 3'-hydroxyls by cleavage

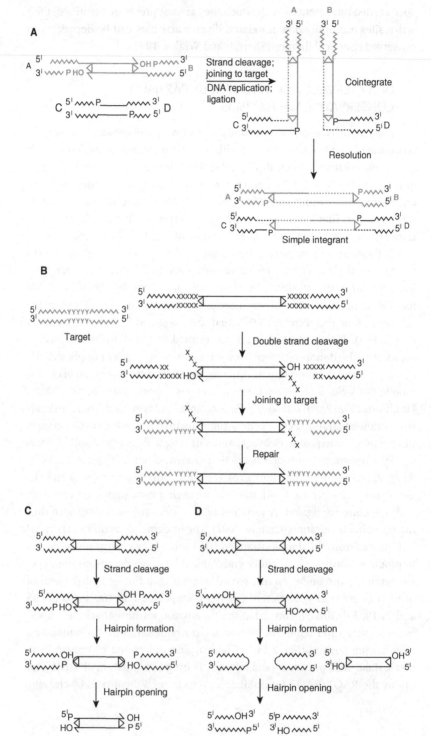

followed by hairpinning occurs at the junction of "coding" and "signal" sequences, analogous to the Tn10 reaction (Fig. 2.14D). The hairpins, though, are formed on the coding sides, and the intervening DNA with the signal ends is removed. Here also, hairpins are opened, often in an off-center mode to produce staggered ends, filled in, and joined in a manner that adds diversity to the junctions between gene segments (Gellert, 2002).

It is quite remarkable that, in several of the systems described previously, a single active site can facilitate multiple phosphoryl transfers, each time using a distinct target phosphodiester and nucleophile. In the Tn10 case, for example, there are four such steps, two carried out by hydrolysis (the first cleavage; hairpin opening) and two by transesterification (hairpinning; strand transfer to target). The conserved catalytic motif in the transposases/integrases is an Asp-Asp-Glu triad (DDE) that is believed to coordinate a divalent cation, usually Mg^{++} or Mn^{++}. Whether catalysis requires one or two metal ions per active site is still under debate. It seems quite likely that the scissile phosphate also contributes to the coordination and positioning of the metal ion(s) in the active site. Results using chiral (phosphorothioate) substrates in Tn10 reactions, together with structural information from Tn5, suggest that the orientation of a scissile phosphate in the multistep reaction differs from that of its counterpart in the previous step (Davies et al., 2000; Haniford, 2002; Kennedy et al., 2000). How do these phosphates enter and exit the active site, and how are the corresponding nucleophiles aligned correctly to attack them? We do not have the answers to

Figure 2.14. Strand exchange by partially conservative mechanisms. **(A)** Hydrolytic single-strand cleavages at the transposon ends (*triangles*), transfer of the cleaved ends to the target by transesterification and replicative repair and ligation result in the cointegrate. In this transposition intermediate, the transposon and the target sites are duplicated. Recombination within the transposon copies releases the donor DNA, leaving a simple integrant as the end product. DNA sequences flanking the transposon and the target are labeled A, B and C, D, respectively. When the donor and target DNA molecules are circular, the cointegrate is a fusion between the two circles. This can be seen by joining the A and B ends and the C and D ends. **(B)** Double-stranded hydrolytic cleavages releases the transposon, and the 3'-hydroxyl ends are joined to the target DNA by transesterification. The Xs are base pairs flanking the transposon; the Ys are base pairs that comprise the target site. **(C)** The transposon is released by single-stranded cleavages (hydrolysis), hairpin formation (transesterification), and hairpin opening (hydrolysis). **(D)** During V(D)J recombination, similar sequences of reactions as in C release the internal signal sequence (bordered by the triangles) and form hairpins at the outside coding ends. See text for further explanations of reactions **A–D**.

these intriguing questions. Nevertheless, there can be little doubt regarding the functional pliability of the transposase active site.

In addition to tranposases, integrases, the RAG1/RAG2 recombinase, and related proteins, metal-assisted transesterification is used almost universally by self-cleaving RNA enzymes (Pyle, 1993; Steitz and Steitz, 1993). One wonders why this widespread and efficient reaction mechanism has not been incorporated into the active sites of conservative site-specific recombinases. Note that transposases or ribozymes must either generate a suitable hydroxyl group in DNA/RNA (3'-OH) by hydrolysis or have one readily available to them, the 3'-OH from a guanine nucleoside or the 2'-OH from RNA, to be able to carry out transesterification. Perhaps the metal ion catalytic strategy is not efficient when the nucleophiles and leaving groups are amino acid side chains from protein enzymes.

BIOLOGICAL CONSEQUENCES OF SITE-SPECIFIC RECOMBINATION

The types of DNA rearrangements promoted by a large number of site-specific recombination systems and their physiological consequences underscores one of the fundamental attributes of life–the capacity to employ the same or similar chemical mechanisms to bring about vastly different end results (Fig. 2.15). It is almost axiomatic in biology that a solution arrived at in the context of a certain biochemical challenge is certain to be adopted and refined by evolution to be deployed in the context of a variety of related challenges.

The integration and excision reactions mediated by the lambda Int protein act as critical developmental switches in the phage's lifestyle (Fig. 2.15A). Integration leads to the quiescent lysogenic state, and excision triggers the multiplicative lytic pathway. Related site-specific recombination systems are responsible for the dissemination of integrons or bacterial genes that are usually associated with antibiotic resistance or pathogenesis (Recchia and Sherratt, 2002). Similarly, the conjugative transposons, Tn916 and Tn1545, and related genetic elements move from the genome of a donor to that of a recipient bacterium by site-specific integration (Churchward, 2002). Apparently, the "transposon" is excised as a circle, and a single strand is transferred to the recipient via conjugation, which is then recircularized in double-stranded form and finally integrated by site-specific recombination. The transposase of a conjugative transposon is in effect an integrase, and these elements are best described by their new name "constins": conjugative, self-transmissible, integrating elements. Programmed DNA rearrangements

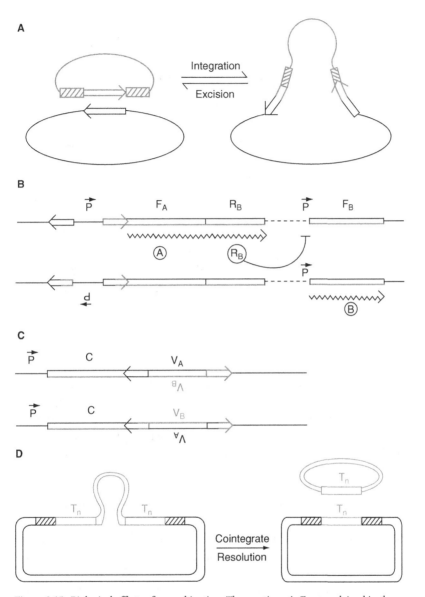

Figure 2.15. Biological effects of recombination. The reactions **A–F** are explained in the text. **(B)** F_A and F_B represent genes coding for two variant forms of a protein; R_B is the repressor for F_B. P stands for transcription promoters. **(C)** V_A and V_B are the variable regions, and C is the constant region of the two alternative forms of a protein. **(D)** T_n denotes the transposon and the hatched bars the duplicated target. **(E)** The replication origin of a bacterial plasmid/chromosome is indicated by the dot; the arrows are the recombination sites for dimer resolution. **(F)** The bars represent inverted recombination sites within the yeast plasmid. The arrows show direction of replication.

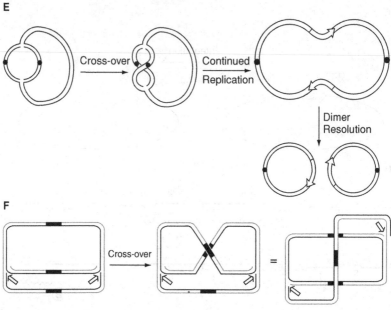

E

F

Figure 2.15 (*cont.*)

MAKKUNI JAYARAM AND IAN GRAINGE

by site-specific recombination have been described in *B. subtilis* and in cyanobacterium (Carrasco et al., 1995; Kunkel et al., 1990). During sporulation of *B. subtilis*, a relatively large element is excised in the terminally differentiating mother cell to form the functional form of the *sigK* gene (coding for a specific sigma factor). Similarly, site-specific deletions of intervening DNA elements occur in cyanobacterial heterocysts to establish the functional configurations of operons involved in nitrogen fixation.

The prokaryotic DNA inversion systems control gene expression by reversing the orientation of a promoter or, by switching protein-coding strands, (Figs. 2.15B and 2.15C). The Hin-mediated DNA inversion controls phase variation in *Salmonella* by promoter inversion. In the "on" state, the promoter drives the expression of one flagellin gene and the repressor for a second flagellin gene that is controlled by its own promoter. In the "off" state, the lack of the repressor turns on the second flagellin gene. The ability to change its flagellar antigen gives an invading *Salmonella* population the opportunity to escape the host immune system. Of course, for the inversion reaction to serve its intended purpose, the frequency of recombination has to be tightly controlled to express only one type of flagellin protein in any one *Salmonella* bacterium. The Gin recombination helps bacteriophage Mu to fuse one of two variable protein domains to a constant domain in its tail fiber protein. The variable domains specify the phage's infectivity toward two distinct set of bacterial hosts. Thus, there is a good chance that a phage population emerging from an infectious cycle will not be faced with extinction, even if one bacterial type is absent in its ecological niche.

The resolvases encoded by transposons are important in reducing the cointegrate intermediate formed during replicative transposition to a simple integrant (Fig. 2.15D). Imagine that a circular plasmid harboring Tn3/$\gamma\delta$ gains access to a bacterial cell. Transposition can coalesce the "donor" plasmid and the "recipient" chromosome into one circle containing two head-to-tail copies of the transposon (the cointegrate). The resolution event releases the donor in its original configuration. It is now free to be potentially transmitted to another bacterium, having already deposited a copy of the transposon in the genome of its current host.

69

The XerC/XerD and related recombination systems ensure the equal segregation of bacterial chromosomes and certain plasmid genomes during cell division (Fig. 2.15E). Due to homologous recombination, there is a finite probability that the duplicated circular chromosomes are present in a dimeric form. The action by XerC/XerD resolves the dimer into monomers that can then be partitioned into the daughter cells. The Cre recombinase is believed to play a similar partitioning role in the propagation of the unit copy bacteriophage/plasmid P1.

The Flp recombinase is central to the copy number control of the 2-μm yeast plasmid. In the event of a drop in copy number, a replication-coupled recombination reaction is believed to trigger the amplification process that restores the steady-state plasmid density (Futcher, 1986). Because of the asymmetric location of the replication origin with respect to the *FRT* sites, the proximal site is duplicated first by the bidirectional fork. A recombination between the unreplicated *FRT* and one of the duplicated *FRT*s inverts one fork with respect to the other (Fig. 2.15F). The unidirectional forks can spin out multiple copies of the plasmid by a bifurcated rolling circle mechanism. The tandem copies of the plasmid in the amplification product can be resolved into monomers by Flp recombination or by homologous recombination.

APPLICATIONS OF SITE-SPECIFIC RECOMBINATION

Site-specific recombination provides an efficient molecular tool for engineering large genomes. The "simple" members of the tyrosine recombinases, namely, Flp and Cre, have been effectively deployed in prokaryotes and eukaryotes to mediate site-specific DNA insertions, targeted DNA deletions,

INTRODUCTION TO SITE-SPECIFIC RECOMBINATION

and expression of proteins from selected chromosomal locales (reviewed in Jayaram et al., 2002, and Sauer, 2002). Recombination-mediated activation or abolition of gene expression has proved valuable in *Drosophila*, plants, and animals for the molecular analysis of development and for the creation of transgenics. The isolation of thermostable versions of recombinases (Buchholz et al., 1998) and the design of ligand-dependent recombination (e.g., by using a hybrid recombinase fused to the steroid hormone ligand-binding domain; Logie and Stewart, 1995) will further enhance the utility of site-specific recombination in the genetic manipulations of higher systems. One serious limitation, though, is that the recombination target site has to be first inserted, or must be fortuitously present (a highly unlikely situation), at the genomic locale of interest. The power of site-specific recombination in genetic engineering can be revolutionized if one is able to preselect a genomic site that resembles a native recombination target and coax the recombinase to acquire this new specificity. The feasibility of obtaining recombinase variants with "made to order" target specificity (Buchholz et al., 1998; Rufer and Sauer, 2002; Santoro and Schultz, 2002; Voziyanov et al., 2002) by directed evolution heralds another potentially significant advance in this direction.

AN EXPANDED AND INCLUSIVE VIEW OF RECOMBINATION

Because biological information flow proceeds from DNA to RNA (or vice versa) and RNA to protein, reshuffling of this information can (and does) occur at all three levels. There are three well-characterized mechanisms for removing intervening sequences from RNA by splicing (Cech, 1990; Fig. 2.16). Group I ribozymes use the 3'-hydroxyl group from guanosine as the nucleophile to attack the 5' splice junction (Fig. 2.16A). The 3'-hydroxyl formed by this cleavage step then attacks the 3'-splice junction to complete the reaction. The second step is mechanistically similar to the strand transfer step of transposition. In the splicing reactions that remove group II introns or introns from premessenger RNAs, the nucleophile in the first transesterification step is the 2'-hydroxyl group from a conserved A within the intron (Fig. 2.16B). Cleavage results in a 2'-5' cyclic phosphate and the 3'-hydroxyl required for the second nucleophilic attack. Completion of splicing causes the release of the intron in the form of a lariat. Introns from eukaryotic tRNAs, as well as archaeal tRNAs and rRNAs, are removed by two endonucleolytic cleavages, followed by joining of the upstream and downstream pieces by ligation (Fig. 2.16C).

Reactions have also been characterized in which RNA-mediated recombination can rearrange information at the DNA level. In one type of reaction,

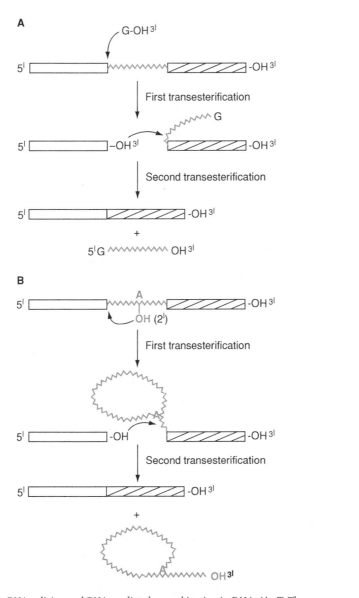

Figure 2.16. RNA splicing and RNA-mediated recombination in DNA. **(A–C)** The different mechanisms for joining interrupted exon sequences in RNA are diagrammed (see text for explanation). **(D)** An intron lariat can be spliced into DNA and then converted into the corresponding DNA sequence. **(E)** The RNA transcript from a mobile element can be incorporated at a DNA site via reverse transcription.

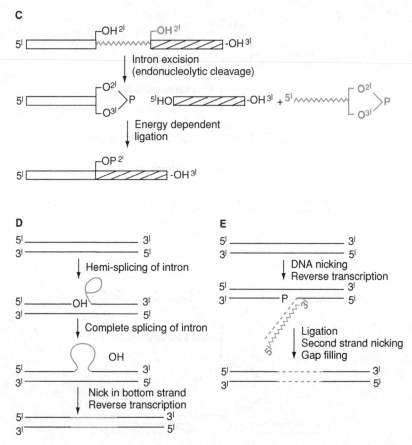

Figure 2.16 (*cont.*)

exemplified by the group II mobile introns, RNA is first spliced into the DNA and then converted into DNA in three steps, by reverse transcription, removal of RNA, and DNA replication (Belfort et al., 2002). In a somewhat related reaction, exemplified by the retrotransposition of LINE elements, the 3′-hydroxyl formed by endonucleolytic cleavage of a DNA strand is used to prime RNA-templated DNA synthesis (Fig. 2.16E). This new information is then copied into the other strand, presumably by a second cleavage event and DNA synthesis.

Finally, posttranslational editing of genetic information can occur by protein splicing, a process in which an intervening polypeptide segment (intein) is removed, and the amino- and carboxyl-terminal flanking polypeptides (exteins) are joined to form the mature protein (Perler, 1998). The

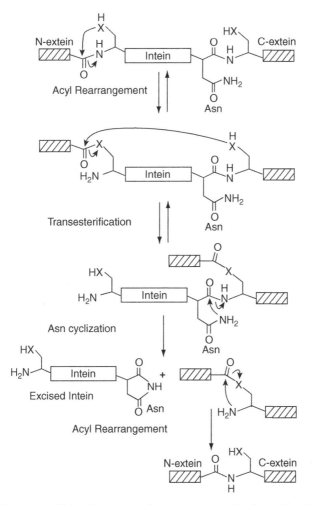

Figure 2.17. Protein splicing. The pathway for protein splicing is redrawn from Perler (1998). The nucleophile X stands for the sulfur or oxygen atom of a cysteine/serine/ threonine side chain.

general reaction mechanism involves four sequential nucleophilic displacements (Fig. 2.17). Although considered an anomaly when it was first discovered, instances of protein splicing have steadily grown into a rather substantial database called InBase (http://www.neb.com/neb/inteins.html; Perler, 2002). Interestingly, the spliced out inteins in several cases are stable proteins with specific biological activities. One example of a functional intein is a homing endonuclease, an enzyme that makes a site-specific double-stranded DNA break in an intronless or inteinless allele (Belfort and Roberts, 1997). The resulting DNA ends initiate a gene conversion event (see Chapter 1) that repairs the break by copying the mobile intron (group I) or the intein gene. Here then is a case of protein recombination serving as the trigger for homologous DNA recombination.

SUMMARY

Pathways for creating genetic diversity likely evolved by the adaptation and modification of reactions devoted to the repair of genome lesions during replication. In contrast to mechanisms aimed at global genetic manipulations, those designed for specific local DNA rearrangements also emerged, primarily among the prokaryotes. Homologous recombination and conservative site-specific recombination represent these two distinct evolutionary paradigms. In addition, a variety of genetic transactions that blend features of site specificity and globalism have been encountered in both prokaryotes and eukaryotes. The chemistry and dynamics of recombination systems reflect the structural intricacy and geometric elegance of the DNA double helix on which they operate. The conservative site-specific recombinases, in particular, have revealed the beauty of the recombination reaction in its most intimate details.

ACKNOWLEDGMENTS

We acknowledge the contributions of numerous laboratories that have advanced our understanding of DNA recombination. One of us, MJ, is grateful to past and present colleagues for their sustained efforts in revealing the secrets of the Flp recombination system. Work in the Jayaram laboratory has been supported over the years by the National Institutes of Health, the National Science Foundation, the Robert F. Welch Foundation, the Human Frontiers in Science Program, the Council for Tobacco Research, and the Texas Higher Education Coordinating Board. We are grateful to Dr. B. R. Knudsen for helpful suggestions and to Drs. N. D. F. Grindley and P. A. Rice for critical comments on selected sections of this chapter.

REFERENCES

Abremski, K., Wierzbicki, A., Frommer, B., and Hoess, R. H. (1986). Bacteriophage P1 Cre-loxP site-specific recombination. Site-specific DNA topoisomerase activity of the Cre recombination protein. J Biol Chem *261*, 391–396.

Arciszewska, L. K., Baker, R. A., Hallet, B., and Sherratt, D. J. (2000). Coordinated control of XerC and XerD catalytic activities during Holliday junction resolution. J Mol Biol *299*, 391–403.

Azaro, M. A., and Landy, A. (2002). Lambda Integrase and the lambda Int family. In N. L. Craig, R. Craigie, M. Gellert, and A. M. Lambowitz, eds. Mobile DNA II (Washington, DC, ASM Press), pp. 118–148.

Baker, T. A., and Mizuuchi, K. (2002). Chemical mechanisms for mobilizing DNA. In N. L. Craig, R. Craigie, M. Gellert, and A. M. Lambowitz, eds. Mobile DNA II (Washington, DC, ASM Press), pp. 12–23.

Barre, F.-X., and Sherratt, D. J. (2002). Xer site-specific recombination: Promoting chromosome segregation. In N. L. Craig, R. Craigie, M. Gellert, and A. M. Lambowitz, eds. Mobile DNA II (Washington, DC, ASM Press), pp. 149–161.

Belfort, M., Derbyshire, V., Parker, M., Cousineau, B., and Lambowitz, A. M. (2002). Mobile introns: Pathways and proteins. In N. L. Craig, R. Craigie, M. Gellert, and A. M. Lambowitz, eds. Mobile DNA II (Washington, DC, ASM Press), pp. 761–783.

Belfort, M., and Roberts, R. J. (1997). Homing endonucleases: Keeping the house in order. Nucleic Acids Res *25*, 3379–3388.

Blakely, G. W., and Sherratt, D. J. (1996). Cis and trans in site-specific recombination. Mol Microbiol *20*, 234–237.

Boocock, M. R., Zhu, X., and Grindley, N. D. (1995). Catalytic residues of gamma delta resolvase act in cis. EMBO J *14*, 5129–5140.

Broach, J. R., and Volkert, F. C. (1991). Circular DNA plasmids of yeasts: The molecular biology of the yeast *Saccharomyces*. Genome dynamics, protein synthesis and energetics. (Cold Spring Harbor Laboratory Press, Cold Spring Harbor, NY), pp. 279–331.

Buchholz, F., Angrand, P. O., and Stewart, A. F. (1998). Improved properties of FLP recombinase evolved by cycling mutagenesis. Nat Biotechnol *16*, 657–662.

Burgin, A. B., Jr. (1997). Can DNA topoisomerases be ribonucleases? Cell *91*, 873–874.

Carrasco, C. D., Buettner, J. A., and Golden, J. W. (1995). Programmed DNA rearrangement of a cyanobacterial hupL gene in heterocysts. Proc Natl Acad Sci U S A *92*, 791–795.

Cech, T. R. (1990). Self-splicing of group I introns. Annu Rev Biochem *59*, 543–568.

Chen, J. W., Lee, J., and Jayaram, M. (1992). DNA cleavage in trans by the active site tyrosine during Flp recombination: Switching protein partners before exchanging strands. Cell 69, 647–658.

Chen, Y., Narendra, U., Iype, L. E., Cox, M. M., and Rice, P. A. (2000). Crystal structure of a Flp recombinase-Holliday junction complex: Assembly of an active oligomer by helix swapping. Mol Cell 6, 885–897.

Cheng, C., and Shuman, S. (1998). A catalytic domain of eukaryotic DNA topoisomerase I. J Biol Chem 273, 11589–11595.

Churchward, G. (2002). Conjugative transposons and related mobile elements. In N. L. Craig, R. Craigie, M. Gellert, and A. M. Lambowitz, eds. Mobile DNA II (Washington, DC, ASM Press), pp. 177–191.

Colloms, S. D., Bath, J., and Sherratt, D. J. (1997). Topological selectivity in Xer site-specific recombination. Cell 88, 855–864.

Cornet, F., Hallet, B., and Sherratt, D. J. (1997). Xer recombination in *Escherichia coli*. Site-specific DNA topoisomerase activity of the XerC and XerD recombinases. J Biol Chem 272, 21927–21931.

Cozzarelli, N. R., Boles, T. C., and White, J. H. (1990). In N. R. Cozzarelli and J. C. Wang, eds. Primer on the topology and geometry of DNA supercoiling. In DNA topology and its biological effects (Cold Spring Harbor Laboratory Press, Cold Spring Harbor, NY), pp. 140–184.

Cozzarelli, N. R., Krasnow, M. A., Gerrard, S. P., and White, J. H. (1984). A topological treatment of recombination and topoisomerases. Cold Spring Harb Symp Quant Biol 49, 383–400.

Craig, N. L. (2002a). Mobile DNA: An introduction. In N. L. Craig, R. Craigie, M. Gellert, and A. M. Lambowitz, eds. Mobile DNA II (Washington, DC, ASM Press), pp. 1–11.

Craig, N. L. (2002b). Tn7. In N. L. Craig, R. Craigie, M. Gellert, and A. M. Lambowitz, eds. Mobile DNA II (Washington, DC, ASM Press), pp. 423–456.

Craigie, R. (2002). Retroviral DNA integration. In N. L. Craig, R. Craigie, M. Gellert, and A. M. Lambowitz, eds. Mobile DNA II (Washington, DC, ASM Press), pp. 613–629.

Crisona, N. J., Weinberg, R. L., Peter, B. J., Sumners, D. W., and Cozzarelli, N. R. (1999). The topological mechanism of phage lambda integrase. J. Mol. Biol. 289, 747–775.

Davies, D. R., Goryshin, I. Y., Reznikoff, W. S., and Rayment, I. (2000). Three-dimensional structure of the Tn5 synaptic complex transposition intermediate. Science 289, 77–85.

Feng, J. A., Johnson, R. C., and Dickerson, R. E. (1994). Hin recombinase bound to DNA: The origin of specificity in major and minor groove interactions. Science 263, 348–355.

Fersht, A. (1977). Enzyme structure and mechanism (San Francisco, W. H. Freeman).

Futcher, A. B. (1986). Copy number amplification of the 2 micron circle plasmid of *Saccharomyces cerevisiae*. J Theor Biol *119*, 197–204.

Gellert, M. (2002). V(D)J recombination. In N. L. Craig, R. Craigie, M. Gellert, and A. M. Lambowitz, eds. Mobile DNA II (Washington, DC, ASM Press), pp. 705–729.

Gillette, W. K., Rhee, S., Rosner, J. L., and Martin, R. G. (2000). Structural homology between MarA of the AraC family of transcriptional activators and the integrase family of site-specific recombinases. Mol Microbiol *35*, 1582–1583.

Gopaul, D. N., Guo, F., and Van Duyne, G. D. (1998). Structure of the Holliday junction intermediate in Cre-loxP site-specific recombination. EMBO J *17*, 4175–4187.

Gopaul, D. N., and Van Duyne, G. D. (1999). Structure and mechanism in site-specific recombination. Curr Opin Struct Biol *9*, 14–20.

Grainge, I., Buck, D., and Jayaram, M. (2000). Geometry of site alignment during Int family recombination: Antiparallel synapsis by the Flp recombinase. J Mol Biol *298*, 749–764.

Grainge, I., Pathania, S., Vologodskii, A., Harshey, R. M., and Jayaram, M. (2002). Symmetric DNA sites are functionally asymmetric within Flp and Cre site-specific DNA recombination synapses. J Mol Biol *320*, 515–527.

Grindley, N. D. F. (2002). The movement of Tn3-like elements: Transposition and cointegrate resolution. In N. L. Craig, R. Craigie, M. Gellert, and A. M. Lambowitz, eds. Mobile DNA II (Washington, DC, ASM Press), pp. 272–304.

Grishin, N. V. (2000). Two tricks in one bundle: Helix-turn-helix gains enzymatic activity. Nucleic Acids Res *28*, 2229–2233.

Guo, F., Gopaul, D. N., and van Duyne, G. D. (1997). Structure of Cre recombinase complexed with DNA in a site-specific recombination synapse. Nature *389*, 40–46.

Guo, F., Gopaul, D. N., and Van Duyne, G. D. (1999). Asymmetric DNA bending in the Cre-loxP site-specific recombination synapse. Proc Natl Acad Sci U S A *96*, 7143–7148.

Hallet, B., Arciszewska, L. K., and Sherratt, D. J. (1999). Reciprocal control of catalysis by the tyrosine recombinases XerC and XerD: An enzymatic switch in site-specific recombination. Mol Cell *4*, 949–959.

Haniford, D. (2002). Transposon Tn10. In N. L. Craig, R. Craigie, M. Gellert, and A. M. Lambowitz, eds. Mobile DNA II (Washington, DC, ASM Press), pp. 457–483.

Huai, Q., Colandene, J. D., Chen, Y., Luo, F., Zhao, Y., Topal, M. D., and Ke, H. (2000). Crystal structure of NaeI–An evolutionary bridge between DNA endonuclease and topoisomerase. EMBO J 19, 3110–3118.

Jayaram, M., Grainge, I., and Tribble, G. D. (2002). Site-specific recombination by the Flp protein of Saccharomyces cerevisiae. In N. L. Craig, R. Craigie, M. Gellert, and A. M. Lambowitz, eds. Mobile DNA II (Washington, DC, ASM Press), pp. 192–218.

Jayaram, M., Yang, X. M., Mehta, S., Voziyanov, Y., and Velmurugan, S. (2004). The 2 micron plasmid of Saccharomyces cerevisiae. In B. Funnell, ed. The biology of plasmids (Washington, DC, ASM Press), pp. 303–324.

Johnson, R. C. (2002). Bacterial site-specific DNA inversion systems. In N. L. Craig, R. Craigie, M. Gellert, and A. M. Lambowitz, eds. Mobile DNA II (Washington, DC, ASM Press), pp. 230–271.

Kanaar, R., Klippel, A., Shekhtman, E., Dungan, J. M., Kahmann, R., and Cozzarelli, N. R. (1990). Processive recombination by the phage Mu Gin system: Implications for the mechanisms of DNA strand exchange, DNA site alignment, and enhancer action. Cell 62, 353–366.

Kennedy, A. K., Haniford, D. B., and Mizuuchi, K. (2000). Single active site catalysis of the successive phosphoryl transfer steps by DNA transposases: Insights from phosphorothioate stereoselectivity. Cell 101, 295–305.

Kilbride, E., Boocock, M. R., and Stark, W. M. (1999). Topological selectivity of a hybrid site-specific recombination system with elements from Tn3 res/resolvase and bacteriophage P1 loxP/Cre. J Mol Biol 289, 1219–1230.

Krogh, B. O., and Shuman, S. (2000). Catalytic mechanism of DNA topoisomerase Ib. Mol Cell 5, 1035–1041.

Kunkel, B., Losick, R., and Stragier, P. (1990). The Bacillus subtilis gene for the development transcription factor sigma K is generated by excision of a dispensable DNA element containing a sporulation recombinase gene. Genes Dev 4, 525–535.

Landy, A. (1989). Dynamic, structural, and regulatory aspects of lambda site-specific recombination. Annu Rev Biochem 58, 913–949.

Lee, J., and Jayaram, M. (1995). Role of partner homology in DNA recombination. Complementary base pairing orients the 5′-hydroxyl for strand joining during Flp site- specific recombination. J Biol Chem 270, 4042–4052.

Lee, J., Jayaram, M., and Grainge, I. (1999). Wild-type Flp recombinase cleaves DNA in trans. EMBO J 18, 784–791.

Lee, J., Tonozuka, T., and Jayaram, M. (1997). Mechanism of active site exclusion in a site-specific recombinase: Role of the DNA substrate in conferring half-of-the-sites activity. Genes Dev 11, 3061–3071.

Logie, C., and Stewart, A. F. (1995). Ligand-regulated site-specific recombination. Proc Natl Acad Sci U S A. *92*, 5940–5944.

Lusetti, S. L., and Cox, M. M. (2002). The bacterial RecA protein and the recombinational DNA repair of stalled replication forks. Annu Rev Biochem *71*, 71–100.

Mizuuchi, K. (1992a). Polynucleotidyl transfer reactions in transpositional DNA recombination. J Biol Chem *267*, 21273–21276.

Mizuuchi, K. (1992b). Transpositional recombination: Mechanistic insights from studies of Mu and other elements. Annu Rev Biochem *61*, 1011–1051.

Mizuuchi, K., and Adzuma, K. (1991). Inversion of the phosphate chirality at the target site of Mu DNA strand transfer: Evidence for a one-step transesterification mechanism. Cell *66*, 129–140.

Nunes-Duby, S., Tirumalai, R. S., Dorgai, L., Yagil, E., Weisberg, R., and Landy, A. (1994). Lambda integrase cleaves DNA in cis. EMBO J *13*, 4421–4430.

Nunes-Duby, S. E., Azaro, M. A., and Landy, A. (1995). Swapping DNA strands and sensing homology without branch migration in lambda site-specific recombination. Curr Biol *5*, 139–148.

Pan, G., Luetke, K., and Sadowski, P. D. (1993). Mechanism of cleavage and ligation by FLP recombinase: Classification of mutations in FLP protein by in vitro complementation analysis. Mol Cell Biol *13*, 3167–3175.

Parsons, R. L., Prasad, P. V., Harshey, R. M., and Jayaram, M. (1988). Step-arrest mutants of FLP recombinase: Implications for the catalytic mechanism of DNA recombination. Mol Cell Biol *8*, 3303–3310.

Pathania, S., Jayaram, M., and Harshey, R. M. (2002). Path of DNA within the Mu transpososome. Transposase interactions bridging two Mu ends and the enhancer trap five DNA supercoils. Cell *109*, 425–436.

Perler, F. B. (1998). Protein splicing of inteins and hedgehog autoproteolysis: Structure, function, and evolution. Cell *92*, 1–4.

Perler, F. B. (2002). InBase: The intein database. Nucleic Acids Res *30*, 383–384.

Pyle, A. M. (1993). Ribozymes: A distinct class of metalloenzymes. Science *261*, 709–714.

Recchia, G. D., and Sherratt, D. J. (2002). Gene acquisition in bacteria by integron mediated site-specific recombination. In N. L. Craig, R. Craigie, M. Gellert, and A. M. Lambowitz, eds. Mobile DNA II (Washington, DC, ASM Press), pp. 162–176.

Redinbo, M. R., Stewart, L., Kuhn, P., Champoux, J. J., and Hol, W. G. (1998). Crystal structures of human topoisomerase I in covalent and noncovalent complexes with DNA. Science *279*, 1504–1513.

Reznikoff, W. (2002). Tn5 transposition. In N. L. Craig, R. Craigie, M. Gellert, and A. M. Lambowitz, eds. Mobile DNA II (Washington, DC, ASM Press), pp. 403–422.

Rice, P. A. (2002). Theme and variation in tyrosine recombinases: Structure of a Flp-DNA complex. In N. L. Craig, R. Craigie, M. Gellert, and A. M. Lambowitz, eds. Mobile DNA II (Washington, DC, ASM Press), pp. 219–229.

Rice, P. A., and Steitz, T. A. (1994a). Model for a DNA-mediated synaptic complex suggested by crystal packing of gamma delta resolvase subunits. EMBO J *13*, 1514–1524.

Rice, P. A., and Steitz, T. A. (1994b). Refinement of gamma delta resolvase reveals a strikingly flexible molecule. Structure *2*, 371–384.

Rufer, A. W., and Sauer, B. (2002). Non-contact positions impose site selectivity on Cre recombinase. Nucleic Acids Res *30*, 2764–2771.

Sanderson, M. R., Freemont, P. S., Rice, P. A., Goldman, A., Hatfull, G. F., Grindley, N. D., and Steitz, T. A. (1990). The crystal structure of the catalytic domain of the site-specific recombination enzyme gamma delta resolvase at 2.7 Å resolution. Cell *63*, 1323–1329.

Santoro, S. W., and Schultz, P. G. (2002). Directed evolution of the site specificity of Cre recombinase. Proc Natl Acad Sci U S A *99*, 4185–4190.

Sau, A. K., DeVue Tribble, G., Grainge, I., Frohlich, R. F., Knudsen, B. R., and Jayaram, M. (2001). Biochemical and kinetic analysis of the RNase active sites of the integrase/tyrosine family site-specific DNA recombinases. J Biol Chem *276*, 46612–46623.

Sauer, B. (2002). Chromosome manipulation by Cre-lox recombination. In N. L. Craig, R. Craigie, M. Gellert, and A. M. Lambowitz, eds. Mobile DNA II (Washington, DC, ASM Press), pp 38–58.

Sekiguchi, J., and Shuman, S. (1997). Site-specific ribonuclease activity of eukaryotic DNA topoisomerase I. Mol Cell *1*, 89–97.

Sherratt, D. J., and Wigley, D. B. (1998). Conserved themes but novel activities in recombinases and topoisomerases. Cell *93*, 149–152.

Stark, W. M., Sherratt, D. J., and Boocock, M. R. (1989). Site-specific recombination by Tn3 resolvase: Topological changes in the forward and reverse reactions. Cell *58*, 779–790.

Steitz, T. A., and Steitz, J. A. (1993). A general two-metal-ion mechanism for catalytic RNA. Proc Natl Acad Sci U S A *90*, 6498–6502.

Stewart, L., Redinbo, M. R., Qiu, X., Hol, W. G., and Champoux, J. J. (1998). A model for the mechanism of human topoisomerase I. Science *279*, 1534–1541.

Stivers, J. T., Jagadeesh, G. J., Nawrot, B., Stec, W. J., and Shuman, S. (2000). Stereochemical outcome and kinetic effects of Rp- and Sp-phosphorothioate substitutions at the cleavage site of vaccinia type I DNA topoisomerase. Biochemistry 39, 5561–5572.

Sumners, D. W., Ernst, C., Spengler, S. J., and Cozzarelli, N. R. (1995). Analysis of the mechanism of DNA recombination using tangles. Q Rev Biophys 28, 253–313.

Van Duyne, G. (2002). A structural view of tyrosine recombinase site-specific recombination. In N. L. Craig, R. Craigie, M. Gellert, and A. M. Lambowitz, eds. Mobile DNA II (Washington, DC, ASM Press), pp. 93–117.

Voziyanov, Y., Pathania, S., and Jayaram, M. (1999). A general model for site-specific recombination by the integrase family recombinases. Nucleic Acids Res 27, 930–941.

Voziyanov, Y., Stewart, A. F., and Jayaram, M. (2002). A dual reporter screening system identifies the amino acid at position 82 in Flp site-specific recombinase as a determinant for target specificity. Nucleic Acids Res 30, 1656–1663.

Wang, J. C. (2002). Cellular roles of DNA topoisomerases: A molecular perspective. Nat Rev Mol Cell Biol 3, 430–440.

Wasserman, S. A., and Cozzarelli, N. R. (1986). Biochemical topology: Applications to DNA recombination and replication. Science 232, 951–960.

Wittschieben, J., Petersen, B. O., and Shuman, S. (1998). Replacement of the active site tyrosine of vaccinia DNA topoisomerase by glutamate, cysteine or histidine converts the enzyme into a site- specific endonuclease. Nucleic Acids Res 26, 490–496.

Xu, C. J., Grainge, I., Lee, J., Harshey, R. M., and Jayaram, M. (1998). Unveiling two distinct ribonuclease activities and a topoisomerase activity in a site-specific DNA recombinase. Mol Cell 1, 729–739.

Yang, W., and Steitz, T. A. (1995). Crystal structure of the site-specific recombinase gamma delta resolvase complexed with a 34 bp cleavage site. Cell 82, 193–207.

CHAPTER 3

Site-specific recombination by the serine recombinases

Sally J. Rowland and W. Marshall Stark

83

It is a curious fact that two unrelated families of enzymes have evolved that promote conservative site-specific recombination. Elsewhere in this volume (see Chapter 2), a large family of "tyrosine recombinases" is described, of which phage lambda integrase is the most famous and senior member. The subject of this chapter is a second large family, the "serine recombinases." The names come from the conserved residue of the recombinase that provides the nucleophile to attack and break the DNA phosphodiester backbone (see "Tn3 and $\gamma\delta$ resolvases: cointegrate resolution" section). Although the serine and tyrosine recombinases are unrelated in sequence, structure, or mechanism, there is no obvious distinction between their biological functions. Why two very different types of site-specific recombinases have survived eons of natural selection, yet continue to play similar roles, remains a mystery. Serine recombinases are widespread in the Eubacteria and Archea, but not in Eukarya, where the few examples found so far may be of recent bacterial origin.

A serine recombinase can be identified by similarity of parts or all of its primary amino acid sequence to that of one of the archetypal members of the family [e.g. Tn3 resolvase (Fig. 3.7)]. Several hundreds of such proteins can now be predicted from available DNA sequences (reviewed by Smith and Thorpe, 2002). The relations of the members of the family are discussed later in this chapter. Each recombinase is a key player in a site-specific recombination *system*, which also includes the sites that are to recombine, and sometimes additional "accessory" sites and proteins, whose roles are also discussed later.

Investigation of many serine recombinase systems is still at a quite early stage, and prediction of their properties is largely based on analogy with a few archetypes that have been studied in detail. These include the resolution systems of the bacterial transposons Tn*3*, γδ, and Tn*21*, and the DNA inversion systems involving the recombinases Gin and Hin. To introduce concepts simply and logically, we adopt a "bottom-up" approach, first discussing in some detail a specific well-characterized serine recombinase system (Tn*3*/γδ resolution), then proceeding to make analogies with other transposon resolution systems, DNA inversion systems, and other more divergent systems, before finally addressing issues relevant to all serine recombinases.

TN3 AND γδ RESOLVASES: COINTEGRATE RESOLUTION

Background

Tn*3* and γδ resolvases are very similar to each other, and perform the same function for their respective transposons (Fig. 3.1a; Grindley, 2002). Replicative transposition, initiated by the transposase enzyme and involving host replication functions, results in a cointegrate molecule, in which two copies of the transposon, in direct repeat, link the donor and target DNA molecules. The transposon-encoded 185-residue resolvase protein acts to split the cointegrate into two DNA circles, each with one copy of the transposon; the overall result being reconstitution of the donor DNA sequence and insertion of a new copy of the transposon into the target DNA. The cutting and rejoining of the DNA strands, required for resolution, is catalyzed by resolvase, at specific chemical bonds within a 114 bp sequence called *res*, to which resolvase binds (Fig. 3.1b; Arthur and Sherratt, 1979; Heffron et al., 1979; Kostriken et al., 1981; Reed, 1981a; Wells and Grindley, 1984). Resolvase also acts as a repressor of transcription of the transposase and resolvase genes *tnpA* and *tnpR*, whose divergent promoters are within *res* (Reed et al., 1982; Wells and Grindley, 1984). Resolution can be reconstituted *in vitro*, using a supercoiled plasmid substrate with two well-separated *res* sequences in direct repeat (head-to-tail) orientation, purified resolvase protein, and a simple reaction buffer (Reed, 1981a). No other proteins or specific components (e.g. metal ions) are required. Early *in vitro* and *in vivo* studies demonstrated the peculiar selectivity of resolvase-mediated recombination. The sites that recombine must be in the same, supercoiled DNA molecule and in direct, not inverted, repeat — that is, arranged as they are in the cointegrate transposition intermediate. Furthermore, resolution yields not free circles or random tangles of the two product circles, but a specific simple (2-noded) catenane (Fig. 3.1c; Krasnow and Cozzarelli, 1983; Reed, 1981a; Wasserman and Cozzarelli, 1985).

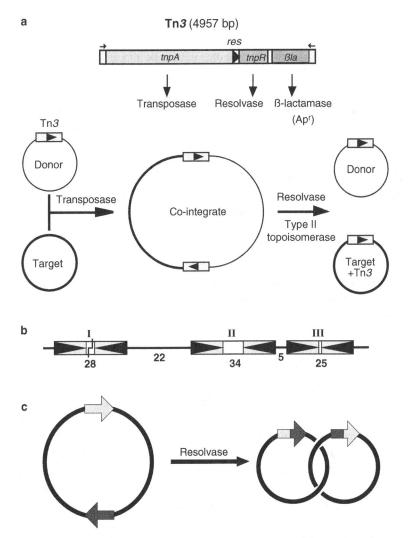

Figure 3.1. Tn3 resolvase action at *res*. **(a)** Structure of Tn3 and outline of transposition. Resolvase acts to divide the large cointegrate plasmid into the two product circles. **(b)** The Tn3 (or $\gamma\delta$) recombination site *res*. The three binding sites for resolvase dimers (I, II, and III) comprise inverted, imperfect copies of a 12 bp sequence motif with different spacings. The lengths of the sites and the linkers between them are given. Resolvase cuts the DNA at the center of site I with a 2 bp stagger, as shown. **(c)** *In vitro* resolution by Tn3 or $\gamma\delta$ resolvase. Resolvase recombines directly repeated copies of *res* in a plasmid molecule to give a simple (2-noded) catenane. For simplicity, the supercoiling in substrate and product circles is not shown.

Three resolvase dimers bind to each *res*, at three binding sites, designated I, II, and III (Fig. 3.1b; Grindley et al., 1982; Kitts et al., 1983). Each binding site comprises two 12 bp motifs in inverted repeat, flanking a central sequence of variable length. The 12 bp motifs are similar but not identical (Grindley et al., 1982). The DNA strands are cut and rejoined at the centre of binding site I, the "crossover site" (Reed, 1981b), and the catalytic steps involved in breaking and rejoining the DNA are performed exclusively by the resolvase subunits bound at site I (Boocock et al., 1995; Dröge et al., 1990; Grindley, 1993). However, the other two "accessory" resolvase-binding sites II and III, in their proper positions relative to site I and each other, are essential for full recombination activity (Bednarz et al., 1990; Kitts et al., 1983; Salvo and Grindley, 1988; Wells and Grindley, 1984). This requirement for accessory sites highlights a ubiquitous characteristic of serine recombinases (and, indeed, of nearly all site-specific recombinases): that the catalytic activity at a "core" site (e.g. *res* site I) is subject to regulation involving additional DNA sequences and/or DNA-bound protein subunits. The function of this regulatory apparatus ensures that recombination occurs only at the appropriate times and between appropriate pairs of sites. For resolvase, regulation manifests as specificity for intramolecular reaction, sites in direct repeat, negative supercoiling, and a simple catenane product, as stated previously. Refer to "The synaptic complex" section for a discussion on how the accessory sites bring about this specificity.

By definition, recombination involves two *res* sites. Therefore, after binding, resolvase must promote the interaction of two *res* sites in a manner suitable for catalysis of strand exchange. The resulting two-*res* structure is usually called the synapse (or synaptic complex). Proper construction of the synapse is a prerequisite for catalytic activity at site I. Elucidation of the structure and function of the synapse is the central aim of much research on resolvase systems; current ideas are discussed in "The synaptic complex" section. Resolvase thus has multifarious roles during the recombination process. It must recognize and bind specifically to each of the three dimer-binding sites in *res*. It must then promote interaction of two *res* sites to form the synapse. It must catalyze the precise breakage and rejoining of four DNA strands to carry out the actual recombination reaction. It must also be adapted to make its catalytic activity conditional on proper synapse formation. That all this functionality can be encompassed in one small (20 kDa) protein is quite astonishing.

Resolvase makes double-stranded breaks at the center of site I, the protein becoming covalently linked to the 5′ end of the DNA (Reed and Grindley, 1981; Fig. 3.2). The protein–DNA linkage is a phosphodiester with the hydroxyl group of the conserved and essential resolvase residue Ser10 (Hatfull

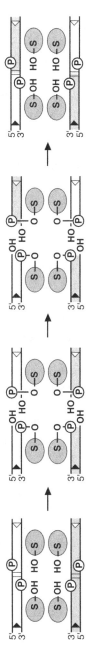

Figure 3.2. Catalysis of strand exchange by serine recombinases. A phosphodiester linkage is formed between the catalytic domain (*shaded oval*) and the 5′ end of the DNA; the nucleophile is the hydroxyl group of a conserved serine residue. The half-sites are rearranged, the left end of one site becoming juxtaposed to the right half of the other site, and then the DNA strands are rejoined, releasing the enzyme.

and Grindley, 1986; Reed and Moser, 1984). The inference that at an intermediate stage in catalysis there are double-stranded breaks in both recombining sites, with a resolvase subunit attached to each "half-site" (Fig. 3.2), has been fully supported in subsequent work (see, for example, Boocock et al., (1995). These results (mirrored in similar studies on the Gin and Hin DNA invertases; see "Gin and Hin: DNA invertases" section) made it clear that the mechanism of strand exchange by serine recombinases is completely different from that of the tyrosine recombinases, where the protein becomes linked to the 3' end of the DNA, strands are exchanged two at a time, with an intermediate Holliday junction, and there are no double-stranded cleaved intermediates (Chapter 2). Current models for the mechanism of strand exchange by resolvase are discussed in the "Mechanism of strand exchange" section.

The resolvase protein and binding to *res*

The resolvase protomer includes two domains, which are separable by limited proteolysis (Abdel-Meguid et al., 1984). The 43-residue $\gamma\delta$ resolvase C-terminal proteolytic fragment, which was predicted to contain a "helix-turn-helix" DNA-binding domain, binds to the outer inverted repeats of site I in the major groove (Abdel-Meguid et al., 1984; Rimphanitchayakit et al., 1989), whereas the N-terminal fragment does not bind detectably to site I. A model for the intact $\gamma\delta$ resolvase dimer bound to site I, based on the differences in protection of the DNA from cleavage by DNase I by the C-terminal fragment and by intact resolvase (Abdel-Meguid et al., 1984), and on other footprinting studies (Falvey and Grindley, 1987; Mazzarelli et al., 1993) was gratifyingly confirmed and refined when the structure of the same entity was solved by X-ray crystallography (Yang and Steitz, 1995; Fig. 3.3). In the crystal, a 37-residue C-terminal domain binds to each end of site I in the major groove. The domain does indeed belong to the helix-turn-helix/homeodomain class of DNA-binding domains. The larger N-terminal domain (about 136 residues) contains all the residues known to contribute to catalytic function. At its core is a five-stranded mixed parallel and antiparallel β-sheet, which is covered by four α-helices. The C-terminal 33 residues of the domain form a fifth long α-helix (the E-helix), which plays an important role in the dimer interface. The domain makes a mainly hydrophobic dimer interaction with the partner subunit. The N- and C-terminal domains are linked by a 10-residue extended peptide sequence that associates with the site I DNA in the minor groove. The crystallized dimer deviates from perfect 2-fold symmetry, in a way that is proposed to be relevant to the mechanism of catalysis (Yang and

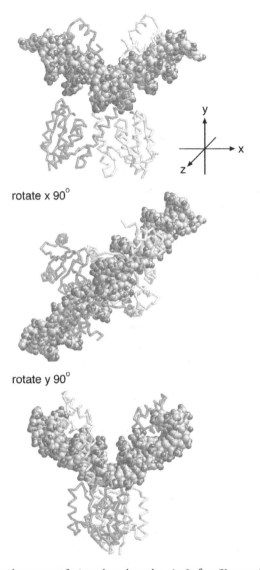

rotate x 90°

rotate y 90°

Figure 3.3. Crystal structure of $\gamma\delta$ resolvase bound to site I of *res* (Yang and Steitz, 1995). The DNA is in spacefill representation (CPK colors). The phosphates that are attacked by resolvase during recombination are in magenta. The resolvase subunits are in yellow and green (backbone representation).

Steitz, 1995). The paired N-terminal domains of the dimer contact the minor groove at the center of the site I DNA. Although the dimer in the co-crystal structure does not make any functionally significant interactions with other dimers, previously solved structures of the $\gamma\delta$ resolvase N-terminal domain, in the absence of DNA (Rice and Steitz, 1994a, 1994b; Sanderson et al., 1990) reveal dimer–dimer interactions that may be crucial for construction of the recombination synapse (Hughes et al., 1990; see also "Mechanism of strand exchange" section). The solution structures of the $\gamma\delta$ resolvase C-terminal domain, and part of the N-terminal domain, have been solved by NMR spectroscopy (Liu et al., 1993, 1994; Pan et al., 1997, 2001) and concur with the crystallographic data, showing also that the C-terminal domain may be largely disordered when not bound to DNA.

The X-ray structure of the DNA-bound $\gamma\delta$ resolvase dimer (Yang and Steitz, 1995) is fully consistent with other studies of resolvase binding at site I (references cited previously). However, the structure did not provide all the hoped-for insights into the mechanism of synapsis and catalysis. As stated previously, the crystallized DNA-bound dimer does not make any function-ally significant interactions with other dimers. Furthermore, the residues implicated in catalysis are not closely engaged with their DNA substrate (Fig. 3.3). In particular, the Ser10 hydroxyl groups are 17.4 Å and 12.1 Å away from their target phosphates. Therefore, the nature of the interaction of the resolvase catalytic site with the DNA during strand exchange itself is still uncertain. Binding of resolvase to site I in a state that is incompetent for catalysis may be a design feature of the system, ensuring that activation by the regulatory components is strictly required before catalysis can take place. To obtain high-resolution structures of species that are more informative about the mechanism of catalysis, modification of the components may be required (e.g. using mutants of resolvase).

No high-resolution structural data are as yet available for resolvase bound at the accessory sites. The interactions of resolvase with sites II and III must be substantially different from the interaction at site I because of the different lengths of the sites. Site II (34 bp) is more than half a turn of DNA longer than site I (28 bp), so the resolvase C-terminal domains binding near the ends of site II must be about 20 Å further apart, and rotated relative to each other because of the helical nature of the DNA. Site III (25 bp) is shorter than site I, and almost a full turn of DNA helix shorter than site II, implying another variation on the binding theme. A model for resolvase binding at site II has been proposed, based on footprinting and related data, in which the linker between the N- and C-terminal domains of one subunit takes a different path over the DNA from that seen at site I (Blake et al., 1995; see also Mazzarelli et al., 1993). The fact that resolvase binds to such disparate sites with similar affinity suggests that it is a remarkably flexible protein (see also Rice and Steitz, 1994b). This flexibility may be an attribute of many serine recombinases (see, for example, Rowland et al., 2002).

The synaptic complex

The synapse formed by the interaction of two *res* sequences is presumed to contain 12 resolvase subunits, as three dimers bind to each *res*. Its structure is still the subject of much speculation. A detailed discussion of current ideas is beyond the scope of this chapter, but the topic is thoroughly covered by Grindley (1994, 2002). The two *res* sequences must be intertwined in the way shown in Fig. 3.4, in order to account for the DNA topologies of the resolution product, and of products formed by multiple rounds of strand exchange (Benjamin and Cozzarelli, 1990; Bliska et al., 1991; Boocock et al., 1987; Krasnow and Cozzarelli, 1983; Stark et al., 1989; Wasserman and Cozzarelli, 1985; Wasserman et al., 1985; reviewed by Stark and Boocock, 1995). This intertwining is brought about primarily by the resolvase-bound accessory sites (II and III), which can synapse (and intertwine) efficiently even when site I has been deleted (Watson et al., 1996), and can impose specific product topology in hybrid recombination systems where resolvase at site I is replaced by the tyrosine recombinases Cre or FLP, bound to their cognate recombination sites (Grainge et al., 2000; Kilbride et al., 1999). Catalysis at site I involves interaction of the two resolvase dimers at site I to make a catalytic tetramer, the nature of which is discussed later in this chapter. It is still unclear whether the synaptic interactions at sites II and III are also between dimers, or whether some more complex arrangement of subunits is formed. A recent model proposes that the synapse involves three similar dimers of dimers, pairing binding sites I-I, II-II, and III-III, such that the DNA is wrapped around the outside (Sarkis et al., 2001; Fig. 3.4b). Adjacent resolvase tetramers are proposed to be united via 2,3′ interactions, as were observed in crystal structures of $\gamma\delta$ resolvase without DNA (Rice and Steitz, 1994a; Sanderson et al., 1990). Mutation of residues at the 2,3′ interface showed that the contacts were essential for recombination activity (Hughes et al., 1990) and for proper synapse construction (Murley and Grindley, 1998). Currently, it seems likely that 2,3′ interactions are involved in the process that activates the resolvase at site I to catalytic competence, as well as having "architectural" roles.

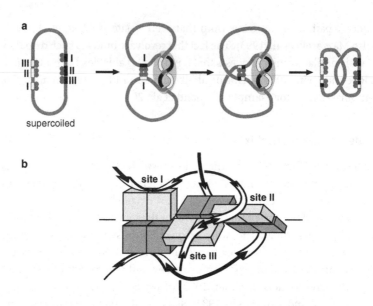

Figure 3.4. (a) Outline of synapsis and strand exchange by Tn3/γδ resolvase. Three resolvase dimers (•) are shown binding to each *res* site in the substrate and product. Synapsis of two directly repeated *res* sites traps three negative interdomainal supercoils; supercoiling other than that associated with synapsis is not shown. The two dimers at sites I synapse; the interactions between accessory sites II and III are not specified. Following strand exchange, the synapse dissociates; the product is a 2-noded catenane. (b) Cartoon of a model for the Tn3/γδ *res*-resolvase synapse (Sarkis et al., 2001; from Rowland et al., 2002). Rectangles with arrowheads represent the binding motifs in the DNA. Dimers of resolvase catalytic domains are represented by dominoes; the DNA-binding domains are not shown. The site I resolvase dimers synapse back-to-back, with the DNA on the outside, as do the site II dimers and site III dimers (see also Fig. 3.5a; 'DNA-out'); this non-crystallographic dimer–dimer interface is hypothetical but structurally plausible. Contacts between corners of adjacent dominoes (e.g. light and dark dominoes at sites I and III, respectively, or dark dominoes at sites III and II) correspond to the crystallographic 2,3′ interface between γδ resolvase dimers.

Many serine recombinases use analogous synaptic structures to regulate their activity. Variations in synapse design are discussed in the "Other resolvases" section.

Mechanism of strand exchange

Catalysis of strand exchange is by the four resolvase subunits bound at the two site Is, which interact to make a tetramer. Recently, some new evidence on the structure of the catalytic tetramer has emerged. Biochemical, structural, and topological data support the idea that the tetramer, and the catalytic mechanism, have functional D_2 symmetry (Sarkis et al., 2001; Stark et al., 1989). Assuming no gross structural changes from that seen in the site I–resolvase co-crystal, we must therefore consider two ways of docking the dimers to make a tetramer with this symmetry. Either the N-terminal domains come together along their approximate two-fold axis, leaving the site I DNA on the outside of the tetramer ("DNA-out"; Fig. 3.5a), or the interaction involves residues nearer the C-terminus, and the two site Is are close to the center of the complex ("DNA-in"). Recent mutagenesis studies, which identify activating mutations allowing resolvase-mediated catalysis in the absence of accessory sites or supercoiling, strongly support the DNA-out hypothesis. Some of these mutations are close to the interface between dimers that would be required for the DNA-out tetramer (Arnold et al., 1999; Burke et al., 2004), and mutant (but catalytically active) resolvases containing three or more of such mutations make a stable complex comprising a resolvase tetramer and two site Is (Sarkis et al., 2001; J.He and Marshall Stark, in preparation). A recent *in vitro* recombination study also indicates a DNA-out tetramer (Leschziner and Grindley, 2003).

Following double-stranded cleavage of the two site Is bound to the catalytic tetramer, the four resolvase-linked "half-sites" must be rearranged into a recombinant configuration, before re-ligation of the cleaved DNA. Complete analysis of the topological changes accompanying strand exchange led to a model for the mechanism of this rearrangement called "subunit rotation" (Stark et al., 1989). In this model, two resolvase subunits rotate though 180° relative to the other two in the catalytic tetramer, along with their attached DNA (Fig. 3.5b). This simple model continues to account for all the hard facts known about strand exchange by serine recombinases, but difficulties in accommodating it within the framework of the published high-resolution crystallographic data on resolvase, and in explaining how rotation could be accomplished without catastrophic dissociation of the parts of the intermediate that rotate relative to each other, have led to alternative proposals. The various models are discussed in detail elsewhere (for a review, see Grindley, 2002); here we consider only one robust alternative to subunit rotation called "domain rotation." In domain rotation, the DNA and N-terminal parts of two resolvase subunits move just as in subunit rotation, but the integrity of the catalytic tetramer is maintained by static interaction of the E-helices of the rotating subunits with the other two subunits (Fig. 3.5c). An additional attractive feature of the domain rotation model is that it can provide an interpretation of some "activating" resolvase mutations (see "Regulation

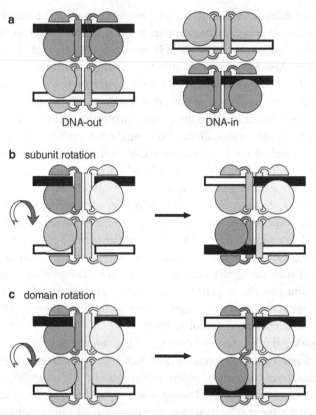

SALLY J. ROWLAND AND W. MARSHALL STARK

Figure 3.5. Models for strand exchange by serine recombinases. In these cartoons, a resolvase subunit comprises a large N-terminal catalytic domain, a small C-terminal helix-turn-helix binding domain, and a long E-helix at the dimer interface. **(a)** The two alternative ways of synapsing resolvase dimer–DNA complexes: 'DNA-out' and 'DNA-in'. **(b)** Subunit rotation model for strand exchange. The resolvase N-terminal domains are covalently attached to the DNA at the center of each site I. The left half-sites, with attached subunits, exchange places; the rotation has a right-handed sense *(arrow)*. **(c)** Domain rotation model for strand exchange (M. Boocock et al., in preparation). Only the N-terminal catalytic domain, with the attached half-site, rotates; the C-terminal domain dissociates from the half-site that moves. Crucially, the dimer–dimer interface is preserved.

of strand exchange by synapse formation" section). However, we still await incontrovertible evidence for any particular model for strand exchange.

Although the simple catenane recombinant is by far the most abundant product from an *in vitro* recombination reaction, minor products that have more complex topologies (multiply interlinked knots and catenanes) have been detected (Krasnow et al., 1983; Wasserman et al., 1985). These products

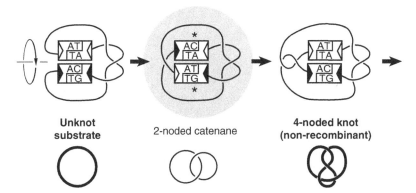

Figure 3.6. Iteration of strand exchange mechanism in a substrate with mismatched sequences at the center of site I. The central 2 bp of each site I and the resolvase-mediated cleavages are shown. Synapsis of two *res* sites traps three interdomainal negative supercoils (sites II and III are not shown). A double round of strand exchange (360° right-handed rotation) generates a non-recombinant 4-noded knot. A 180° round of strand exchange would give a recombinant 2-noded catenane with two mismatched base pairs (marked by asterisks, shaded panel, not detected).

are interpreted as being formed by iteration of the strand exchange mechanism, and provided important evidence on the disposition of the DNA in the synaptic intermediate (reviewed in Stark and Boocock, 1995). Iteration of strand exchange can be induced artificially by using a substrate in which one *res* has a mutation of one of the two base pairs at the center of binding site I. Resolution, which would lead to recombinant sites with mismatched base pairs, is not observed; instead, a second round of strand exchange reconstitutes the non-recombinant sites and causes knotting of the plasmid DNA (Stark et al., 1991; Fig. 3.6). A related phenomenon has been observed in invertase-catalyzed reactions (see "Gin and Hin: DNA invertases" section). Models for the mechanism of strand exchange must account for this facile iteration of the strand exchange mechanism (McIlwraith et al., 1997; Stark and Boocock, 1994). Subunit rotation easily accommodates iteration, whereas the situation is more complicated for domain rotation (Boocock et al., manuscript in preparation), but current data cannot decisively distinguish between the two models.

Alignment of the known serine recombinase amino acid sequences reveals a number of highly conserved residues (Fig. 3.7), several of which may contribute to formation of the recombinase active site. Only Ser10 (numbering residues according to the Tn$3/\gamma\delta$ resolvase sequence) has been clearly shown to have a catalytic function, acting as the nucleophile that attacks

96

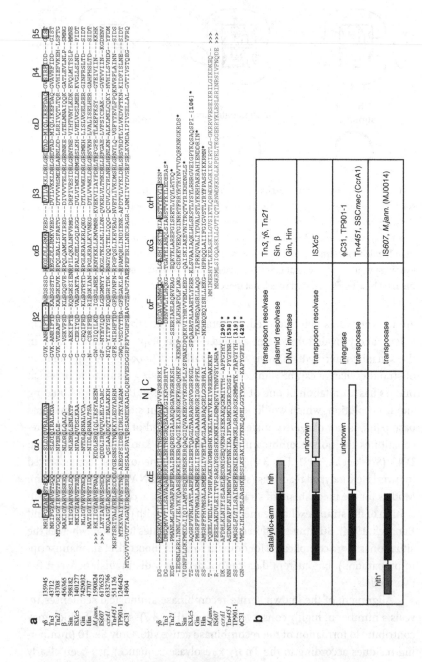

the target phosphodiester. Other residues are implicated in catalysis by their proximity to Ser10 in the crystal structures and/or by loss of catalytic activity when they are mutated; these include Tyr6, Arg8, Gln14, Arg45, and Arg68 (reviewed by Grindley, 2002). Studies on the DNA invertases Hin and Gin have provided complementary results (see "Gin and Hin: DNA invertases" section). It would be expected that, as for other enzymes that break DNA strands, such as tyrosine recombinases, topoisomerases, and endonucleases, active site residues would be involved in withdrawing electron density from the phosphorus targeted by the nucleophile, and acting as general acids and bases; however, at present, assigning these roles to specific resolvase residues would be speculative.

Regulation of strand exchange by synapse formation

How is resolvase catalytic activity made conditional on formation of the full synapse of two *res* sites? There is still no clear answer, although some evidence suggests that contacts at the 2,3' interface are important (Murley and Grindley, 1998; Sarkis et al., 2001). In a recent structural model of the synapse (Sarkis et al., 2001; see "Mechanism of strand exchange" section), resolvase subunits bound at site III contact those at site I via 2,3'-like interactions, and could perhaps thereby provide a stimulus to catalytic activity. The failure of isolated wild-type resolvase-site I complexes to make a stable synapse suggests that one factor in activation might be the juxtaposition of the two site Is in a catalytically productive geometry, imposed by the structure of the complex formed by the paired accessory sites and resolvase. Alternatively, it might be that activation of resolvase at site I requires a protein conformational change that is stimulated by contacts with the rest of the synapse. Resolvase mutants

Figure 3.7. Alignment of serine recombinase sequences. (a) An alignment of the sequences of selected serine recombinases (with accession numbers). The secondary structure elements of $\gamma\delta$ resolvase, for which the crystal structure is known, are shown. An arrow marks the junction between the N- and C-terminal fragments of $\gamma\delta$ resolvase obtained by proteolysis. Conserved residues in or near the active site are highlighted (shaded); S10 (Tn3/$\gamma\delta$ numbering) is marked (o). The number of residues in a C-terminal extension to a sequence (not shown) is in brackets. The C-terminus is indicated by an asterisk. For the *Methanococcus jannaschii* ("*M. jann.*") and IS607 transposase sequences, the N- and C-terminal domains are aligned with the C- and N-terminal domains, respectively, of $\gamma\delta$ resolvase. (b) A cartoon showing the domain structures of the recombinases in A.

that catalyze strand exchange at site I in the absence of the accessory sites have been described (Arnold et al., 1999; Burke et al., 2004). Some of these mutants (unlike wild-type resolvase) can promote the formation of a stable synapse of two resolvase dimer-site I complexes, as mentioned previously (Sarkis et al., 2001; J. He and Marshall Stark, manuscript in preparation). Furthermore, additional mutations at the 2,3' interface do not reduce the strand exchange activity of these mutants, showing that requirement for 2,3' interactions has somehow been bypassed. At present, it is unclear whether the activating mutations simply stabilize an interface between two dimers whose structure is like that seen in the co-crystal, or whether they promote a protein conformational change that is required for tetramer stabilization.

To summarize, our current view of the crucial synaptic intermediate in resolvase-mediated recombination is that it comprises a catalytic module (a resolvase tetramer holding together the two site Is where strand exchange is to occur), linked to a regulatory module (the intertwined accessory sites with their bound resolvase) that interacts with the catalytic module in such a way as to stimulate catalysis.

What is the regulation for? The requirement for a synapse involving intertwining of the *res* sites ensures that recombination is between two sites in the same molecule, and furthermore, that the two sites are in direct repeat, as they are in the natural cointegrate transposition intermediate. The principle by which this topological selectivity operates has been established and is reviewed elsewhere (Grindley, 2002; Stark and Boocock, 1995). We currently view the dependence on negative supercoiling, and the specific formation of a catenane resolution product, as incidental consequences of the mechanism for selectivity, rather than having any biological significance per se. Other resolvases and DNA invertases of the serine recombinase family use very similar topological selectivity strategies (see "Other resolvases" and "Gin and Hin: DNA invertases" sections), as do some tyrosine recombinases (see, for example, Colloms et al., 1997; reviewed by Hallet and Sherratt, 1997). The biological *raison d'être* of the selectivity of Tn3/$\gamma\delta$ resolution might be to ensure that recombination is strictly between the two transposon copies formed by replicative transposition, and not between transposons on separate DNA molecules, or elsewhere within the same DNA molecule. These illegitimate reactions would rearrange the DNA with potentially damaging consequences for the viability of the transposon. It is also possible that the selectivity prevents undesirable recombination between copies of the transposon in replication intermediates; for example, when replication of the cointegrate precedes its resolution.

OTHER RESOLVASES

Many transposons have serine recombinase-based cointegrate resolution systems similar to those of Tn3 and $\gamma\delta$ (reviewed in Grindley, 2002; Smith and Thorpe, 2002). Also, some bacterial plasmids use related systems to promote monomerization of multimers. These plasmid systems may have been acquired via ancestral transposon insertions. In all cases, the resolvase sequences can be aligned with Tn3/$\gamma\delta$ resolvase along almost their entire length; in a few cases, small insertions or deletions are required in regions predicted to be loops between secondary structure elements (Fig. 3.7; Smith and Thorpe, 2002). Some resolvases have extended sequences (up to about 100 extra residues) at their C-terminus. This extra sequence is of unknown function, and in at least one case (ISXc5 resolvase) it can be deleted without affecting recombination activity (Schneider et al., 2000).

Of special importance in the development of our ideas about synapsis and topological selectivity has been the resolution system of Tn21, which has been studied in depth by Stephen Halford's group (e.g. Halford et al., 1985; Parker and Halford, 1991; Soultanas et al., 1995). Tn21 res, like Tn3 res, has three binding sites for resolvase dimers, but the lengths and spacing of the sites are subtly different (Hall and Halford, 1993). The topology and selectivity of Tn21 resolvase-mediated recombination are essentially identical to those of Tn3 and $\gamma\delta$ resolvases (Stark et al., 1994), suggesting common principles in synapse formation and reactivity.

The res "design concept" is quite general; that is, a "core" recombinase dimer-binding site ("crossover site"), at the center of which DNA cleavage and rejoining occurs, is adjacent to "accessory sites," which bind additional recombinase subunits. However, there are a number of interesting variations on this theme (Fig. 3.8). Many res sites, like that of Tn21, are clearly of similar design to Tn3/$\gamma\delta$ res, having two accessory dimer-binding sites adjacent to site I. Intriguingly, the variation in site I–site II distance is restricted to approximately integral numbers of DNA helix turns (Fig. 3.8; Rowland et al., 2002). It is notable that the $\gamma\delta$ res site I–site II distance can be increased by integral numbers of DNA turns with retention of activity, whereas insertion of other lengths of DNA abolish activity (Salvo and Grindley, 1988). In some other systems, the arrangement of binding motifs (presumed to be recognized by the C-terminal domain) is quite complex, and a "site II" or "site III" cannot be identified by sequence inspection (Rowland et al., 2002). Thus, alternative arrangements of resolvase subunits at the accessory sites may exist. A group of recombinases, exemplified by the Sin resolvase from *Staphylococcus aureus* and the β-recombinase from *Streptococcus pyogenes*, is

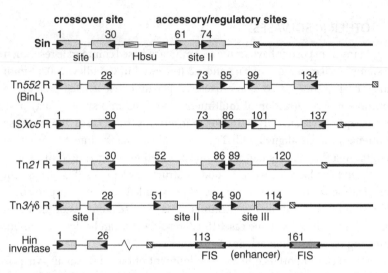

Figure 3.8. Variations in arrangement of binding sites in *res* (from Rowland et al., 2002). Boxes represent imperfect copies of a 12 bp sequence motif (10 bp for Hin); white boxes represent poorer matches to the consensus. Hatched boxes are start codons of recombinase genes (*thick lines*). The first base pair of each repeat motif is numbered, beginning with the left end of each recombination site.

highly diverged from Tn3/γδ resolvases (Petit et al., 1995; Rojo and Alonso, 1994; Rowland et al., 2002). Their accessory sequences have a very different organization from those of Tn3 or γδ *res*. Footprinting analysis has indicated that the recombinase binds to only one accessory site (site II; Petit et al., 1995; Rojo and Alonso, 1995; Rowland et al., 2002), in which the two binding motifs are proposed to be in tandem repeat, not inverted as in site II of Tn3 *res* (Fig. 3.8). Furthermore, the Sin and β systems require another protein for recombination, a DNA-bending "architectural" protein (e.g. HU or HMG-1; Alonso et al., 1995; Petit et al., 1995; Rowland et al., 2002). The *Bacillus sub-tilis* HU homologue, Hbsu, was shown to bind between sites I and II when they are occupied by Sin, and is believed to stabilize a tight bend between them (Rowland et al., 2002). Despite these striking structural differences, the Tn3/γδ and Sin/β systems share a similar selectivity for intramolecular re-combination, directly repeated sites, a negatively supercoiled substrate, and simple catenane resolution products. Is there, therefore, some core structural unit that is common to all serine recombinase resolution systems, and can be assembled from subunits bound at diverse *res* sites? Or do some synapses have an architecture unlike that of Tn3/γδ? A model for the Sin synapse (Fig. 3.9; Rowland et al., 2002; see also Canosa et al., 2003) proposes the former option, with a core structure that corresponds to the interacting

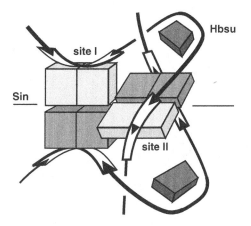

Figure 3.9. Cartoon of a model for the Sin synapse (from Rowland et al., 2002). For explanation, see legend to Fig. 3.4b (a similar representation of the Sarkis et al. model for the Tn*3*/γδ resolvase synapse). Note that a DNA bend in each *res* site is stabilized by Hbsu in the Sin synapse, and by resolvase at site II in the Tn*3*/γδ synapse.

tetramers at sites I and III in the γδ synapse structure of Sarkis et al. (2001) (Fig. 3.4b).

Note that some type II transposons have tyrosine recombinase-based resolution systems (see Grindley, 2002).

GIN AND HIN: DNA INVERTASES

The DNA invertases comprise a compact group of serine recombinases, whose biological role is to invert a defined DNA segment, thereby changing gene expression or protein sequence (for a recent comprehensive review, see Johnson, 2002). The Gin and Hin invertases, and the systems of which they are the protagonists, are the most fully characterized of the group. Gin is encoded by bacteriophage Mu. Inversion of the Mu "G segment," promoted by Gin, switches between two alternative coding sequences for the C-terminus of a tail fiber protein. In *Salmonella typhimurium*, Hin-mediated inversion of the "H segment" switches the expression of a bacterial surface protein on or off. In both cases, the inversion system is a strategem to evade host defenses (reviewed by Johnson, 2002).

The components of the Gin or Hin inversion systems differ considerably from those of the resolution systems. Strand exchange takes place at the centers of two synapsed recombinase dimer-binding sites (*gix* or *hix*, respectively), which can be identified with site I of Tn*3 res*, but there are no essential adjacent "accessory" recombinase-binding sites (Glasgow et al.,

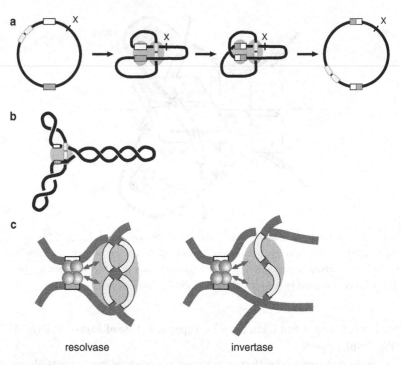

Figure 3.10. (a) Synapsis and recombination by DNA invertases. The enhancer is shown as a pair of narrow boxes, and the inversion sites are white or grey. Invertase and FIS are represented by shaded ovals. (b) Inversion synapse (equivalent to the second panel in a), showing how this complex might assemble at a 3-way junction in supercoiled DNA (Heichman and Johnson, 1990). (c) Comparison of models for the Tn3/γδ resolvase-*res* resolution synapse and the inversion synapse to illustrate how the function of the intertwined accessory sequences of *res*, bound by resolvase (diagram on the left), might be analogous to the function of the Fis-bound enhancer (diagram on the right). The arrows indicate possible interactions between the catalytic recombinase subunits and the proteins bound at the accessory sequences.

1989; Mertens et al., 1988). However, efficient inversion requires a third site, the "enhancer" or "*sis*," which may be at an indeterminate distance from the two *hix* or *gix* sites, but must be in the same DNA molecule (Huber et al., 1985; Johnson and Simon, 1985; Kahmann et al., 1985; Fig. 3.10a). In the natural systems, the 60 bp enhancer is within the coding region of the recombinase gene, overlapping the codon for the conserved serine that attacks the DNA. This sequence binds two dimers of a bacterially encoded DNA-binding protein Fis (factor for inversion stimulation), 48 bp apart (Bruist et al., 1987; Hubner and Arber, 1989). Fis is an abundant "architectural" DNA-binding protein of *Escherichia coli*, with some sequence selectivity, which has other

Within the figure region: resolvase   invertase

102

SALLY J. ROWLAND AND W. MARSHALL STARK

roles in the cell (as a transcription activator), and is also involved in the phage λ integration system (Finkel and Johnson, 1992; Travers et al., 2001). The crystal structure of Fis has been solved (Kostrewa et al., 1991; Yuan et al., 1991), but we do not yet have its structure in a complex with DNA. How Fis and the enhancer might act in inversion by Gin/Hin is discussed later in this chapter. The Hin system requires yet another protein, the bacterially encoded HU, which is a non-specific DNA-binding and -bending protein (Rice, 1997). HU is required to facilitate the tight bending of the short segment of DNA between one of the *hix* sites and the enhancer; this requirement is abolished if the spacing between these two elements is increased (Haykinson and Johnson, 1993).

Gin and Hin have been purified and studied extensively *in vitro* (reviewed by Johnson, 2002). To cut a long story short, the mechanism of catalysis of strand exchange by these enzymes seems to be essentially identical to that of Tn*3*/γδ resolvase. Both DNA strands at the center of the two *hix* or *gix* sites are broken by attack of the serine residue homologous to Ser10 of resolvase, making double-strand breaks with invertase attached to the 5′ ends of the DNA (Johnson and Bruist, 1989; Klippel et al., 1988a). The top- and bottom-strand breaks are staggered by 2 bp. Unlike site I of Tn*3* or γδ *res*, the central 2 bp of the inversion sites are not palindromic, so the 2-nucleotide extensions on the cleaved half-sites are non-identical. Strand exchange then takes place by a mechanism equivalent to a simple rotation of one pair of DNA ends relative to the other pair, as for resolvase (Crisona et al., 1994; Kanaar et al., 1988, 1990). There are no high-resolution structures of an intact invertase protein, but a structure of the Hin DNA-binding domain in a complex with its DNA recognition motif has been solved (Feng et al., 1994). The structure is similar to that of the γδ resolvase C-terminal domain.

Like Tn*3*/γδ resolution, inversion *in vitro* shows strong topological selectivity. Recombination is intramolecular, and requires a negatively supercoiled substrate with *gix* or *hix* sites in inverted repeat. The recombination (inversion) product is unknotted (Johnson and Simon, 1985; Kanaar et al., 1988). These observations, and many further biochemical and topological studies, indicated that inversion takes place in a synaptic complex of defined topology, analogous to the synapse of the resolution systems (reviewed by Johnson, 2002; Fig. 3.10). It is proposed that this synapse contains the enhancer with its bound Fis, in contact with the paired *hix*/*gix* sites (Heichman and Johnson, 1990). The *hix*-Hin dimer complex can make a stable synaptic interaction with another identical complex, unlike the site I–resolvase complex, but this synapse is not competent for strand exchange in the absence of the Fis-enhancer complex (Johnson and Bruist, 1989; Kanaar et al., 1989, 1990). In a substrate with *gix* (or *hix*) sites in the "wrong" (direct repeat) orientation, no resolution (or inversion) products are formed, but surprisingly, catalysis of strand exchange is not blocked. Topological analysis revealed extensive knotting of the reaction products, consistent with multiple "double rounds" of strand exchange (Heichman et al., 1991; Kanaar et al., 1990). This phenomenon is analogous to the resolvase-catalyzed knotting of a substrate containing "mismatched" res sites, described in the "Mechanism of strand exchange" section, and is a consequence of the fact that the central 2 bp of gix or hix is not palindromic. If the sites in a direct-repeat substrate are built into a synapse with the same structure as for normal inversion, the gix (or hix) sites will be aligned such that one round of strand exchange would create recombinants with two mismatched base pairs at the center of each site. Presumably, this configuration is unstable, and a second round of strand exchange restores correct base pairing in the non-recombinant configuration, but the rotation of the DNA ends results in knotting (c.f. Fig. 3.6; Moskowitz et al., 1991). Other types of recombination systems do not have this problem; for resolvases, the interactions of the accessory sites preclude recombination of sites in inverted repeat (Bednarz et al., 1990; Rowland et al., 2002), and tyrosine recombinases such as FLP or Cre detect non-homology of the overlap sequence at the center of the recombination site at an early, reversible stage of strand exchange (see Chapter 2).

There are striking similarities between the resolution and inversion synapses. In both, catalysis of strand exchange takes place at a pair of dimer-binding DNA sequences held together by (presumably) a tetramer of recombinase, and this assembly is believed to contact additional protein bound at accessory sites; more resolvase in the case of resolution, or Fis in the case of inversion. The DNA in the synapse is intertwined in such a way that selectivity for the appropriate relation of sites can be achieved (see Fig. 3.10). Catalytic activity by invertases must require something provided by the synaptic complex: a specific juxtaposition of the sites, specific contacts with Fis, or both – like the resolvase subunits bound at res site I, except that their contacts would be with other resolvase subunits. Despite a great deal of work, the surfaces of Hin or Gin that are involved in contacts with Fis remain undefined. In contrast, Fis residues essential for activation of the invertase have been identified and can be mapped onto the Fis crystal structure (Deufel et al., 1997; Safo et al., 1997). Because the Fis–enhancer interaction is not well characterized, proposals on the detailed structure for the Hin/Gin synapse (e.g. Merickel et al., 1998) remain tentative.

An important development in the field was the identification of mutant invertases that recombine in the absence of Fis or the enhancer element (Haffter and Bickle, 1988; Haykinson et al., 1996; Klippel et al., 1988b). Most of the "activating" mutations are at residues predicted (by mapping them onto the $\gamma\delta$ resolvase crystal structures) to be buried within the recombinase dimer. Several of these residues contribute either to the dimer interface or to contacts of the N-terminal part of the subunit (residues 1–100) with the E-helix (Fig. 3.3). There is a strong relation between these mutations and analogous activating mutations in Tn3 resolvase (see "Regulation of strand exchange by synapse formation" section), and in several cases, invertase and resolvase activating mutations have been found at homologous residues (Burke et al., 2004). The importance of the enhancer for recombination selectivity was highlighted by the loss of selectivity in the mutants; they no longer require supercoiling, and recombine sites in direct or inverted repeat orientations or on separate molecules. Furthermore, the product topology is no longer specified, and mixtures of knots or catenanes are obtained (Crisona et al., 1994; Klippel et al., 1993).

Why do invertases require bacterially encoded proteins for activity? The Hin system requires both Fis and HU. Host proteins are also required in some resolution systems (e.g. the Sin system; see "Other resolvases" section), and in some systems that use tyrosine recombinases (e.g. phage λ integration/excision, which uses Fis and the HU-like protein IHF). It has been proposed that dependence on Fis might allow recombination to be coupled to cell development, as levels of Fis vary substantially at different phases of the bacterial cell cycle (reviewed by Johnson, 2002). Whether these "architectural" bacterial proteins (Nash, 1996; Rice, 1997) are used simply "because they're there" or whether they are indeed used to couple recombination to other cellular activities is still uncertain for serine recombinases.

TRANSPOSASES AND PHAGE INTEGRASES OF THE SERINE RECOMBINASE FAMILY

Many temperate bacteriophages use serine recombinases to integrate and excise their DNA from the host. The recombinase proteins (integrases) are diverse in sequence, and universally are much bigger than the resolvases and invertases discussed previously (known examples are between 441 and 772 residues; Groth and Calos, 2004; Smith and Thorpe, 2002). The sequence homologous to resolvase lies in the N-terminal region of the integrases (the first 150 residues, approximately) and corresponds to the N-terminal (catalytic) domain of resolvase, including the E-helix. Nearer the C-terminus, there is no further sequence similarity, so there is no region corresponding to the helix-turn-helix DNA-binding domain of resolvases and invertases. The sequence following the catalytic domain contains several conserved residues, including an arrangement of cysteine residues reminiscent of zinc finger DNA-binding proteins (Smith and Thorpe, 2002). However, the functions of these large extensions to the basic recombinase module are still unclear. A few of these systems have been reconstituted *in vitro*, the most extensively characterized being from the *Streptomyces* bacteriophage φC31 (Ghosh et al., 2003; Groth and Calos, 2004; Smith et al., 2004a, 2004b; Thorpe and Smith, 1998; Thorpe et al., 2000). Purified φC31 integrase promotes recombination between the phage-encoded 50 bp *attP* site and the 51 bp host attachment site *attB*. Somehow, the sequence of the sites induces selectivity; *attP* does not recombine with *attP*, nor *attB* with *attB*. Neither do the recombinant sites, *attL* and *attR*, recombine with each other, as they would have to in an excision reaction (Thorpe et al., 2000). The factor(s) required for stimulation of *attL* × *attR* activity is not yet known. The site sequence selectivity of φC31 integrase implies that the mechanisms involved in regulation of catalytic activity must be very different from those described for resolvase (see "Regulation of strand exchange by synapse formation" section), possibly involving the C-terminal extension of the enzyme. Nevertheless, catalysis of recombination by this enzyme has features in common with resolvase. The "crossover sites" each bind a recombinase dimer, and the mode of breaking and rejoining the strands seems to be similar; a conserved serine residue provides the nucleophile that attacks the target phosphodiester of the DNA backbone, and double-stranded cleavage occurs at the center of the dimer-binding site, with the strand breaks staggered by 2 bp in the same way as at Tn*3 res* site I (Breüner et al., 2001; Thorpe and Smith, 1998).

Two other members of this class of recombinases, the SpoIVCA recombinase from *Bacillus subtilis* and XisF from *Anabaena*, reported as being involved in developmentally regulated DNA rearrangements, are now believed to be the integrases of prophage-like elements whose excision are necessary to allow differentiation (Kunkel et al., 1990; Ramaswamy et al., 1997). Related recombinases (transposases), encoded by various large transposons, such as Tn*4451*, Tn*5397*, and the *Staphylococcus* methicillin-resistance element SSC*mec* (Katayama et al., 2000; Lyras and Rood, 2000; Wang et al., 2000), also belong to this family. The mechanism of transposition seems to be essentially equivalent to that of phage integration. A circularized excised version of the transposon is an intermediate, which is inserted into the target DNA site by conservative recombination. So, unlike "traditional" transposases, these enzymes do not cause any target site duplication; when they excise, the original sequence of the target is restored. Usually the integration site has a 2 bp central "overlap" sequence that is identical to that of the transposon's crossover site (Crellin and Rood, 1997), so that integration does not create mismatched base pairs.

Some transposons have a different type of serine recombinase as their transposase. In these cases, the enzyme is similar in size to resolvase, and almost the entire sequence can be aligned, except that the helix-turn-helix motif, which in resolvase is at the C-terminus, is shifted to the N-terminus (Fig. 3.7). The N-terminal, presumed DNA-recognition domain containing the helix-turn-helix motif resembles that of the mercury repressor family (Heldwein and Brennan, 2001). The best studied example of this type is the IS607 transposase from *Helicobacter pylori*, but others have been identified by sequence analysis, including many examples from the Archea (Kersulyte et al., 2000; M.R. Boocock and Marshall Stark, unpublished analysis). The catalytic mechanism of these "back to front" recombinases is currently unknown.

EVOLUTION OF THE SERINE RECOMBINASES

The serine recombinase family is surprisingly homogeneous. A large proportion of the known members can be aligned with each other quite easily and are of similar size (Fig. 3.7). This may reflect some strong functional constraints on structure. Another conserved feature is that the catalytic domain is always at the extreme N-terminus of the protein, except in transposases such as that of IS607, which have an N-terminal DNA-binding domain. In this context, it is intriguing that domains similar to the C-terminal DNA-binding domain of the serine resolvases and invertases are found at the N-terminus of many "classical" transposases (e.g. Tc1: Iszvák et al., 2002, and references therein; Tn552: Rowland and Dyke, 1990). Creating "evolutionary trees" of serine recombinases is difficult, as might perhaps be expected from proteins that are quintessentially associated with mobile DNA. However, several sub-groups can be identified (Fig. 3.7; Smith and Thorpe, 2002). It has been speculated that the DNA invertases evolved relatively recently from one or more transposon resolvases, following insertions of Tn3-like transposons into bacteriophage or bacterial chromosomes (Kamp et al., 1984). Strangely, there are no bona fide examples of serine recombinases in higher eukaryotes. It is possible that we just cannot recognize them because of sequence divergence. Alternatively, some aspect of their structure or mechanism might be incompatible with the eukaryotic genome lifestyle.

Where did serine recombinases originate? It has recently been appreciated that the tyrosine recombinases are structurally related to the type IB topoisomerases (reviewed by Sherratt and Wigley, 1998). Serine recombinases invariably have topoisomerase activity too (Falvey et al., 1988; Krasnow and Cozzarelli, 1983), and structural similarity has been noted between the resolvase catalytic domain and the type IA topoisomerases (M.R. Boocock, personal communication). However, it is possible that the similarity is coincidental or convergent. A notable difference is that the type IA topoisomerases use the hydroxyl group of a tyrosine, rather than a serine residue as the DNA-cleaving nucleophile. There is also some structural similarity of the serine recombinase catalytic domain with parts of the purine and *lac* repressors.

A fascinating recent development has been the discovery that *res* sites are targets for insertion of some types of transposons (Kamali-Moghaddam and Sundstrom, 2000; Minakhina et al., 1999; Rowland et al., 2002). This targeting requires the cognate resolvase protein, and the positions of the insertions suggest that the transposase–transposon end complex recognizes some specific structural feature(s) of the resolvase–DNA complex. It is not yet clear whether the resolvases are specifically adapted to attract the transposition complex to the *res* site. The phenomenon could be important for transposon evolution, by promoting novel combinations of genes useful for mobile genetic elements (e.g. Liebert et al., 1999).

HYBRID/CHIMERIC SERINE RECOMBINASES

In research on DNA-binding proteins, a very common aim has been to change their sequence recognition properties to probe the basis of sequence-specific binding, create useful tools for further research, and more recently, create proteins that may have uses in biotechnology. Sequence recognition by resolvase-like serine recombinases has been shown to involve the small C-terminal domain, so a possible strategy is to substitute this domain with the homologous region of a related recombinase that recognizes a different sequence. Several "hybrid" recombinases have been made in this way, or by substitution of smaller parts of the C-terminal domain (Ackroyd et al., 1990; Avila et al., 1990; Schneider et al., 2000). The hybrid recombinases retain catalytic activity, but generally still prefer to act at the sites proper to the N-terminal domain, suggesting that this domain plays a significant role in recognition of the DNA. An alternative approach to alteration of sequence recognition, which also achieved partial conversion to new specificity, was to select mutants that bound to a chosen altered sequence (Grindley, 1993).

The identification of activated mutants of resolvase and invertases, which act at simple dimer-binding sites (site I of *res* or *gix/hix*) without the need for accessory sites (see "Tn*3* and $\gamma\delta$ resolvases: cointegrate resolution" and "Gin and Hin: DNA invertases" sections), has opened up the possibility of a much more radical approach, in which the C-terminal domain is replaced by an unrelated DNA-binding domain. The isolated N-terminal domain of an activated Tn*3* resolvase variant did not bind to site I with high affinity, nor catalyze any recombination, but when it was coupled to the DNA-binding zinc finger domain of the mouse transcription factor Zif268, the chimeric protein recognized and recombined at novel sites comprising 13 bp from the center of site I, flanked by the motifs recognized by Zif268 (Akopian et al., 2003). It is currently hoped that the Zif268 DNA-binding domain can be mutated systematically to create proteins that recognize any specific DNA sequence (Choo and Isalan, 2000; Isalan et al., 2001), so the resolvase-Zif chimeras may allow site-specific recombination to be directed to chosen pairs of sequences occurring naturally in a genome of interest, thereby providing powerful tools for therapeutic genetic alteration and biotechnology.

SERINE RECOMBINASES IN BIOTECHNOLOGY

Site-specific recombination is now used extensively to promote artificial controlled rearrangements of DNA in useful organisms, for experimental, genetic engineering, and potential gene therapy uses (Gorman and Bullock, 2000; Kilby et al., 1993; Nagy, 2000). Nearly all applications to date have used the tyrosine recombinases Cre or FLP. The dearth of application of serine recombinases may reflect their more tightly regulated activities, which has limited their transferability to novel situations. However, this may change soon, following identification of resolvase and invertase variants that no longer require regulatory sequences or factors, and the development of the bacteriophage integrases as integration-specific recombinases.

Activated mutants of $\gamma\delta$ resolvase, and wild-type β-recombinase from *S. pyogenes,* have been shown to be active in mammalian cell lines (Diaz et al., 1999; Schwikardi and Dröge, 2000), and an activated mutant of Gin has been shown to be active in plant protoplasts (Maeser and Kahmann, 1991). Highly activated Tn3 resolvase mutants recombine substrates containing two copies of the 28 bp *res* site I in mammalian cell lines (T. Brown and Marshall Stark, unpublished results). There has been much interest in the ϕC31 and other phage integrases because of their selectivity for integration, which is desirable for many biotechnology applications. Integration has been demonstrated in mammalian cells, and experiments to identify natural sequences in the human genome that might be targeted by ϕC31 Int, or mutants selected for improved activity, have been carried out (Groth and Calos, 2004; Groth et al., 2000; Sclimenti et al., 2001; Stoll et al., 2002).

CONCLUDING REMARKS

The feature shared by all serine recombinases is the "catalytic domain" homologous to the N-terminal part of Tn3 resolvase. This domain catalyzes conservative recombination by a mechanism that has some well-conserved features: the use of a serine nucleophile; linkage of the recombinase to the 5' ends of the cleaved DNA, double-stranded breaks with a 2 bp stagger at the center of a recombinase dimer-binding site, and a strand exchange mechanism equivalent to a rotation of the DNA ends, mediated by a catalytic domain tetramer. This basic catalytic module has been used in the natural recombinases, along with a domain conferring specific DNA recognition. The resolvases and DNA invertases have a helix-turn-helix domain attached at the C-terminus. The transposases related to that of IS607 have a similar domain, but at the N-terminus. The phage integrases and related transposases have a domain (or domains) of unknown type, C-terminal to the catalytic domain. Protein sequences with unknown function are attached at the C-terminus in some resolvases. Catalysis of strand exchange is universally subject to regulation that restricts the types, or geometrical relations, of pairs of sites that can recombine. This regulation often involves accessory DNA sequences and protein subunits. For the resolvases and DNA invertases, the structures of the synaptic complexes required for activation of catalysis are becoming better understood (see "Tn3 and $\gamma\delta$ resolvases: cointegrate resolution" and "Other resolvases" sections).

Every class of site-specific recombination (resolution, inversion, integration, and transposition) is now known to be catalyzed by serine recombinases specific to that role. We still have much to learn about this important family of enzymes.

REFERENCES

Abdel-Meguid, S. S., Grindley, N. D. F., Templeton, N. S., and Steitz, T. A. (1984). Cleavage of the site-specific recombination protein $\gamma\delta$ resolvase: The smaller of the two fragments binds DNA specifically. *Proc. Natl. Acad. Sci. USA*, **81**, 2001–2005.

Ackroyd, A. J., Avila, P., Parker, C. N., and Halford, S. E. (1990). Site-specific recombination by mutants of Tn21 resolvase with DNA recognition functions from Tn3 resolvase. *J. Mol. Biol.*, **216**, 633–643.

Akopian, A., He, J., Boocock, M. R., and Stark, W. M. (2003). Chimeric site-specific recombinases with designed DNA sequence recognition. *Proc. Natl. Acad. Sci. USA*, **100**, 8688–8691.

Alonso, J. C., Weise, F., and Rojo, F. (1995). The *Bacillus subtilis* histone-like protein Hbsu is required for DNA resolution and DNA inversion mediated by the β recombinase of pSM19035. *Mol. Microbiol.*, **18**, 471–478.

Arnold, P. H., Blake, D. G., Grindley, N. D. F., Boocock, M. R., and Stark, W. M. (1999). Mutants of Tn3 resolvase which do not require accessory binding sites for recombination activity. *EMBO J.*, **18**, 1407–1414.

Arthur, A., and Sherratt, D. J. (1979). Dissection of the transcription process: A transposon-encoded site-specific recombination system. *Mol. Gen. Genet.*, **175**, 267–274.

Avila, P., Ackroyd, A. J., and Halford, S. E. (1990). DNA binding by mutants of Tn21 resolvase with DNA recognition functions from Tn3 resolvase. *J. Mol. Biol.*, **216**, 645–655.

Bednarz, A. L., Boocock, M. R., and Sherratt, D. J. (1990). Determinants of correct *res* site alignment in site-specific recombination by Tn3 resolvase. *Genes Dev.*, **4**, 2366–2375.

Benjamin, H. W., and Cozzarelli, N. R. (1990). Geometric arrangements of Tn3 resolvase sites. *J. Biol. Chem.*, **265**, 6441–6447.

Blake, D. G., Boocock, M. R., Sherratt, D. J., and Stark, W. M. (1995). Cooperative binding of Tn3 resolvase monomers to a functionally asymmetric binding site. *Curr. Biol.*, **5**, 1036–1046.

Bliska, J. B., Benjamin, H. W., and Cozzarelli, N. R. (1991). Mechanism of Tn3 resolvase recombination *in vivo*. *J. Biol. Chem.*, **226**, 2041–2047.

Boocock, M. R., Brown, J. L., and Sherratt, D. J. (1987). Topological specificity in Tn3 resolvase catalysis. *In* T. J. Kelly and R. McMacken, eds. *DNA replication and recombination* (Alan R. Liss, New York), pp. 703–718.

Boocock, M. R., Zhu, X., and Grindley, N. D. F. (1995). Catalytic residues of γδ resolvase act in cis. *EMBO J.*, **14**, 5129–5140.

Breüner, A., Brøndsted, L., and Hammer, K. (2001). Resolvase-like recombination performed by the TP901-1 integrase. *Microbiology*, **147**, 2051–2063.

Bruist, M. F., Glasgow, A. C., Johnson, R. C., and Simon, M. I. (1987). Fis binding to the recombinational enhancer of the Hin DNA inversion system. *Genes Dev.*, **1**, 762–772.

Burke, M. E., Arnold, P. H., He, J., Wenwieser, S. V. C. T., Rowland, S. J., Boocock, M. R., and Stark, W. M. (2004). Activating mutations of Tn3 resolvase marking interfaces important in recombination catalysis and its regulation. *Mol. Microbiol.*, **51**, 937–948.

Canosa, I., Lopez, G., Rojo, F., Boocock, M. R., and Alonso, J. C. (2003). Synapsis and strand exchange in the resolution and DNA inversion reactions catalysed by the β recombinase. *Nucleic Acids Res.*, **31**, 1–7.

Choo, Y., and Isalan, M. (2000). Advances in zinc finger engineering. *Curr. Opin. Struct. Biol.*, **10**, 411–416.

Colloms, S. D., Bath, J., and Sherratt, D. J. (1997). Topological selectivity in Xer site-specific recombination. *Cell*, **88**, 855–864.

Crellin, P. K., and Rood, J. I. (1997). The resolvase/invertase domain of the site-specific recombinase TnpX is functional and recognizes a target sequence that resembles the junction of the circular form of the *Clostridium perfringens* transposon Tn*4451*. *J. Bacteriol.*, **179**, 5148–5156.

Crisona, N.J., Kanaar, R., Gonzalez, T. N., Zechiedrich, E. L., Klippel, A., and Cozzarelli, N. R. (1994). Processive recombination by wild-type Gin and an enhancer-independent mutant. Insight into the mechanisms of recombination site selectivity and strand exchange. *J. Mol. Biol.*, **243**, 437–457.

Deufel, A., Hermann, T., Kahmann, R., and Muskhelishvili, G. (1997). Stimulation of DNA inversion by FIS: Evidence for enhancer-independent contacts with the Gin-gix complex. *Nucleic Acids Res.*, **25**, 3832–3839.

Diaz, V., Rojo, F., Martinez, A. C., Alonso, J. C., and Bernad, A. (1999). The prokaryotic β-recombinase catalyses site-specific recombination in mammalian cells. *J. Biol. Chem.*, **274**, 6634–6640.

Dröge, P., Hatfull, G. F., Grindley, N. D. F., and Cozzarelli, N. R. (1990). The two functional domains of $\gamma\delta$ resolvase act on the same recombination site: Implications for the mechanism of strand exchange. *Proc. Natl. Acad. Sci. USA*, **87**, 5336–5340.

Falvey, E., and Grindley, N. D. F. (1987). Contacts between $\gamma\delta$ resolvase and the $\gamma\delta$ res site. *EMBO J.*, **6**, 815–821.

Falvey, E., Hatfull, G. F., and Grindley, N. D. F. (1988). Uncoupling of the recombination and topoisomerase activities of the $\gamma\delta$ resolvase by a mutation at the crossover point. *Nature*, **332**, 861–863.

Feng, J. A., Johnson, R. C., and Dickerson, R. E. (1994). Hin recombinase bound to DNA: The origin of specificity in major and minor groove interactions. *Science*, **263**, 348–355.

Finkel, S. E., and Johnson, R. C. (1992). The FIS protein: It's not just for DNA inversion anymore. *Mol. Microbiol.*, **6**, 3257–3265.

Ghosh, P., Kim, A. I., and Hatfull, G. F. (2003). The orientation of mycobacteriophage Bxb1 integration is solely dependent on the central dinucleotide of *attP* and *attB*. *Mol. Cell*, **12**, 1101–1111.

Glasgow, A. C., Bruist, M. F., and Simon, M. I. (1989). DNA-binding properties of the Hin recombinase. *J. Biol. Chem.*, **264**, 10072–10082.

Gorman, C., and Bullock, C. (2000). Site-specific gene targeting for gene expression in eukaryotes. *Curr. Opin. Biotech.*, **11**, 455–460.

Grainge, I., Buck, D., and Jayaram, M. (2000). Geometry of site alignment during Int family recombination: Antiparallel synapsis by the Flp recombinase. *J. Mol. Biol.*, **298**, 749–764.

Grindley, N. D. F. (1993). Analysis of a nucleoprotein complex: The synaptosome of $\gamma\delta$ resolvase. *Science*, **262**, 738–740.

SALLY J. ROWLAND AND W. MARSHALL STARK

Grindley, N. D. F. (1994). Resolvase-mediated site-specific recombination. *In* F. Eckstein and D. M. J. Lilley, eds. *Nucleic acids and molecular biology, vol. 8* (Springer-Verlag, Berlin), pp. 236–267.

Grindley, N. D. F. (2002). The movement of Tn3-like elements: Transposition and cointegrate resolution. *In* N. Craig, R. Craigie, M. Gellert, and A. Lambowitz, eds. *Mobile DNA II* (ASM Press, Washington, DC), pp. 272–302.

Grindley, N. D. F., Lauth, M. R., Wells, R. G., Wityk, R. J., Salvo, J. J., and Reed, R. R. (1982). Transposon-mediated site-specific recombination: Identification of three binding sites for resolvase at the *res* site of $\gamma\delta$ and Tn3. *Cell*, **30**, 19–27.

Groth, A. C., and Calos, M. P. (2004). Phage integrases: Biology and applications. *J. Mol. Biol.*, **335**, 667–678.

Groth, A. C., Olivares, E. C., Thyagarajan, B., and Calos, M. P. (2000). A phage integrase directs efficient site-specific integration in human cells. *Proc. Natl. Acad. Sci. USA*, **97**, 5995–6000.

Haffter, P., and Bickle, T. A. (1988). Enhancer-independent mutants of the Cin recombinase have a relaxed topological specificity. *EMBO J.*, **7**, 3991–3996.

Halford, S. E., Jordan, S. L., and Kirkbride, E. A. (1985). The resolvase protein from the transposon Tn21. *Mol. Gen. Genet.*, **200**, 169–175.

Hall, S. C., and Halford, S. E. (1993). Specificity of DNA recognition in the nucleoprotein complex for site-specific recombination by Tn21 resolvase. *Nucleic Acids Res.*, **21**, 5712–5719.

Hallet, B., and Sherratt, D. J. (1997). Transposition and site-specific recombination: Adapting DNA cut-and-paste mechanisms to a variety of genetic rearrangements. *FEMS Microbiol. Rev.*, **21**, 157–178.

Hatfull, G. F., and Grindley, N. D. F. (1986). Analysis of $\gamma\delta$ resolvase mutants *in vitro*: Evidence for an interaction between serine-10 of resolvase and site I of *res*. *Proc. Natl. Acad. Sci. USA*, **83**, 5429–5433.

Haykinson, M. J., Johnson, L. M., Soong, J., and Johnson, R. C. (1996). The Hin dimer interface is critical for Fis-mediated activation of the catalytic steps of site-specific DNA inversion. *Curr. Biol.*, **6**, 163–177.

Haykinson, M. J., and Johnson, R. C. (1993). DNA looping and the helical repeat *in vitro* and *in vivo*: Effect of HU protein and enhancer location on Hin invertasome assembly. *EMBO J.*, **12**, 2503–2512.

Heffron, F., McCarthy, B. J., Ohtsubo, H., and Ohtsubo, E. (1979). DNA sequence analysis of the transposon Tn3: Three genes and three sites involved in transposition of Tn3. *Cell*, **18**, 1153–1163.

Heichman, K. A., and Johnson, R. C. (1990). The Hin invertasome: Protein-mediated joining of distant recombination sites at the enhancer. *Science*, **249**, 511–517.

Heichman, K. A., Moskowitz, I. P. G., and Johnson, R. C. (1991). Configuration of DNA strands and mechanism of strand exchange in the Hin invertasome as revealed by analysis of recombinant knots. *Genes Dev.*, **5**, 1622–1634.

Heldwein, E. E., and Brennan, R. G. (2001). Crystal structure of the transcription activator BmrR bound to DNA and a drug. *Nature*, **409**, 378–382.

Huber, H. E., Iida, S., Arber, W., and Bickle, T. A. (1985). Site-specific DNA inversion is enhanced by a DNA sequence element in *cis*. *Proc. Natl. Acad. Sci. USA*, **82**, 3776–3780.

Hubner, P., and Arber, W. (1989). Mutational analysis of a prokaryotic recombinational enhancer with two functions. *EMBO J.*, **8**, 577–585.

Hughes, R. E., Hatfull, G. F., Rice, P. A., Steitz, T. A., and Grindley, N. D. F. (1990). Cooperativity mutants of the $\gamma\delta$ resolvase identify an essential interdimer interaction. *Cell*, **63**, 1331–1338.

Isalan, M., Klug, A., and Choo, Y. (2001). A rapid, generally applicable method to engineer zinc fingers illustrated by targeting the HIV-1 promoter. *Nat. Biotechnol.*, **19**, 656–660.

Iszvák, Z., Khare, D., Behlke, J., Heinemann, U., Plasterk, R. H., and Ivics, Z. (2002). Involvement of a bifunctional, paired-like DNA-binding domain and a transpositional enhancer in *Sleeping Beauty* transposition. *J. Biol. Chem.*, **277**, 34581–34588.

Johnson, R. C. (2002). Bacterial site-specific DNA inversion systems. In N. Craig, R. Craigie, M. Gellert, and A. Lambowitz, eds. *Mobile DNA II* (ASM Press, Washington, DC), pp. 230–271.

Johnson, R. C., and Bruist, M. F. (1989). Intermediates in Hin-mediated DNA inversion: A role for FIS and the recombinational enhancer in the strand exchange reaction. *EMBO J.*, **8**, 1581–1590.

Johnson, R. C., and Simon, M. I. (1985). Hin-mediated site-specific recombination requires two 26 bp recombination sites and a 60 bp enhancer. *Cell*, **41**, 781–791.

Kahmann, R., Rudt, F., Koch, C., and Mertens, G. (1985). G inversion in bacteriophage Mu DNA is stimulated by a site within the invertase gene and a host factor. *Cell*, **41**, 771–780.

Kamali-Moghaddam, M., and Sundstrom, L. (2000). Transposon targeting determined by resolvase. *FEMS Microbiol. Lett.*, **186**, 55–59.

Kamp, D., Kardas, W., Ritthaler, W., Sandulache, R., Schmucker, R., and Stern, B. (1984). Comparative analysis of invertible DNA in phage genomes. *Cold Spring Harb. Symp. Quant. Biol.*, **49**, 301–311.

Kanaar, R., Klippel, A., Shekhtman, E., Dungan, J. M., Kahmann, R., and Cozzarelli, N. R. (1990). Processive recombination by the phage Mu Gin system: Implications for the mechanism of DNA exchange, DNA site alignment, and enhancer action. *Cell*, **62**, 353–366.

Kanaar, R., van de Putte, P., and Cozzarelli, N. R. (1988). Gin-mediated DNA inversion: Product structure and the mechanism of strand exchange. *Proc. Natl. Acad. Sci. USA*, **85**, 752–756.

Kanaar, R., van de Putte, P., and Cozzarelli, N. R. (1989). Gin-mediated recombination of catenated and knotted DNA substrates: Implications for the mechanism of interaction between cis-acting sites. *Cell*, **58**, 147–159.

Katayama, Y., Ito, T., and Hiramatsu, K. (2000). A new class of genetic element, *Staphylococcus* cassette chromosome *mec*, encodes methicillin resistance in *Staphylococcus aureus*. *Antimicrob. Agents Chemother.*, **44**, 1549–1555.

Kersulyte, D., Mukhopadhyay, A. K., Shirai, M., Nakazawa, T., and Berg, D. E. (2000). Functional organization and insertion specificity of IS *607*, a chimeric element of *Helicobacter pylori*. *J. Bacteriol.*, **182**, 5300–5308.

Kilbride, E., Boocock, M. R., and Stark, W. M. (1999). Topological selectivity of a hybrid site-specific recombination system with elements from Tn*3 res*/resolvase and bacteriophage P1 *loxP*/Cre. *J. Mol. Biol.*, **289**, 1219–1230.

Kilby, N.J., Snaith, M. R., and Murray, J. A. H. (1993). Site-specific recombinases: Tools for genetic engineering. *Trends Genet.*, **9**, 413–421.

Kitts, P. A., Symington, L. S., Dyson, P., and Sherratt, D. J. (1983). Transposon-encoded site-specific recombination: Nature of the Tn*3* DNA sequences which constitute the recombination site *res*. *EMBO J.*, **2**, 1055–1060.

Klippel, A., Cloppenborg, K., and Kahmann, R. (1988b). Isolation and characterisation of unusual *gin* mutants. *EMBO J.*, **7**, 3983–3989.

Klippel, A., Kanaar, R., Kahmann, R., and Cozzarelli, N. R. (1993). Analysis of strand exchange and DNA binding of enhancer-independent Gin recombinase mutants. *EMBO J.*, **12**, 1047–1057.

Klippel, A., Mertens, G., Patschinsky, T., and Kahmann, R. (1988a). The DNA invertase Gin of phage Mu: Formation of a covalent complex with DNA via a phosphoserine at amino acid position 9. *EMBO J.*, **7**, 1229–1237.

Kostrewa, D., Granzin, J., Koch, C., Choe, H. W., Raghunathan, S., Wolf, W., et al. (1991). Three-dimensional structure of the *E. coli* DNA-binding protein FIS. *Nature*, **349**, 178–180.

Kostriken, R., Morita, C., and Heffron, F. (1981). Transposon Tn*3* encodes a site-specific recombination system: Identification of essential sequences, genes, and actual site of recombination. *Proc. Natl. Acad. Sci. USA*, **78**, 4041–4045.

Krasnow, M. A., and Cozzarelli, N. R. (1983). Site-specific relaxation and recombination by the Tn*3* resolvase: Recognition of the DNA path between oriented *res* sites. *Cell*, **32**, 1313–1324.

Krasnow, M. A., Stasiak, A., Spengler, S. J., Dean, F., Koller, T., and Cozzarelli, N. R. (1983). Determination of the absolute handedness of knots and catenanes of DNA. *Nature*, **304**, 559–560.

Kunkel, B., Losick, R., and Stragier, P. (1990). The *Bacillus subtilis* gene for the development transcription factor sigma K is generated by excision of a dispensable DNA element containing a sporulation recombinase gene. *Genes Dev.*, **4**, 525–535.

Leschziner, A. E., and Grindley, N. D. F. (2003). The architecture of the $\gamma\delta$ resolvase crossover site synaptic complex revealed by using constrained DNA substrates. *Mol. Cell*, **12**, 775–781.

Liebert, C. A, Hall, R. M., and Summers, A. O. (1999). Transposon Tn*21*, flagship of the floating genome. *Microbiol. Mol. Biol. Rev.*, **63**, 507–522.

Liu, T., DeRose, E. F., and Mullen, G. P. (1994). Determination of the structure of the DNA binding domain of $\gamma\delta$ resolvase in solution. *Protein Sci.*, **3**, 1286–1295.

Liu, T., Liu, D., DeRose, E. F. and Mullen, G. P. (1993). Studies of the dimerization and domain structure of $\gamma\delta$ resolvase. *J. Biol. Chem.*, **268**, 16309–16315.

Lyras, D., and Rood, J. I. (2000). Transposition of Tn*4451* and Tn*4453* involves a circular intermediate that forms a promoter for the large resolvase, TnpX. *Mol. Microbiol.*, **38**, 588–601.

Maeser, S., and Kahmann, R. (1991). The Gin recombinase of phage Mu can catalyse site-specific recombination in plant protoplasts. *Mol. Gen. Genet.*, **230**, 170–176.

Mazzarelli, J. M., Ermácora, M. R., Fox, R. O., and Grindley, N. D. F. (1993). Mapping interactions between the catalytic domain of resolvase and its DNA substrate using cysteine-coupled EDTA-Iron. *Biochemistry*, **32**, 2979–2986.

McIlwraith, M. J., Boocock, M. R., and Stark, W. M. (1997). Tn*3* resolvase catalyses multiple recombination events without intermediate rejoining of DNA ends. *J. Mol. Biol.*, **266**, 108–121.

Merickel, S. K., Haykinson, M. J., and Johnson, R. C. (1998). Communication between Hin recombinase and Fis regulatory subunits during coordinate activation of Hin-catalyzed site-specific recombination. *Genes Dev.*, **12**, 2803–2816.

Mertens, G. A., Klippel, A., Fuss, H., Blocker, H., Frank, R., and Kahmann, R. (1988). Site-specific recombination in bacteriophage Mu: Characterization of binding sites for the DNA invertase Gin. *EMBO J.*, **7**, 1219–1227.

Minakhina, S., Kholodii, G., Mindlin, S., Yurieva, O., and Nikiforov, V. (1999). Tn5053 family transposons are *res* site hunters sensing plasmidal *res* sites occupied by cognate resolvases. *Mol. Microbiol.*, **33**, 1059–1068.

Moskowitz, I. P., Heichman, K. A., and Johnson, R. C. (1991). Alignment of recombination sites in Hin-mediated site-specific recombination. *Genes Dev.*, 5, 1635–1645.

Murley, L. L., and Grindley, N. D. F. (1998). Architecture of the $\gamma\delta$ resolvase synaptosome: Oriented heterodimers identify interactions essential for synapsis and recombination. *Cell*, 95, 553–562.

Nagy, A. (2000). Cre recombinase: The universal reagent for genome tailoring. *Genesis* 26, 99–109.

Nash, H. A. (1996). The HU and IHF proteins: Accessory factors for complex protein-DNA assemblies. In E. C. C. Lin and A. S. Landes, eds. *Regulation of gene expression in* Escherichia coli (RG Landes Company, Austin, TX), pp. 149–179.

Pan, B., Deng, Z., Liu, D., Ghosh, S., and Mullen, G. P. (1997). Secondary and tertiary structural changes in gammadelta resolvase: Comparison of the wild-type enzyme, the I110R mutant, and the C-terminal DNA binding domain in solution. *Protein Sci.*, 6, 1237–1247.

Pan, B., Maciejewski, M. W., Marintchev, A., and Mullen, G. P. (2001). Solution structure of the catalytic domain of $\gamma\delta$ resolvase: Implications for the mechanism of catalysis. *J. Mol. Biol.*, 310, 1089–1107.

Parker, C. N., and Halford, S. E. (1991). Dynamics of long range interactions on DNA: The speed of synapsis during site-specific recombination by resolvase. *Cell*, 66, 781–791.

Petit, M. A., Ehrlich, D., and Janniere, L. (1995). pAMβ1 resolvase has an atypical recombination site and requires a histone-like protein HU. *Mol. Microbiol.*, 18, 271–282.

Ramaswamy, K. S., Carrasco, C. D., Fatma, T., and Golden, J. W. (1997). Cell-type specificity of the Anabaena *fdxN*-element rearrangement requires *xisH* and *xisI*. *Mol. Microbiol.*, 23, 1241–1249.

Reed, R. R. (1981a). Transposon-mediated site-specific recombination: A defined *in vitro* system. *Cell*, 25, 713–719.

Reed, R. R. (1981b). Resolution of cointegrates between transposons $\gamma\delta$ and Tn3 defines the recombination site. *Proc. Natl. Acad. Sci. USA*, 78, 3428–3432.

Reed, R. R., and Grindley, N. D. F. (1981). Transposon-mediated site-specific recombination *in vitro*: DNA cleavage and protein-DNA linkage at the recombination site. *Cell*, 25, 721–728.

Reed, R. R., and Moser, C. D. (1984). Resolvase-mediated recombination intermediates contain a serine residue covalently linked to DNA. *Cold Spring Harb. Symp. Quant. Biol.*, 49, 245–249.

Reed, R. R., Shibuya, G. I., and Steitz, J. A. (1982). Nucleotide sequence of the $\gamma\delta$ resolvase gene and demonstration that its gene product acts as a repressor of transcription. *Nature*, **300**, 381–383.

Rice, P. A. (1997). Making DNA do a U-turn: IHF and related proteins. *Curr. Opin. Struct. Biol.*, **7**, 86–93.

Rice, P. A., and Steitz, T. A. (1994a). Model for a DNA-mediated synaptic complex suggested by crystal packing of $\gamma\delta$ resolvase subunits. *EMBO J.*, **13**, 1514–1524.

Rice, P. A., and Steitz, T. A. (1994b). Refinement of $\gamma\delta$ resolvase reveals a strikingly flexible molecule. *Structure*, **2**, 371–384.

Rimphanitchayakit, V., Hatfull, G. F., and Grindley, N. D. F. (1989). The 43 residue DNA-binding domain of $\gamma\delta$ resolvase binds adjacent major and minor grooves of DNA. *Nucleic Acids Res.*, **17**, 1035–1050.

Rojo, F., and Alonso, J. C. (1994). A novel site-specific recombinase encoded by the *Streptococcus pyogenes* plasmid pSM19035. *J. Mol. Biol.*, **238**, 159–172.

Rojo, F., and Alonso, J. C. (1995). The β recombinase of plasmid pSM19035 binds to two adjacent sites, making different contacts at each of them. *Nucleic Acids Res.*, **23**, 3181–3188.

Rowland, S-J., and Dyke, K. G. H. (1990). Tn552, a novel transposable element from *Staphylococcus aureus*. *Mol. Microbiol.*, **4**, 961–975.

Rowland, S-J., Stark, W. M., and Boocock, M. R. (2002). Sin recombinase from *Staphylococcus aureus*: Synaptic complex architecture and transposon targeting. *Mol. Microbiol.*, **44**, 607–619.

Safo, M. K., Yang, W. Z., Corselli, L., Cramton, S. E., Yuan, H. S., and Johnson, R. C. (1997). The transactivation region of the Fis protein that controls site-specific DNA inversion contains extended mobile beta-hairpin arms. *EMBO J.*, **16**, 6860–6873.

Salvo, J. J., and Grindley, N. D. F. (1988). The $\gamma\delta$ resolvase bends the *res* site into a recombinogenic complex. *EMBO J.*, **11**, 3609–3616.

Sanderson, M. R., Freemont, P. S., Rice, P. A., Goldman, A., Hatfull, G. F., Grindley, N. D. F., and Steitz, T. A. (1990). The crystal structure of the catalytic domain of the site-specific recombination enzyme $\gamma\delta$ resolvase at 2.7 Å resolution. *Cell*, **63**, 1323–1329.

Sarkis, G. J., Murley, L. L., Leschziner, A. E., Boocock, M. R., Stark, W. M., and Grindley, N. D. F. (2001). A model for the $\gamma\delta$ resolvase synaptic complex. *Mol. Cell*, **8**, 623–631.

Schneider, F., Schwikardi, M., Muskhelishvili, G., and Dröge, P. (2000). A DNA-binding domain swap converts the invertase Gin into a resolvase. *J. Mol. Biol.*, **295**, 767–775.

Schwikardi, M., and Dröge, P. (2000). Site-specific recombination in mammalian cells catalyzed by $\gamma\delta$ resolvase mutants: Implications for the topology of episomal DNA. *FEBS Lett.*, **471**, 147–150.

Sclimenti, C. R., Thyagarajan, B., and Calos, M. P. (2001). Directed evolution of a recombinase for improved genomic integration at a native human sequence. *Nucleic Acids Res.*, **29**, 5044–5051.

Sherratt, D. J., and Wigley, D. B. (1998). Conserved themes but novel activities in recombinases and topoisomerases. *Cell*, **93**, 149–152.

Smith, M. C. A., Till, R., Brady, K., Soultanos, P., Thorpe, H., and Smith, M. C. M. (2004b). Synapsis and DNA cleavage in ϕc31 integrase-mediated site-specific recombination. *Nucleic Acids Res.*, **32**, 2607–2617.

Smith, M. C. A., Till, R., and Smith, M. C. M. (2004a). Switching the polarity of a bacteriophage integration system. *Mol. Microbiol.*, **51**, 1719–1728.

Smith, M. C. M., and Thorpe, H. M. (2002). Diversity in the serine recombinases. *Mol. Microbiol.*, **44**, 299–307.

Soultanas, P., Oram, M., and Halford, S. E. (1995). Site-specific recombination at *res* sites containing DNA-binding sequences for both Tn*21* and Tn*3* resolvases. *J. Mol. Biol.*, **245**, 208–218.

Stark, W. M., and Boocock, M. R. (1994). The linkage change of a knotting reaction catalysed by Tn*3* resolvase. *J. Mol. Biol.*, **239**, 25–36.

Stark, W. M., and Boocock, M. R. (1995). Topological selectivity in site-specific recombination. *In* D. J. Sherratt, ed. *Mobile genetic elements* (Frontiers in Molecular Biology series) (Oxford University Press), pp. 101–129.

Stark, W. M., Grindley, N. D. F., Hatfull, G. F., and Boocock, M. R. (1991). Resolvase-catalysed reactions between *res* sites differing in the central dinucleotide of subsite I. *EMBO J.*, **10**, 3541–3548.

Stark, W. M., Parker, C. N., Halford, S. E., and Boocock, M. R. (1994). Stereoselectivity of DNA catenane fusion by resolvase. *Nature*, **368**, 76–78.

Stark, W. M., Sherratt, D. J., and Boocock, M. R. (1989). Site-specific recombination by Tn*3* resolvase: Topological changes in the forward and reverse reactions. *Cell*, **58**, 779–790.

Stoll, S. M., Ginsburg, D. S., and Calos, M. P. (2002). Phage TP901-1 site-specific integrase functions in human cells. *J. Bacteriol.*, **184**, 3657–3663.

Thorpe, H. M., and Smith, M. C. M. (1998). *In vitro* site-specific integration of bacteriophage DNA catalysed by a recombinase of the resolvase/invertase family. *Proc. Natl. Acad. Sci. USA*, **95**, 5505–5510.

Thorpe, H. M., Wilson, S. E., and Smith, M. C. M. (2000). Control of directionality in the site-specific recombination system of the *Streptomyces* phage phiC31. *Mol. Microbiol.*, **38**, 232–241.

Travers, A., Schneider, R., and Muskhelishvili, G. (2001). DNA supercoiling and transcription in *Escherichia coli*: The FIS connection. *Biochimie*, **83**, 213–217.

Wang, H., Roberts, A. P., Lyras, D., Rood, J. I., Wilks, M., and Mullany, P. (2000). Characterization of the ends and target sites of the novel conjugative transposon Tn*5397* from *Clostridium difficile*: Excision and circularization are mediated by a large resolvase, TndX. *J. Bacteriol.*, **182**, 3775–3783.

Wasserman, S. A., and Cozzarelli, N. R. (1985). Determination of the stereostructure of the product of Tn3 resolvase by a general method. *Proc. Natl. Acad. Sci. USA*, **82**, 1079–1083.

Wasserman, S. A., Dungan, J. M., and Cozzarelli, N. R. (1985). Discovery of a predicted DNA knot substantiates a model for site-specific recombination. *Science*, **229**, 171–174.

Watson, M. A., Boocock, M. R., and Stark, W. M. (1996). Rate and selectivity of synapsis of *res* recombination sites by Tn3 resolvase. *J. Mol. Biol.*, **257**, 317–329.

Wells, R. G., and Grindley, N. D. F. (1984). Analysis of the $\gamma\delta$ *res* site: Sites required for site-specific recombination and gene expression. *J. Mol. Biol.*, **179**, 667–687.

Yang, W., and Steitz, T. A. (1995). Crystal structure of the site-specific recombinase $\gamma\delta$ resolvase complexed with a 34 bp cleavage site. *Cell*, **82**, 193–207.

Yuan, H. S., Finkel, S. E., Feng, J. A., Kaczor-Grzeskowiak, M., Johnson, R. C., and Dickerson, R. E. (1991). The molecular structure of wild-type and a mutant Fis protein: Relationship between mutational changes and recombinational enhancer function or DNA binding. *Proc. Natl. Acad. Sci. USA*, **88**, 9558–9562.

CHAPTER 4

Mobile introns and retroelements in bacteria

Steven Zimmerly

Introns and retroelements are hallmarks of eukaryotic genomes, but they are also found in bacteria. Here the different types of bacterial introns and retroelements are summarized, including group I introns, group II introns, archaeal bulge-helix-bulge (BHB) introns, intervening sequences (IVSs) in rRNAs, retrons, and diversity generating elements (DGRs). Except for the retroelements, these elements are evolutionarily unrelated, but nevertheless share intriguing properties. The elements all appear mobile within and among bacterial genomes, and in general, do not have clear phenotypic consequence to their host cells. It is possible that introns and retroelements spread from bacteria to eukaryotes as selfish DNAs or were present in the common ancestor of bacteria and eukaryotes.

INTRODUCTION

Introns and retroelements are typically considered components of eukaryotic genomes, because they were discovered in eukaryotes and are particularly abundant in higher eukaryotes. The human genome, for example, contains roughly 200,000 introns and nearly 3 million retroelements (including SINEs), dwarfing the number of functional genes, which are estimated at 30,000 (International Human Genome Sequencing Consortium, 2001). Together, introns and retroelements make up nearly half of the human genome and constitute the major types of "junk DNAs."

In bacteria, the major types of "junk DNAs" are transposons and prophages, which can constitute anywhere from <1% to 7% of a genome (e.g., Glaser et al., 2001). Introns and retroelements are comparatively rare in bacteria, but they have generated substantial interest because of their parallels to eukaryotic introns and retroelements. This chapter summarizes the varieties of introns and retroelements currently known in bacteria. The assemblage is somewhat miscellaneous because the elements are unrelated, but their similarities suggest similar biological niches and evolutionary pressures.

Introns are interrupting sequences within genes that are excised post-transcriptionally, with the flanking exons ligated together to generate a functional RNA. Mechanistically, there are three types of introns, which in bacteria are exemplified by group I introns, group II introns, and BHB introns. An additional intron-like element is included here, the IVSs of rRNAs, which are not true introns because their flanks are not ligated together after IVS excision. However, they are included because their mobility characteristics have implications for other types of introns.

Retroelements are genetic units that encode reverse transcriptases (RTs). Here the term "retroelement" is used regardless of whether the element is known to replicate itself by reverse transcription, or to be independently mobile. Of the three types of bacterial retroelements currently known (group II introns, retrons, DGRs), only group II introns have been shown to be independently mobile.

Apart from their connections to eukaryotic elements, these elements are similar in that they appear to be "silent parasites" that do not obviously benefit or damage their hosts, with the exception of DGRs, which have a clear function. In addition, the elements appear to be mobile, capable of disseminating to other sites within a genome or spreading to other genomes. These properties may be connected: If the elements are neutral phenotypically, then mobility may be necessary for their long-term survival.

INTRONS

Group I introns

Structure and distribution

Group I introns are self-splicing RNAs of roughly 250 to 500 nucleotides, which fold into a conserved secondary structure of 10 paired regions (P1–P10) (Fig. 4.1A). The intron RNA structures are classified into five subclasses based on structural variations (IA, IB, IC, ID, and IE), and the subclasses are subdivided further (e.g., A1, A2, etc.; Jaeger et al., 1996; Michel and Westhof, 1990). The conserved structures allow the introns to self-splice *in vitro* without the help of protein. Many group I introns encode ORFs, which loop out from one of the peripheral regions (e.g., the loop of P2, P8, or P9). The ORFs fall into four families of nucleases: LAGLIDADG, GIY-YIG, HNH, and His-Cys

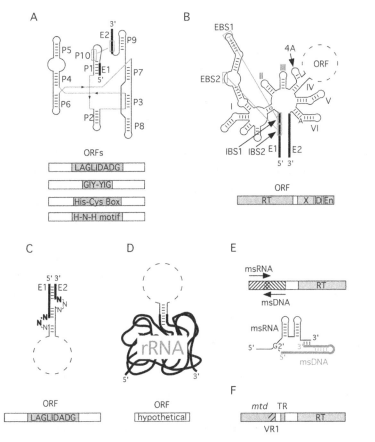

Figure 4.1. RNA and ORF structures of introns and retroelements. Exons are indicated by thick solid segments and are labeled E1 and E2. **(A)** Generalized group I intron RNA structure with 10 pairing regions (P1-P10). The four classes of intron-encoded ORFs are diagrammed below. Arrows show polarity for the connecting lines. The ORFs can be in multiple locations including loops of P2, P5, P6, P8, and P9. Most intron-encoded LAGLIDADG proteins have two LAGLIDADG motifs although only one is drawn. **(B)** Generalized group II intron RNA structure with 6 domains, and with the ORF looped out of domain 4. The high affinity binding site for the protein is domain 4A. The ORF domains are RT (reverse transcriptase), X (maturase, or splicing) D (DNA binding), and En (nuclease). **(C)** Generalized structure of the archaeal bulge-helix-bulge (BHB) motif. **(D)** Generalized structure of IVSs in ribosomal rRNAs. The IVS consists of an extension of an existing rRNA stem with a loop of variable length and sometimes a hypothetical ORF. **(E)** Retron structure, consisting of msRNA, msDNA, and RT genes. The msDNA produced from the retron is an RNA linked to a DNA by a branched 2'-5' phosphate bond. **(F)** *Bordetella* phage retroelement structure, consisting of the *mtd* gene (putative tail protein) with the VR1 (variable region) at its end, followed by the TR (template repeat) gene, and the reverse transcriptase gene.

(Belfort et al., 2002). The intron-encoded proteins have two functions. First, they are nucleases that function in intron mobility. Second, some intron-encoded proteins of the LAGLIDADG family aid in the intron-specific splicing of their host introns, a function known as *maturase activity*.

Hundreds of group I introns have been identified in the nuclear, mitochondrial, and chloroplast genomes of lower eukaryotes (Cannone et al., 2002). In bacteria, in contrast, there are comparatively few group I introns, approximately 20 in all, although the number is increased if one considers essentially identical introns in related species (Edgell et al., 2000). Interestingly, the smallest known group I introns are bacterial (*Azoarcus* and *Anabaena* tRNAs [Leu], 205 and 237 bp respectively; Reinhold-Hurek and Shub, 1992). Bacterial group I introns do not reflect the full range of structural diversity known in eukaryotic sources, and are represented only by A2, B4, and C3 classes (Everett et al., 1999; Landthaler and Shub, 1999; Tanner and Cech, 1996). Their homing endonucleases include only LAGLIDADG, GIY-YIG, and HNH families, but not the His-Cys class (Edgell et al., 2000). One bacterial intron encodes a novel nuclease not yet found in organellar introns (Bonocora and Shub, 2001). Group I introns are present in gram-negative, gram-positive, chlamydial, and cyanobacterial species, and in phages of gram-negative and gram-positive bacteria. It has been observed that group I introns in the chromosome are located exclusively in tRNA and rRNA genes, whereas introns in phages and prophages are located in protein genes, suggesting a functional difference (Edgell et al., 2000). However, this distinction was blurred more recently by the discovery of a group I intron in the chromosomal *recA* gene of *Bacillus anthracis*, which splices *in vivo* and does not appear to interfere with *recA* gene function (Ko et al., 2002).

Splicing

The splicing mechanism of group I introns is well established and proceeds through two transesterifications (Lambowitz et al., 1999). In the first transesterification, the 3' OH of a free guanosine attacks the 5' junction, becoming linked to the 5' end of the intron by a 5'-3' phosphate bond (Fig. 4.2A). The 3' OH of the 5' exon then attacks the 3' intron–exon junction to ligate the exons together and release linear intron with the added G residue.

Although some group I introns are capable of self-splicing *in vitro*, most do not, and *in vivo*, additional splicing factors are required. In yeast and *Neurospora* mitochondria, such splicing factors include the nuclear-encoded proteins CYT-18, CYT-19, CBP2, and MSS116 (Lambowitz et al., 1999; Mohr et al., 2002; Seraphin et al., 1989; Weeks and Cech, 1995). The splicing

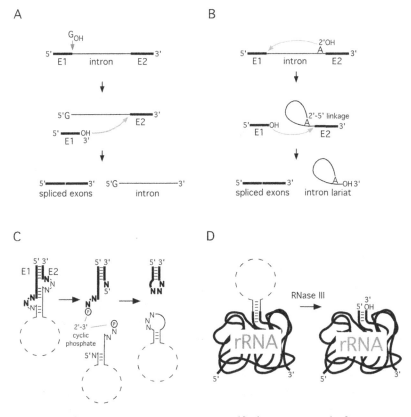

Figure 4.2. Splicing reactions. **(A)** Group I intron self-splicing reaction. The first transesterification is initiated by the 3′ OH of a free guanosine. In the second transesterification, the exons are ligated and intron released in a linear form. **(B)** Group II intron self-splicing reaction. Like group I introns, splicing occurs through two transesterifications, but the initiating nucleophile is the 2′ OH of a bulged A in domain 6 (see Fig. 4.1B), and the intron is released in a lariat form. **(C)** Bulge-helix-bulge (BHB) intron splicing reaction. Splicing is catalyzed by several protein enzymes. The initial cleavages generate 5′ OH and 2′-3′-cyclic phosphate ends. Both intron and exons are circularized during the ligation step. **(D)** IVSs in rRNAs are removed by RNase III, but flanking RNA sequences are not ligated together.

co-factors can act through multiple mechanisms, either by forming stable complexes between the protein and intron RNA to help the intron RNAs achieve a catalytic structure, or by transient interactions that resolve misfolded RNAs. In addition to host-encoded proteins, some intron-encoded LAGLIDADG proteins have also been shown to have maturase activity and to aid in splicing through direct binding to the intron RNA (Bolduc et al.,

2003). However, many intron-encoded proteins probably do not have a role in splicing and only function in mobility (Lambowitz et al., 1999).

In bacteria, unlike the mitochondrial introns, little is known about splicing cofactors for group I introns. So far, no bacterial intron-encoded proteins are known to have maturase activity. Of host-encoded proteins, StpA of *Escherichia coli* has been shown to suppress a group I intron splicing defect *in vivo* (Zhang et al., 1995), and ribosomal protein S12 has been shown to aid splicing *in vitro* (Coetzee et al., 1994). Because these proteins are not stably associated with the RNAs, they have been called RNA chaperones. However, it is not clear whether they are required for *in vivo* splicing under normal conditions.

Mobility

About two-thirds of bacterial group I introns encode ORFs for endonucleases and are putatively mobile. Mobility occurs via homing, in which the intron inserts site specifically into intronless homing sites. The homing site typically extends for 20 to 30 bp and includes sequence from both exon flanks. Thus, the homing endonucleases cleave exon sites lacking intron but not sites interrupted by an intron. Homing occurs when cells expressing intron are invaded by homing site DNA during processes such as conjugation or phage infection.

Homing has been studied in most molecular detail in *E. coli* for the *td* (thymidylate synthase) intron of phage T4. When phage lacking the *td* intron infects a cell expressing the *td* intron from a plasmid, the intron inserts into the phage's intronless *td* gene. First, the intron-encoded homing endonuclease (of the GIY-YIG class) cleaves the phage *td* gene, after which endogenous nucleases degrade the exposed DNA ends by tens to hundreds of base pairs. The exposed ends stimulate a gene conversion process, in which the cleaved DNA is repaired using intron-containing DNA as a template (Fig. 4.3A). For this process, there must be sequence homology between the donor allele and recipient homing site, but the intron's catalytic RNA activity does not play a role. Gene conversion can proceed through the classic double-stranded break repair pathway with a Holliday junction intermediate, or the related synthesis-dependent strand annealing pathway (Belfort et al., 2002). Other introns are expected to follow the same overall pathway, but there may be variations dependent on the repair and recombination enzymes available in different host cells.

A variation of homing occurs for the *Bacillus subtilis* phages SP01 and SP82 (Goodrich-Blair and Shub, 1996). Each phage contains a related but nonidentical group I intron with an HNH ORF. The introns are located in

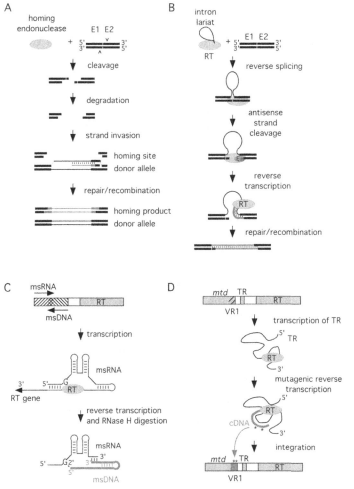

Figure 4.3. Mobility-related reactions of bacterial introns and retroelements. **(A)** Group I intron mobility. The intron-encoded endonuclease cleaves the homing site DNA target. The exposed ends are degraded by nucleases, and the 3′ strands invade an intron-containing allele. Repair activities result in gene conversion. **(B)** Group II intron mobility. The RNP particle consisting of an intron lariat and RT initiates mobility. The intron reverse splices into the sense strand of the DNA target. The En domain cleaves the bottom strand, and the RT reverse transcribes the intron RNA. Repair and recombination activities complete insertion. **(C)** The retron reverse transcription reaction. The msRNA, msDNA, and RT genes are transcribed in a single transcript. The RT binds to the transcript, and the 2′ OH of a G is the primer for partial reverse transcription of the RNA. **(D)** Mutagenic homing by the *Bordetella* phage retroelement. The TR RNA template is reverse transcribed by the RT and the cDNA is integrated into the VR1 region of the *mtd* gene. Reverse transcription of TR is mutagenic and results in insertion of all four nucleotides at positions coded by **A**, resulting in a partially randomized *mtd* sequence.

the phage DNA polymerase gene, and each HNH protein nicks the bottom strand of both intronless and intron-containing alleles, but with a preference for the heterologous intron. The *in vivo* consequence is that, upon coinfection of the two phages, the resident intron is replaced with its counterpart from the other phage. Thus, the two introns compete for the same position within a single gene, rather than filling an intronless site.

In addition to homing, group I introns can spread to ectopic sites (transposition) at much lower frequencies. The transposition process is important because it allows introns to spread to new locations, as has occurred in nature. However, because these events are rare, the mechanism is unknown for group I introns. One possibility is illegitimate integrations during homing reactions, which might rely on cryptic stretches of very short homology between the donor and recipient flanks (Belfort et al., 2002). An alternative is reverse splicing of the intron into noncognate RNAs, followed by reverse transcription and integration of the sequence into a new chromosomal site through homologous recombination. Such reverse splicing into noncognate RNAs has been observed *in vivo* for the *Tetrahymena* group I intron into the *E. coli* 23S rRNA (Roman and Woodson, 1998). However, the sequence was not subsequently integrated into the genomic DNA, and hence, this mechanism remains hypothetical.

In contrast to such mobility, at least one intron, the group I intron of tRNA ^{Leu} (UAA), has been predicted to be immobile because the split between eukaryotes and eubacteria (Kuhlsel et al., 1990). This conclusion is based on the observation that homologous introns are found in tRNA ^{Leu} (UAA) of cyanobacteria and chloroplasts, and the introns do not encode homing endonucleases. However, the situation became more complicated when a closer inspection of cyanobacteria revealed a nonuniform distribution, with many strains lacking the intron, and with some strains containing a second, less related tRNA ^{Leu} intron that appears to have been horizontally transferred into cyanobacteria (Costa et al., 2002; Rudi and Jakobsen, 1999). Still, the data are most consistent with primarily vertical transmission of the tRNA ^{Leu} (UAA) intron (Simon et al., 2003).

Group II introns

Group II introns share many properties with group I introns, including self-splicing activity, intron-encoded maturase proteins that aid in splicing and mobility, and mobility through homing and transposition. However, because group II introns act as retroelements, they are described in the "Retroelement" section.

Archaebacterial bulge-helix-bulge introns

Archaebacterial BHB introns are structurally the simplest of introns. Their conserved motif consists of a four base pair helix flanked by two three-nucleotide bulges (Fig. 4.1C). The distal loop is variable in length to accommodate different sizes of introns and sometimes ORFs (Lykke-Andersen et al., 1997; Trotta and Abelson, 1999). BHB introns are quite common in archaeal tRNA genes, and usually the introns are located 1 nt after the anticodon residue, with the anticodon paired in the four base pair stem (Fig. 4.1C). The introns can also occur in other locations within tRNAs, or in several locations within rRNAs. More recently, a BHB intron was found in an mRNA gene (Cbf5p) in *Aeropyrum* and *Sulfolobus* (Watanabe et al., 2002), and so this intron type is not restricted to structural RNA genes. Interestingly, BHB motifs also occur within the intergenic spacers between 16S and 23S rRNAs (Tang et al., 2002), and here their function is in processing rRNA transcripts, with the BHB motifs directing cleavages of precursor transcripts.

Splicing of BHB introns is accomplished by several protein enzymes. The introns are not self-splicing, although one BHB intron has been reported to have traces of self-cleavage activity *in vitro* (Weber et al., 1996). The most important splicing enzyme is the endonuclease that recognizes the BHB motif and makes the initial two cleavages, producing $2'$-$3'$ cyclic phosphates and $5'$ hydroxyls (Fig. 4.2C). Subsequent steps to accomplish exon ligation have not been worked out biochemically in archaebacteria, but involve resolution of the cyclic phosphate and ligation of the two exon pieces. Interestingly, exon ligation results in circularization of the intron, at least for some longer introns (Dalgaard and Garrett, 1992; Lykke-Andersen et al., 1997). It is interesting that for rRNA processing events at BHB motifs, exon ligation and intron circularization also occur (Tang et al., 2002). The significance of intron circularization is not clear for either intron splicing or rRNA processing.

Evolutionarily, the archaeal introns are related to tRNA introns in nuclear genomes. Nuclear tRNA introns are common in lower eukaryotes (e.g., yeast) and even occur in humans (Trotta and Abelson, 1999). The eukaryotic introns are similar to archaeal BHB introns in structure, but occur exclusively in tRNAs and always one nucleotide past the anticodon. The definitive evolutionary connection came after purification and cloning of the splicing endonucleases from yeast and archaeal sources, which showed the endonucleases to be homologous (Kleman-Leyer et al., 1997; Trotta et al., 1997). In yeast, the enzymes acting downstream of the endonuclease have been identified (Trotta and Abelson, 1999), but these enzymes are not present in archea, suggesting that either the enzymes are too divergent to identify or that other enzymes serve the equivalent functions in archaebacteria.

Like group I and group II introns, some archaebacterial introns encode ORFs and are mobile. Most archaeal intron ORFs belong to the LAGLIDADG family that is common in group I introns. There are also other putative intron-encoded ORFs with no similarity to known proteins (Burggraf et al., 1993). By analogy with group I introns, the LAGLIDADG proteins are expected to be homing endonucleases (Fig. 4.3A), and indeed, two BHB-encoded LAGLIDADG proteins have been expressed and shown to have endonuclease activities specific for the predicted homing sites (Dalgaard et al., 1993; Lykke-Andersen et al., 1994). Mobility was demonstrated *in vivo* by transforming a nonreplicating plasmid expressing intron into an intronless strain (Aagaard et al., 1995). In a small proportion of cells, a homing event was detected, with the intron inserted into the chromosomal tRNA homing site. Then, the intron-containing cells dominated the cell culture. It was concluded that homing initially occurred at a low level and that cells with intron had a replicative advantage, but also that the intron actively spread to intronless cells secondarily.

Intervening sequences in rRNAs of eubacteria

IVSs are found in the rRNAs of many eubacteria, including at least 17 species of proteobacteria and spirochetes (Pronk and Sanderson, 2001). The IVSs interrupt rRNA genes and are excised from mature rRNAs; however, they are not introns because the two rRNA halves are not ligated together after excision. The IVSs are tens to hundreds of nucleotides in length and do not have a conserved structure, except that their 5′ and 3′ ends are paired to each other, and form an extension of an existing helix in the rRNA (Fig. 4.1D). The extended stem is cleaved by RNase III, which also cleaves other dsRNA targets (Burgin et al., 1990). Ribosomes containing the cleaved rRNAs remain functional and are not structurally perturbed. Even when an IVS is not excised (e.g., in an RNase III-minus cell), the ribosomes are still functional, suggesting that the IVSs are on the periphery of the ribosome and that the presence or absence of IVS is phenotypically silent (Gregory et al., 1996). The IVSs can occur in at least three locations in the rRNA sequence, in helix 9, 25, and 45.

Like introns, IVSs appear to be mobile. The IVS of the spirochete *Leptospira* has a sporadic occurrence in *Leptospira* species, as well as among related strains, and phylogenetic evidence has inferred horizontal transfers (Ralph and McClelland, 1994). Intriguingly, the *Leptospira* IVS encodes a conserved, hypothetical ORF of approximately 120 amino acids, of unknown significance (Ralph and McClelland, 1993). IVSs in other species also show sporadic distributions, even when not encoding a putative ORF. These include the helix 9 IVS in alpha-proteobacteria (Evguenieva-Hackenberg and Klug, 2000), and helix-25 and helix-45 IVSs in enterobacteria (Miller et al., 2000; Pronk and Sanderson, 2001). A specific example of a horizontal transfer was inferred from the IVSs in *Salmonella* and *Yersinia*, where the IVSs are >80% identical, whereas host genomes are <20% identical (Skurnik and Toivanen, 1991). Finally, a survey of IVS composition of 88 *Salmonella* natural isolates showed widely heterogeneous content of the 25 and 45 helix IVSs (Pabbaraju et al., 2000). Of the seven *rrn* operons within the genome, some strains contain no IVSs, whereas others contain IVSs in all seven operons. The IVS distribution has surprisingly little correlation with strain groupings based on multilocus enzyme electrophoresis data, which led to the conclusion that these IVSs are mobile and horizontally transferred among closely related *Salmonella* strains. IVS gain and loss is inferred to be fairly inefficient because most strains have a mixture of filled and unfilled *rrn* sites. However, IVS movement must occur frequently enough for related strains to have different IVS compositions.

The mechanism of mobility has not been demonstrated experimentally. Based on the simplicity of the IVS structure, the mechanism is expected to rely wholly on gene conversion events, accomplished by host recombination and repair pathways. The point here is that the IVSs achieve sporadic distributions similar to those seen for group I and group II introns, even though IVSs do not appear to have intrinsic mobility activities. Therefore, host recombination activities may be sufficient to give sporadic distributions of introns or other elements. It is possible that some dispersal of group I, group II, and BHB introns in nature may also occur through cellular recombination mechanisms rather than the autonomous mobility activities of the introns.

RETROELEMENTS

Group II introns

Structure and distribution

The group II intron RNA structure consists of approximately 600 nucleotides with six domains (Fig. 4.1B). RNA structural subtypes are A1, A2, B1, B2, and C (Lambowitz and Zimmerly, 2004) but unlike group I introns, the structural variety is greater in bacteria than in organelles (Toor et al., 2001). Also unlike organellar introns, virtually all bacterial group II introns encode reverse transcriptase ORFs, which are located in domain IV (Fig. 4.1B). The proteins function in both intron splicing and mobility. The ORF contains an RT domain, domain X, domain D, and an En domain. Domain X is probably the "thumb" domain of the RT, analogous to thumb domains of other polymerases, but has acquired a role in splicing. Domain D is a DNA-binding domain, and the En domain is a nuclease domain important for mobility.

Like group I introns, group II introns were discovered in organellar genomes, where they are relatively abundant, and were subsequently found in bacteria. By now, barely 10 years after their discovery in bacteria (Ferat and Michel, 1993), more than 125 bacterial group II introns are known (Dai et al., 2003), and the number is growing steadily as more genomes are sequenced. Approximately one-fourth of sequenced bacterial genomes encode a full-length group II intron, and the introns are found across the bacterial spectrum, including gram-positive bacteria, proteobacteria, cyanobacteria, and archaebacteria.

Group II introns in bacteria appear to behave mainly as retroelements rather than introns, judging from their distribution (Dai and Zimmerly, 2002a). Bacterial group II introns are primarily located within mobile elements, such as plasmids, IS elements, prophages, or integrons. Many introns are located outside of ORFs, and virtually none are located in conserved genes. Some introns are inserted in the reverse orientation in genes, thereby inactivating the genes because the introns cannot be spliced out. In *E. coli*, group II intron content is highly heterogeneous among natural isolates (Dai and Zimmerly, 2002b), resembling the natural distribution of IVSs or retrons. In contrast, in *Sinorhizobium*, the intron content is comparatively homogeneous but still with variation (Muñoz et al., 2001). Finally, some introns are organized in nested configurations, which are formed when an intron inserts into another group II intron. Up to four levels of intron insertions can occur, and there are no flanking exons that would encode a functional gene product in these examples (Dai and Zimmerly, 2003). Of the bacterial retroelements described in this chapter, group II introns appear to have the most prominent characteristics of retroelements.

Splicing

Like group I introns, splicing of group II introns occurs in two transesterifications (Fig. 4.2B). In the first step, the 2′ OH of a bulged A of intron domain 6 attacks the 5′ splice, forming a 2′-5′ phosphate bond and a lariat structure. In the second step, the 3′ OH of the 5′ exon attacks the 3′ splice site, ligating together the exons and releasing the intron as a lariat. During this reaction, the exon-binding sequences 1 and 2 (EBS1, EBS2) of the intron base pair with the intron-binding sequences 1 and 2 (IBS1, IBS2), and this base pairing anchors the 5' exon during the two steps (Fig. 4.1B; Lambowitz and Zimmerly, 2004).

The only splicing factor known for group II introns in bacteria is the maturase encoded internally in the introns. Maturase activity has been demonstrated for several bacterial intron ORFs, including *Lactococcus* Ll.ltrB, *Sinorhizobium* RmInt1, and the *Clostridium* intron (Martínez-Abarca et al., 1998; Matsuura et al., 1997; Roberts et al., 2001). Biochemically, maturase function has been dissected for Ll.ltrB. The protein binds to domain 4A (Fig. 4.1B), the protein's high-affinity binding site, and then makes secondary contacts in domains 1, 2, and 6 (Matsuura et al., 2001), together inducing a conformation that precipitates splicing (Wank et al., 1999). Chemical protection experiments showed that the structural changes upon protein binding are similar to the structure achieved by the intron RNA under self-splicing conditions (Matsuura et al., 2001). After splicing, the protein remains tightly bound to the intron lariat, forming an RNP particle that carries out a mobility reaction.

Retrohoming

Again like group I introns, group II introns are mobile by homing into specific target sites of approximately 30 bp. The process is known as retrohoming because it is RT mediated, and the mechanism is termed target-primed reverse transcription (TPRT). Retrohoming has been studied intensively for Ll.ltrB. In the first step, the RNP particle (RT protein complexed to intron lariat) recognizes the homing site through specific recognition of nucleotides in the upstream exon target (−25 to −12 for Ll.ltrB; Mohr et al., 2000; Singh and Lambowitz, 2001), and also by RNA base pairing between the intron's EBS1 and 2 and the target's IBS1 and 2 (−12 to −1; Guo et al., 1997). Next, the intron RNA reverse splices into the sense strand of the DNA target, which is essentially the same reaction as splicing, but in the reverse direction. The En domain cleaves the antisense strand at position +9 (other group II introns cleave at position +10), and the exposed 3' OH of the bottom strand is the primer for reverse transcription of the intron (Matsuura et al., 1997; Zimmerly et al., 1995a, 1995b). Presumably, repair and recombination activities complete insertion, but the full details have not been reproduced *in vitro*. Coconversion studies indicate that in bacteria, flanking markers are not coconverted during bacterial homing (Cousineau et al., 1998), a difference from homing of organellar group II introns or group I introns. In addition, *recA* function is not required for homing of Ll.ltrB in either *E. coli* or *Lactococcus lactis* (Cousineau et al., 1998; Mills et al., 1997). Together, these differences suggest that in bacteria the intron completely reverse splices into the DNA target and is completely reverse transcribed, whereas in yeast mitochondria, both steps are partial, and thus repair functions are needed to finish the insertion reaction (Eskes et al., 2000).

Interestingly, whereas Ll.ltrB requires the En domain for mobility, most bacterial group II intron ORFs do not contain an En domain. One of these introns, RmInt1 of *Sinorhizobium meliloti*, is efficiently mobile nevertheless (Martínez-Abarca et al., 2000). Recent work indicates that mobility can occur without the En domain by a process in which an intron reverse splices into one strand of a target, and then a nascent strand at a DNA replication fork is used as the primer for reverse transcription (Zhong and Lambowitz, 2003; Lambowitz and Zimmerly, 2004).

Retrotransposition

In addition to homing, group II introns can insert into ectopic sites (retrotransposition). Retrotransposition has been experimentally addressed for Ll.ltrB in *L. lactis*. A genetic system was devised that selects for Ll.ltrB insertion into noncognate sites in the chromosome from a donor plasmid (Cousineau et al., 2000; Ichiyanagi et al., 2002). Although the mechanism of retrotransposition was initially suggested to involve reverse splicing into noncognate RNAs, followed by reverse transcription and homologous recombination (Cousineau et al., 2000), further characterization showed that the mechanism is reverse splicing into DNA targets (Ichiyanagi et al., 2002). The reaction is largely independent of RecA function and the En domain, and strand biases suggested reverse splicing into single-stranded regions of a replication fork (Ichiyanagi et al., 2002, Ichiyanagi et al., 2003).

Evolution

Group II introns are of great evolutionary interest because they are probably the ancestors of spliceosomal introns in eukaryotes. This idea is supported by essentially identical splicing pathways for the two types of introns (Fig. 4.2B), by RNA structural similarities between group II introns and the snRNAs of the spliceosome, and by evidence for catalytic activity by snRNAs (Valadkhan and Manley, 2001). According to this idea, ancestral self-splicing introns migrated to the nucleus and fragmented into the snRNA components of the spliceosome, thus becoming dependent on many proteins. Although this is an attractive idea, it is not proved because of many differences, or "missing links," between the two types of introns.

Another evolutionary connection for group II introns is that their RTs are related to RTs of non-LTR retroelements, a prominent class of retroelements in eukaryotes (e.g., LINE elements). Non-LTR element transposition parallels group II intron mobility because the cleaved target site is used to prime reverse transcription, and in fact, group II introns and non-LTR elements are the only known retroelements that transpose by a TPRT mechanism (Eickbush, 1999). Therefore, it is likely that either group II introns and non-LTR elements had a common ancestor that transposed by TPRT, or that group II introns themselves were ancestors of non-LTR elements.

RETRONS

Retrons were the first retroelements identified in bacteria, yet their function remains mysterious. It is still unknown whether they are independently mobile or whether they provide a useful function to their host cells. Retrons consist of a *ret* (RT) gene, a *msr* (RNA-encoding) gene, and a *msd* (DNA-encoding) gene (Fig. 4.1E). The *msr* and *msd* genes overlap in opposite orientations, and all three genes are transcribed as a single RNA transcript. The RNA structure corresponding to the *msr* and *msd* regions folds into a secondary structure containing two or three stem loops. The 2' OH of a conserved G residue is the primer for reverse transcription, forming a 2'-5' linkage between the RNA and the first polymerized DNA nucleotide (Fig. 4.3C). (Superficially, the G resembles the bulged A in domain 6 of group II introns, whose 2' OH also forms a branched 2'-5' linkage. The similarity is reinforced by the fact that yeast debranching enzyme can hydrolyze both 2'-5' linkages (Nam et al., 1994); however, the similarities may be coincidental.) Reverse transcription of the RNA template proceeds through the *msd* gene portion and halts at a region midway through the RNA template, leaving a 5 to 8 base pair helix connecting the RNA and DNA portions of the molecule. The RNA template strand is degraded by RNase H, which is most often provided by the cell's endogenous RNase H activity, although some retron RTs have an RNase H domain. (Retron RTs are the only bacterial RTs known to possess an RNase H domain.) After reverse transcription, the RT remains bound to the RNA–DNA branched hybrid molecule, which is known as an msDNA (multicopy single-stranded DNA), and can accumulate to levels of 500 to 700 copies in the cell (Lampson et al., 2001; Yamanaka et al., 2002).

Retrons are somewhat less common in bacteria than group II introns, with approximately forty currently known. They are found mainly in proteobacteria, including myxomycetes, *E. coli*, *Salmonella*, *Rhizobium*, *Klebsiella*, *Proteus*, *Melittangium*, and *Vibrio*. More recently, they have even been found in the archaebacterium *Methanococcus acetivorans* (Rest and Mindell, 2003), suggesting they will continue to be found in diverse bacterial species.

There are two patterns of retron distribution and spread, exemplified by the retrons in myxobacteria and in *E. coli*. In myxobacteria, the retron composition is widespread and fairly homogeneous, suggesting vertical inheritance and a long history within the organism. For example, all strains of *Myxococcus xanthus* and 95% of other myxobacterial species contain the retron Mx162 (the retron name is based on the species abbreviation and the length of the msDNA in nucleotides). Some strains contain an additional retron Mx65 (Lampson et al., 1991). Vertical inheritance is inferred from the uniform retron content and the retron codon usage, which is very similar to that of the host genome, suggesting a long association with the myxobacterial species (Inouye and Inouye, 1991; Rice and Lampson, 1995).

In contrast, retrons in proteobacteria such as *E. coli* are sporadic in distribution. Most *E. coli* strains do not contain retrons, but there are at least seven different retrons scattered in various strains (Ec48, Ec67, Ec73, Ec78, Ec83, Ec86, and Ec107). These seven retrons are not closely related to each other, suggesting horizontal transfers into *E. coli* from other sources (Herzer et al., 1990). Supporting this scenario, the codon usage of the retrons differs from that of *E. coli* (Inouye and Inouye, 1991).

Independent mobility of retrons has not yet been demonstrated, and it remains unclear how retrons are dispersed. The only observed mobility event in the laboratory was for Ec73, which is contained in a cryptic prophage, ϕR73 (Inouye et al., 1991). By co-infecting the strain with a helper phage, ϕR73 was induced to excise from the genome and infect and lysogenize a new host strain, resulting in msDNA formation in the new strain. Although this provides a mechanism for retron dispersal among species, it does not settle the question of whether the retron itself is independently mobile. In another example, the insertion site of retron Ec107 was compared in different *E. coli* strains, and it was concluded that insertion of Ec107 resulted in the replacement of a 38-bp segment in the *E. coli* chromosome with the retron; however, the precise mechanism of insertion remains unclear (Herzer et al., 1992).

One might expect that retrons should be independently mobile because their RTs are clearly active for reverse transcription, and at least some retrons have sporadic distributions. However, in the absence of clear evidence for independent mobility, it is also possible that their dispersal relies wholly on host mobile DNAs (e.g., a prophage) or general recombination functions. According to this possibility, it would be necessary that the retrons provide a function beneficial to the host organism in order to maintain their survival and propagation. This situation does in fact appear to be the case for the *Bordetella* phage retroelement (below), which provides a useful function for its host phage.

Although it is not clear whether retrons benefit their hosts, two functions have been suggested. First, it has been observed that some *E. coli* retrons are mutagenic to their host strains. The retrons Ec83 and Ec86 have stem loops with mismatched base pairs. It was shown that the mismatches in the retron RNA sequester MutS protein, which is required for mismatch repair, and thus the repair machinery is less available to repair DNA damage (Maas et al, 1994, 1996). However, not all retrons contain mismatched bases (Yamanaka et al., 2002).

A second possible function is speculated for *Vibrio* msDNA Vp96. In *Vibrio*, there is an exact correlation between pathogenic *Vibrio* strains and the presence of retron Vp96 in 12 of 21 isolates tested (Yamanaka et al., 2002). This correlation does not hold for *E. coli* retrons, and furthermore, myxobacteria always contain a retron but are nonpathogenic. So if there is a contribution to pathogenicity, the effect would seem to be localized to certain situations and genetic backgrounds. Therefore, the question remains open as to whether retrons have a function or are independently mobile.

Diversity-generating retroelements

The most recently identified retroelements in bacteria are the DGRs, first discovered in the bacteriophage BPP-1 of *Bordetella* (Liu et al., 2002). The *Bordetella* DGR consists of a reverse transcriptase gene (*brt*) preceded by an RNA template gene (TR) and a putative tail protein gene (*mtd*; Fig. 4.1F). The retroelement is not actively mobile, but instead serves an important role in phage tropism switching. *Bordetella* species are pathogenic and have two growth phases, Bvg+ and Bvg−. The Bvg+ phase occurs during pathogenic growth, whereas in the Bvg− phase virulence genes are repressed. The infecting phage similarly has three states, which preferentially infect either Bvg+ or Bvg− cells. BPP phages infect Bvg+ cells, BMP phages infect Bvg− cells, and BIP phages infect either Bvg+ or Bvg− cells at lower efficiencies. The three forms of phage can interconvert, and when their genomes were sequenced, the differences were located in the VR1 (variable region) of the *mtd* (tail) sequence, implying that a change in tail protein sequence determines whether the phage can infect Bvg+ or Bvg− cells. Remarkably, directly downstream of the *mtd* gene lies a 134-bp sequence (TR) that is 90% identical to the tail sequence that changes during interconversion of phage states. Genetically, it was shown that phage interconversion occurs through reverse transcription of the TR sequence and incorporation of the variant sequence into *mtd*. Because the TR sequence is constant, the variation in *mtd* sequence is caused by inherent inaccuracy of the RT. For all positions encoded by A in the TR template, all four nucleotides are incorporated in the cDNA. Thus, a partially randomized tail gene sequence is produced, from which emerges the new specificity toward Bvg+ or Bvg− cells. The overall process is called "mutagenic homing" because of similarities to group II intron homing.

The molecular mechanism of mutagenic homing is not yet clear. By analogy with other retroelements, it is expected that the TR template region folds into an RNA structure to which the RT binds, thus positioning the RT to reverse transcribe the proper sequence. Incorporation of the resulting cDNA into the *mtd* gene might occur by TPRT or another mechanism. One difference from group II introns is that the integration event is the replacement of old sequence with new, rather than simply being an insertion reaction. More recent characterization points to a remarkable intermediate containing a heteroduplex in the VR1 sequence. The intermediate appears to consist of one strand of "old" VR1 sequence and one strand of mutagenized cDNA, which then is resolved by recombination and repair machinery. The evidence for this is that there is not a fixed requirement for direction or continuity of mutagenesis (i.e. 3′-to-5′ as expected for cDNA synthesis), but rather, the mutagenesis can occur anywhere in the VR1 region, and in multiple patterns that are selected for genetically (Doulatov et al., 2004).

Interestingly, the *Bordetella* element is just one of many in a class, as revealed by homologs in the databases (Doulatov et al., 2004). There are at least nine other DGR examples in various bacteria, including pathogens and nonpathogens and in both phage and chromosomal locations. Each DGR has the same basic components of an RT, TR, and VR1 in a neighboring gene, and surprisingly, four examples have two neighboring genes with VR1 sequences.

DGRs are most closely related to group II introns based on phylogenetic analysis of RT sequences (Doulatov et al., 2004), and it would seem that either group II introns and DGRs descended from a common ancestor, or that group II introns are the direct ancestor of DGRs. DGRs cannot be ancestral, however, because reverse transcription is mutagenic and it is highly unlikely that the elements can replicate themselves. Instead they appear to be derived from a mobile retroelement, and adapted a useful function that perpetuates its survival. The discovery of DGRs is exciting not only for its novelty and obvious useful applications in biotechnology for *in vivo* mutagenesis of protein cassettes, but also because the discovery points to additional bacterial retroelements that yet may be discovered hidden away in selfish DNA elements.

CONCLUSION

This chapter has summarized the mobile introns and retroelements currently known in bacteria. It is striking that so much information has come recently from genome projects, and it is expected that the picture will be defined further as more genomes are sequenced. For example, group I introns may yet be found in archaebacteria, BHB introns may be found in eubacteria, etc., and there may be even entirely new varieties of introns or retroelements to be discovered.

Although the elements are not related, they share properties that suggest adaptation to similar biological niches. All the introns and retroelements appear to be phenotypically neutral (except DGRs) and are mobile in some way, being found in variable distributions in related strains or species. The simplest mechanism for spread is exemplified by the IVSs of rRNAs, which appear to move by host-mediated gene conversion events. This mode may also occur for other elements found in conserved genes or conserved sites (BHB, group I, and group II introns). Another possibly shared mechanism of spread is "hitchhiking" via host mobile DNAs (group I, group II introns, retrons, DGRs). In addition to these general mechanisms, group I and BHB introns share a specific mobility mechanism mediated by homing endonucleases, and group II introns possess a unique mechanism of dissemination.

Site-specific mobility is the rule for most of the elements, but transposition to new sites has obviously occurred during evolution for nearly all these elements. Transposition mechanisms may occur by illegitimate recombination during homing reactions (group I, BHB introns), by reverse splicing into DNA in cryptic target sites (group II introns) or by reverse splicing into RNA (group I, group II, BHB introns). Although reverse splicing into RNA remains hypothetical, it is also the favored model for how spliceosomal introns move to new sites.

A persistent question is whether these elements have a useful function. It is hard to imagine that such widespread elements have no function at all, and indeed, possible functions have been proposed for nearly all of them. Group I introns have been proposed to be involved in regulation of nucleic acid metabolism because several of the introns are found in nucleic acid metabolism genes, and splicing requires a guanosine nucleotide (Shub, 1991). IVSs have been speculated to be involved in rRNA turnover (Hsu et al., 1994). Some eukaryotic tRNA introns (relatives of BHB introns) are required for specific tRNA nucleotide modifications (Grosjean et al., 1997). Finally, retrons can be mutagenic and may sometimes contribute to pathogenicity. However, all these examples either apply to a subset of elements or are hypothetical. In nearly all cases, there are related strains or species that lack the element, with little apparent effect. Hence, if the elements have an effect on the organism, it would appear to be relatively small or specific to certain situations.Perhaps the most straightforward explanation is that these elements are essentially silent phenotypically, but that in some instances they develop a useful function. This would appear to be the case for the *Bordetella* element, which is a derivative of a mobile retroelement.

Although the reasons for the existence of these elements are not completely obvious, it is likely that understanding their function and perpetuation will shed light on their counterparts in higher eukaryotes. In eukaryotes, the roles of introns and retroelements are likewise not satisfactorily rationalized. Nevertheless, retroelements and introns have spread quite spectacularly and become embedded in eukaryotic biology.

REFERENCES

Aagaard, C., Dalgaard, J. Z., and Garrett, R. A. (1995). Intercellular mobility and homing of an archaeal rDNA intron confers a selective advantage over intron⁻ cells of *Sulfolobus acidocaldarius*. Proc. Natl. Acad. Sci. U S A, 92, 12285–12289.

Belfort, M., Derbyshire, V., Parker, M. M., Cousineau, B., and Lambowitz, A. M. (2002). Mobile introns: Pathways and proteins. In Mobile DNA II (J. L. Craig, R. Craigie, M. Gellert, A. M. Lambowitz, eds.), pp. 761–783. Washington, DC: ASM Press.

Bolduc, J. M., Spiegel, P. C., Chatterjee, P., Brady, K. L., Downing, M. E., Caprara, M. G., Waring, R. B., and Stoddard, B. L. (2003). Structural and biochemical analyses of DNA and RNA binding by a bifunctional homing endonuclease and group I intron splicing factor. Genes Dev., 17, 2875–2888.

Bonocora, R. P., and Shub, D. A. (2001). A novel group I intron-encoded endonuclease specific for the anticodon region of tRNA[fMet] genes. Mol. Microbiol., 39, 1299–1306.

Burggraf, S., Larsen, N., Woese, C. R., and Stetter, K. O. (1993). An intron within the 16S ribosomal RNA gene of the archaeon *Pyrobaculum aerophilum*. Proc. Natl. Acad. Sci. U S A, 90, 2547–2550.

Burgin, A. B., Parodos, K., Lane, D. J., and Pace, N. R. (1990). The excision of intervening sequences from *Salmonella* 23S ribosomal RNA. Cell, 60, 405–414.

Cannone, J. J., Subramanian, S., Schnare, M. N., Collett, J. R., D'Souza, L. M., Du, Y., Feng, B., et al. (2002). The Comparative RNA Web (CRW) site: An online database of comparative sequence and structure information for ribosomal, intron, and other RNAs. BioMed Central Bioinformatics, 3, 2.

Coetzee, T., Herschlag, D., and Belfort, M. (1994). *Escherichia coli* proteins, including ribosomal protein S12, facilitate *in vitro* splicing of phage T4 introns by acting as RNA chaperones. Genes Dev., 8, 1575–1588.

Costa, J. L., Paulsrud, P., and Lindblad, P. (2002). The cyanobacterial tRNA [Leu] (UAA) intron: Evolutionary patterns in a genetic marker. Mol. Biol. Evol., 19, 850–857.

Cousineau, B., Lawrence, S., Smith, D., and Belfort, M. (2000). Retrotransposition of a bacterial group II intron. Nature, 404, 1018–1021.

Cousineau, B., Smith, D., Lawrence-Cavanaugh, S., Mueller, J. E., Yang, J., Mills, D., Manias, D., et al. (1998). Retrohoming of a bacterial group II intron: Mobility via complete reverse splicing, independent of homologous DNA recombination. Cell, 94, 451–462.

Dai, L., Toor, N., Olson, R., Keeping, A., and Zimmerly, S. (2003). Database for mobile group II introns. Nucleic Acids Res., 31, 424–426.

Dai, L., and Zimmerly, S. (2002a). Compilation and analysis of group II intron insertions in bacterial genomes: Evidence for retroelement behavior. Nucleic Acids Res., 30, 1091–1102.

Dai, L., and Zimmerly, S. (2002b). The dispersal of five group II introns among natural populations of *E. coli*. RNA, 8, 1294–1307.

Dai, L., and Zimmerly, S. (2003). ORF-less and RT-encoding group II introns in archaebacteria, with a pattern of homing into related group II intron ORFs. RNA, 9, 14–19.

Dalgaard, J. Z., and Garrett, R. A. (1992). Protein-coding introns from the 23S rRNA-encoding gene form stable circles in the hyperthermophilic archaeon *Pyrobaculum organotrophum*. Gene, 121, 103–110.

Dalgaard, J. Z., Garrett, R. A., and Belfort, M. (1993). A site-specific endonuclease encoded by a typical archaeal intron. Proc. Natl. Acad. Sci. U S A, 90, 5414–5417.

Doulatov, S., Hodes, A., Dai, L., Mandhana, N., Liu, M., Deora, R., Simons, R. W., Zimmerly, S., and Miller, J. F. (2004). Tropism switching in *Bordetella* bacteriophage defines a family of diversity-generating retroelements. Nature, 431, 476–481.

Edgell, D. R., Belfort, M., and Shub, D. A. (2000). Barriers to intron promiscuity in bacteria. J. Bacteriol., 182, 5281–5289.

Eickbush, T. H. (1999). Mobile introns: Retrohoming by complete reverse splicing. Current Biology, 9, R11–R14.

Eskes, R., Liu, L., Ma, H., Chao, M. Y., Dickson, L., Lambowitz, A. M., and Perlman, P. S. (2000). Multiple homing pathways used by yeast mitochondrial group II introns. Mol. Cell. Biol., 20, 8432–8446.

Everett, K. D. E., Kahane, S., Bush, R. M., and Friedman, M. G. (1999). An unspliced group I intron in 23S rRNA links *Chlamydiales*, chloroplasts and mitochondria. J. Bacteriol., 181, 4734–4740.

Evguenieva-Hackenberg, E., and Klug, G. (2000). RNase III processing of intervening sequences found in helix 9 of 23S rRNA in the alpha subclass of *Proteobacteria*. J. Bacteriol., 182, 4719–4729.

Ferat, J. L., and Michel, F. (1993). Group II self-splicing introns in bacteria. Nature, 364, 358–361.

Glaser, P., Frangeul, L., Buchrieser, C., Rusniok, C., Amend, A., et al. (55 authors). (2001). Comparative genomics of Listeria species. Science, 294, 849–852.

Goodrich-Blair, H., and Shub, D. A. (1996). Beyond homing: Competition between intron endonucleases confers a selective advantage on flanking genetic markers. Cell, 84, 211–221.

Gregory, S. T., O'Connor, M., and Dahlberg, A. E. (1996). Functional *Escherichia coli* 23S rRNAs containing processed and unprocessed intervening sequences from *Salmonella typhimurium*. Nucleic Acids Res., 24, 4918–4923.

Grosjean, H., Szweykowska-Kulinska, Z., Motorin, Y., Fasiolo, F., and Simos, G. (1997). Intron-dependent enzymatic formation of modified nucleosides in eukaryotic tRNAs: A review. Biochimie, 79, 293–302.

Guo, H., Zimmerly, S., Perlman, P. S., and Lambowitz, A. M. (1997). Group II intron endonucleases use both RNA and protein subunits for recognition of specific sequences in double-stranded DNA. EMBO J., 16, 6835–6848.

Herzer, P. J., Inouye, S., and Inouye, M. (1992). Retron-Ec107 is inserted into the *Escherichia coli* genome by replacing a palindromic 34-bp intergenic sequence. Mol. Microbiol., 6, 345–354.

Herzer, P. J., Inouye, S., Inouye, M., and Whittam, T. S. (1990). Phylogenetic distribution of branched RNA-linked multicopy single-stranded DNA among natural isolates of *Escherichia coli*. J. Bacteriol., 172, 6175–6181.

Hsu, D., Shih, L. M., and Zee, Y. C. (1994). Degradation of rRNA in *Salmonella* strains: A novel mechanism to regulate the concentrations of rRNA and ribosomes. J. Bacteriol., 176, 4761–4765.

Ichiyanagi, K., Beauregard, A., and Belfort, M. (2003). A bacterial group II intron favors retrotransposition into plasmid targets. Proc. Natl. Acad. Sci. U S A, 100, 15742–15747.

Ichiyanagi, K., Beauregard, A., Lawrence, S., Smith, D., Cousineau, B., and Belfort, M. (2002). Retrotransposition of the Ll.LtrB group II intron proceeds predominantly via reverse splicing into DNA targets. Mol. Microbiol., 46, 1259–1272.

Inouye, M., and Inouye, S. (1991). msDNA and bacterial reverse transcriptase. Annu. Rev. Microbiol., 45, 163–186.

Inouye, S., Sunshine, M. G., Six, E. W., and Inouye, M. (1991). Retronphage φR73: An E. coli phage that contains a retroelement and integrates into a tRNA gene. Science, 252, 969–971.

International Human Genome Sequencing Consortium (256 co-authors). (2001). Initial sequencing and analysis of the human genome. Nature, 409, 860–921.

Jaeger, L., Michel, F., and Westhof, E. (1996). The structure of group I ribozymes. Nucleic Acids Mol. Biol., 10, 33–51.

Kleman-Leyer, K., Armbruster, D. W., and Daniels, C. J. (1997). Properties of H. volcanii tRNA intron endonuclease reveal a relationship between the archaeal and eucaryal tRNA intron processing systems. Cell, 89, 839–847.

Ko, M., Choi, H., and Park, C. (2002). Group I self-splicing intron in the recA gene of Bacillus anthracis. J. Bacteriol., 184, 3917–3922.

Kuhlsel, M. G., Strickland, R., and Palmer, J. D. (1990). An ancient group I intron shared by eubacteria and chloroplasts. Science, 250, 1570–1573.

Lambowitz, A. M., Caprara, M., Zimmerly, S., and Perlman, P. S. (1999). Group I and group II ribozymes as RNPs: Clues to the past and guides to the future. In The RNA World, 2nd ed. (R. F. Gesteland, T. R. Cech, J. F. Atkins, eds.), pp. 451–485. Cold Spring Harbor, NY: Cold Spring Harbor Laboratory Press.

Lambowitz, A. M., and Zimmerly, S. (2004). Mobile group II introns. Annu. Rev. Genet., 38, 1–35.

Lampson, B. C., Inouye, M., and Inouye, S. (1991). Survey of multicopy single-stranded DNAs and reverse transcriptase genes among natural isolates of Myxococcus xanthus. J. Bacteriol., 173, 5363–5370.

Lampson, B. C., Inouye, M., and Inouye, S. (2001). The msDNAs of bacteria. Prog. Nucleic Acid Res. Mol. Biol., 67, 65–91.

Landthaler, M., and Shub, D. A. (1999). Unexpected abundance of self-splicing introns in the genome of bacteriophage T wort: Introns in multiple genes, a single gene with three introns, and exon skipping by group I ribozymes. Proc. Natl. Acad. Sci. U S A, 96, 7005–7010.

Liu, M., Deora, R., Doulatov, S. R., Gingery, M., Eiserling, F. A., Preston, A., et al. (2002). Reverse transcriptase-mediated tropism switching in Bordetella bacteriophage. Science, 295, 2091–2094.

Lykke-Andersen, J., Aagaard, C., Semionenkov, M., and Garrett, R. A. (1997). Archaeal introns: Splicing, intercellular mobility and evolution. Trends Biochem. Sci., 22, 326–331.

Lykke-Andersen, J., Thi-Ngoc, H. P., and Garrett, R. A. (1994). DNA substrate specificity and cleavage kinetics of an archaeal homing-type endonuclease from Pyrobaculum organotrophum. Nucleic Acids Res., 22, 4583–4590.

Maas, W. K., Wang, C., Lima, T., Hach, A., and Lim, D. (1996). Multicopy single-stranded DNA of *Escherichia coli* enhances mutation and recombination frequencies by titrating MutS protein. Mol. Microbiol., 19, 505–509.

Maas, W. K., Wang, C., Lima, T., Zubay, G., and Lim, D. (1994). Multicopy single-stranded DNAs with mismatched base pairs are mutagenic in *Escherichia coli*. Mol. Microbiol., 14, 437–441.

Martínez-Abarca, F., García-Rodriguez, F. M., and Toro, N. (2000). Homing of a bacterial group II intron with an intron-encoded protein lacking a recognizable endonuclease domain. Mol. Microbiol., 35, 1405–1412.

Martínez-Abarca, F., Zekri, S., and Toro, N. (1998). Characterization and splicing *in vivo* of a *Sinorhizobium meliloti* group II intron associated with particular insertion sequences of the IS630-Tc1/IS3 retroposon superfamily. Mol. Microbiol., 28, 1295–1306.

Matsuura, M., Noah, J. W., and Lambowitz, A. M. (2001). Mechanism of maturase-promoted group II intron splicing. EMBO J., 20, 7259–7270.

Matsuura, M., Saldanha, R., Ma, H., Wank, H., Yang, J., Mohr, G., Cavanagh, S., et al. (1997). A bacterial group II intron encoding reverse transcriptase, maturase, and DNA endonuclease activities: Biochemical demonstration of maturase activity and insertion of new genetic information within the intron. Genes Dev., 11, 2910–2924.

Michel, F., and Westhof, E. (1990). Modelling of the three-dimensional architecture of group I catalytic introns based on comparative sequence analysis. J. Mol. Biol., 216, 585–610.

Miller, W. L., Pabbaraju, K., and Sanderson, K. E. (2000). Fragmentation of 23S rRNA in strains of *Proteus* and *Providencia* results from intervening sequences in the *rrn* (rRNA) genes. J. Bacteriol., 182, 1109–1117.

Mills, D. A., Manias, D. A., McKay, L. L., and Dunny, G. M. (1997). Homing of a group II intron from *Lactococcus lactis* subsp. *lactis* ML3. J. Bacteriol., 179, 6107–6111.

Mohr, G., Smith, D., Belfort, M., and Lambowitz, A. M. (2000). Rules for DNA target-site recognition by a lactococcal group II intron enable retargeting of the intron to specific DNA sequences. Genes Dev., 14, 559–573.

Mohr, S., Stryker, J., and Lambowitz, A. M. (2002). A DEAD-box protein functions as an ATP-dependent RNA chaperone in group I intron splicing. Cell, 109, 769–779.

Muñoz, E., Villadas, P. J., and Toro, N. (2001). Ectopic transposition of a group II intron in natural bacterial populations. Mol. Microbiol., 41, 645–652.

Nam, K., Hudson, R. H. E., Chapman, K. B., Ganeshan, K., Damha, M. J., and Boeke, J. D. (1994). Yeast lariat debranching enzyme: Substrate and sequence specificity. J. Biol. Chem., 269, 20613–20621.

Pabbaraju, K., Miller, W. L., and Sanderson, K. E. (2000). Distribution of intervening sequences in the genes for 23S rRNA and rRNA fragmentation among
strains of the *Salmonella* reference collection B (SARB) and SARC sets.
J. Bacteriol., 182, 1923–1929.

Pronk, L. M., and Sanderson, K. E. (2001). Intervening sequences in *rrl* genes
and fragmentation of 23S rRNA in genera of the family *Enterobacteriaceae*.
J. Bacteriol., 183, 5782–5787.

Ralph, D., and McClelland, M. (1993). Intervening sequence with conserved open
reading frame in eubacterial 23S rRNA genes. Proc. Natl. Acad. Sci. U S A,
90, 6864–6868.

Ralph, D., and McClelland, M. (1994). Phylogenetic evidence for horizontal
transfer of an intervening sequence between species in a spirochete genus.
J. Bacteriol., 176, 5982–5987.

Reinhold-Hurek, B., and Shub, D. A. (1992). Self-splicing introns in tRNA genes
of widely divergent bacteria. Nature, 357, 173–176.

Rest, J. S., and Mindell, D. P. (2003). Retroids in archaea: Phylogeny and lateral
origins. Mol. Biol. Evol., 20, 1134–1142.

Rice, S. A., and Lampson, B. C. (1995). Phylogenetic comparison of retron
elements among the myxobacteria: Evidence for vertical inheritance. J.
Bacteriol., 177, 37–45.

Roberts, A. P., Braun, V., Von Eichel-Streiber, C., and Mullany, P. (2001). Demonstration that the group II intron from the clostridial conjugative transposon
Tn5397 undergoes splicing *in vivo*. J. Bacteriol., 183, 1296–1299.

Roman, J., and Woodson, S. A. (1998). Integration of the *Tetrahymena* group I
intron into bacterial rRNA by reverse splicing *in vivo*. Proc. Natl. Acad. Sci.
U S A, 95, 2134–2139.

Rudi, K., and Jakobsen, K. S. (1999). Complex evolutionary patterns of tRNA [Leu]
(UAA) group I introns in cyanobacterial radiation. J. Bacteriol., 181, 3445–
3451.

Seraphin, B., Simon, M., Boulet, A., and Faye, G. (1989). Mitochondrial splicing
requires a protein from a novel helicase family. Nature, 337, 84–87.

Shub, D. A. (1991). The antiquity of group I introns. Curr. Opin. Genet. Dev., 1,
478–484.

Simon, D., Fewer, D., Friedl, T., and Bhattacharya, D. (2003). Phylogeny and
self-splicing ability of the plastid tRNA-Leu group I intron. J. Mol. Evol., 57,
710–720.

Singh, N. N., and Lambowitz, A. M. (2001). Interaction of a group II intron ribonucleoprotein endonuclease with its DNA target site investigated
by DNA footprinting and modification interference. J. Mol. Biol., 309, 361–
386.

Skurnik, M., and Toivanen, P. (1991). Intervening sequences (IVSs) in the 23S ribosomal RNA genes of pathogenic *Yersinia enterocolitica* strains: The IVSs in *Y. enterocolitica* and *Salmonella typhimurium* have a common origin. Mol. Microbiol., 5, 585–593.

Tang, T. H., Rozhdestvensky, T. S., Clouet D'Orval, B., Bortolin, M. L., Huber, et al. (2002). RNomics in archaea reveals a further link between splicing of archaeal introns and rRNA processing. Nucleic Acids Res., 30, 921–930.

Tanner, M. A., and Cech, T. R. (1996). Activity and thermostability of the small self-splicing group I intron in the pre-tRNA ^{Ile} of the purple bacterium *Azoarcus*. RNA, 2, 74–83.

Toor, N., Hausner, G., and Zimmerly, S. (2001). Coevolution of group II intron RNA structures with their intron-encoded reverse transcriptases. RNA, 7, 1142–1152.

Trotta, C. R., and Abelson, J. (1999). tRNA splicing: An RNA world add-on or an ancient reaction? In The RNA World, 2nd ed. (R. F. Gesteland, T. R. Cech, J. F. Atkins, eds.), pp. 561–584. Cold Spring Harbor, NY: Cold Spring Harbor Laboratory Press.

Trotta, C. R., Miao, F., Arn, E. A., Stevens, S. W., Ho, C. K., Rauhut, R., and Abelson, J. N. (1997). The yeast tRNA splicing endonuclease: A tetrameric enzyme with two active site subunits homologous to the archaeal tRNA endonucleases. Cell, 89, 849–858.

Valadkhan, S., and Manley, J. L. (2001). Splicing-related catalysis by protein-free snRNAs. Nature, 413, 701–707.

Wank, H., Sanfilippo, J., Singh, R. N., Matsuura, M., and Lambowitz, A. M. (1999). A reverse transcriptase/maturase promotes splicing by binding at its own coding segment in a group II intron RNA. Mol. Cell, 4, 239–250.

Watanabe, Y., Yokobori, S., Inaba, T., Yamagishi, A., Oshima, T., Kawarabayasi, Y., Kikuchi, H., et al. (2002). Introns in protein-coding genes in archaea. FEBS Lett., 510, 27–30.

Weber, U., Beier, H., and Gross, H. J. (1996). Another heritage from the RNA world: Self-excision of intron sequences from nuclear pre-tRNAs. Nucleic Acids Res., 24, 2212–2219.

Weeks, K. M., and Cech, T. R. (1995). Protein facilitation of group I intron splicing by assembly of the catalytic core and the 5' slice site domain. Cell, 82, 221–230.

Yamanaka, K., Shimamoto, T., Inouye, S., and Inouye, M. (2002). Retrons. In Mobile DNA II (J. L. Craig, R. Craigie, M. Gellert, A. M. Lambowitz, eds.), pp. 784–795. Washington, DC: ASM Press.

STEVEN ZIMMERLY

Zhang, A., Derbyshire, V., Salvo, J. L. G., and Belfort, M. (1995). *Escherichia coli* protein StpA stimulates self-splicing by promoting RNA assembly *in vitro*. RNA, 1, 783–793.

Zhong, J., and Lambowitz, A. M. (2003). Group II intron mobility using nascent strands at DNA replication forks to prime reverse transcription. EMBO J., 22, 4555–4565.

Zimmerly, S., Guo, H., Eskes, R., Yang, J., Perlman, P. S., and Lambowitz, A. M. (1995a). A group II intron RNA is a catalytic component of a DNA endonuclease involved in intron mobility. Cell, 83, 529–538.

Zimmerly, S., Guo, H., Perlman, P. S., and Lambowitz, A. M. (1995b). Group II intron mobility occurs by target DNA-primed reverse transcription. Cell, 82, 545–554.

Part 2 Horizontal Gene Transfer and Genome Plasticity

CHAPTER 5

The F-plasmid, a paradigm for bacterial conjugation

Michael J. Gubbins, William R. Will, and Laura S. Frost

The F factor is often associated with *Escherichia coli* and appears to have been adapted by the bacterial host to act as an agent of genetic exchange and evolution. F encodes a type IV secretion system (T4SS) that enables bacterial conjugation, the transfer of DNA from a donor F$^+$ to a recipient F$^-$ cell. The delivery of DNA containing either host or foreign genes has important consequences for the bacterium, allowing it to enlarge or modify its genetic content and rapidly adapt to an environmental niche. Unlike other plasmids, the conjugative functions of F and F-like plasmids appear to be controlled by a complex regulatory network that involves many host proteins resulting in a symbiotic relation between F and its host. This chapter outlines the predicted and known functions for all the genes on the F plasmid and its close relatives, and describes our current knowledge about the regulation of F conjugation.

BRIEF HISTORY OF THE F PLASMID

The discovery of horizontal gene transfer between bacteria can be attributed to the work of Lederberg and Tatum (1946), who observed that different strains of *E. coli* K-12 could be genetically and phenotypically altered when mixed together. A series of experiments led to the conclusion that direct contact between bacteria was required in order for genetic material to be transferred between the cells (Davis, 1950). This transfer was found to occur in one direction, from donor to recipient cells, by a mechanism contained within the donor cells (Hayes, 1952). By 1953, it had been determined that the donor cells in fact contained a separate genetic element, the F (fertility) sex factor, which could be transferred to recipient cells lacking this element (Hayes, 1953; Lederberg, Cavalli, and Lederberg, 1952).

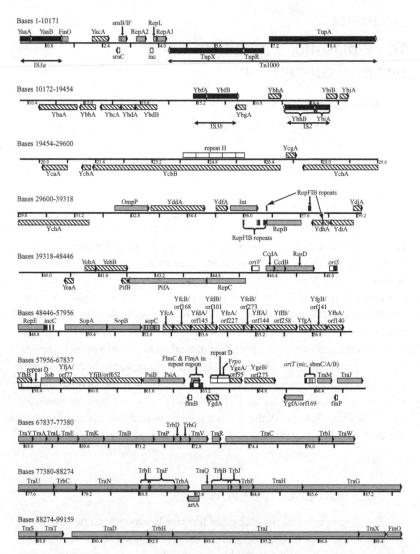

Figure 5.1. Map of the F plasmid. Sequence information for the entire plasmid (bases 1 to 99,159) was obtained from the GenBank sequence submitted by Shimizu et al. (2000; AP001918). The transfer region (U01159; Frost, Ippen-Ihler, and Skurray, 1994) and leading region (AF106329; Manwaring, Skurray, and Firth, 1999) sequences have been compiled and deposited separately. The plasmid map has been divided into ~10-kb segments. The exact region covered in each line is indicated above each section by a start and end base number, and markers (kb) are indicated every 1.6 kb under each region. Black regions represent insertion sequences and transposons, as well as the known and predicted genes encoded within them. Grey regions represent known, characterized

MICHAEL J. GUBBINS, WILLIAM R. WILL, AND LAURA S. FROST

Subsequent analysis of the F sex factor confirmed that it was an autonomously replicating plasmid that encoded an extracellular filament, the pilus, essential for conjugation (Brinton, Gemski, and Carnahan, 1964; Marmur et al., 1961). Other self-transmissible genetic elements that encoded multiple drug resistance genes were subsequently identified, suggesting that conjugation was important for the dissemination of bacterial traits as demonstrated by transfer of the R (resistance) factors between antibiotic resistant *Shigella* spp. (Watanabe, 1963, 1966). Since that time, intergeneric and interkingdom transfer of plasmids has been demonstrated, with replication of the plasmid in the new host cell being the most important barrier to its establishment and maintenance (Davison, 1999; Mazodier and Davies, 1991). We now know that conjugative DNA transfer is a function of a subfamily of the type IV secretion systems (T4SS), with the F-like T4SS representing a group of systems seemingly involved with the transport of nucleic acids into and out of cells (Lawley et al., 2003). The presence of F-like T4SS in many bacterial plasmids and genomes is one way that F contributes to the development of the bacterial genome. It also promotes gene evolution and genetic exchange by mobilizing genes between bacteria either as segments of the chromosome or cargo embedded in the F plasmid itself.

THE STRUCTURE OF THE F PLASMID

The F plasmid of *E. coli* is approximately 100 kb in size (99,159 bp; GenBank accession number: AP001918; NC002483), more than one-third of which is comprised of operons involved in facilitating plasmid transfer (reviewed in Frost, Ippen-Ihler, and Skurray, 1994; Shimizu et al., 2000; Fig. 5.1, Table 5.1). The majority of the known genes lie in two major regions on the plasmid. One contains the transfer (*tra*; coordinates 66.1F to 100/0F; GenBank accession number U01159) and leading (coordinates 53.3F to 66.1F; GenBank accession number AF106329) regions, which contain genes

Figure 5.1 (*cont.*) proteins. Hatched regions indicate predicted protein products. White regions represent miscellaneous features such as *ori* regions, genes encoding antisense RNAs, and repeat sequence regions (e.g., iterons). Only those genes and features that are currently annotated in the available databases are indicated. Genes and features that were previously identified, but are not annotated in the current databases, are discussed in numerous older publications (e.g., see Willetts and Skurray, 1987, for a complete list). Further details of the various features on this map are discussed in the text and at the beginning of this chapter.

Table 5.1. *Annotation of the F plasmid sequence*

Feature Name[a]	Coordinates[b]	Probable/known function	gi #	Notes/sequence references[c]
IS3a	1 .. 1258	Insertion sequence	N.A.	(Hu, Ohtsubo, and Davidson, 1975)
YaaA	65 .. 365	IS3a transposase	8918824	
YaaB	366 .. 1228	IS3a transposase	8918825	
FinO	1259 .. 1426	RNA binding/chaperone	N.A.	Interrupted by IS3 in F (Cheah and Skurray, 1986)
YacA	2110 .. 2550	Unknown	8918826	
srnC	2821 .. 2882	Antisense RNA	8918827	Modulation of srnB expression (Nielsen et al., 1991; Thisted, Nielsen, and Gerdes, 1994).
SrnB	2869 .. 3075	Postsegregational killing	8918828	Toxin protein (Nielsen et al., 1991; Thisted, Nielsen, and Gerdes, 1994)
RepA2	3359 .. 3616	Replication regulatory protein	8918829	Part of RepFIC replicon remnant (Bergquist, Saadi, and Maas, 1986; Saadi et al., 1987; Willetts and Skurray, 1987).
inc	3747 .. 3839	Inhibitor of RepA1 protein synthesis	N.A.	Part of RepFIC replicon remnant (Bergquist, Saadi, and Maas, 1986; Saadi et al., 1987; Willetts and Skurray, 1987).
RepL	3850 .. 3924	Leader peptide of RepA1	8918830	Part of RepFIC replicon remnant (Bergquist, Saadi, and Maas, 1986; Saadi et al., 1987; Willetts and Skurray, 1987).

Gene	Position	Function	Accession	Reference
RepA1	3917 . . 4190	RepFIC replication initiation protein	N.A.	Inactivated by insertion sequence $\gamma\delta$ (Tn1000); part of RepFIC replicon remnant (Bergquist, Saadi, and Maas, 1986; Saadi et al., 1987; Willetts and Skurray, 1987).
Tn1000	4191 . . 10171	Transposon $\gamma\delta$	N.A.	(Guyer, 1978)
TnpX	4303 . . 6399	$\gamma\delta$ (Tn1000) hypothetical protein	8918831	
TnpR	6415 . . 6966	$\gamma\delta$ (Tn1000) resolvase	8918832	
TnpA	7030 . . 10138	$\gamma\delta$ (Tn1000) transposase	8918833	
YbaA	10729 . . 11790	Selenite dissimilatory reduction protein	8918834	SedR (gi852525) of F-like plasmid in *Stenotrophomonas maltophilia*
YbbA	11900 . . 12313	Unknown	8918835	
YbcA	12477 . . 13040	Unknown	8918836	
YbdA	13047 . . 13460	Unknown	8918837	
YbdB	13453 . . 14055	Unknown	8918838	
IS3b	15093 . . 16350	Insertion sequence	N.A.	(Hu, Ohtsubo, and Davidson, 1975)
YbfA	15158 . . 15457	IS3b hypothetical protein	8918839	
YbfB	15454 . . 16320	IS3b putative transposase	8918840	
YbgA	16315 . . 16571	Unknown	8918841	
YbhA	17216 . . 17548	Unknown	8918842	

(cont.)

Table 5.1 (cont.)

Feature Name[a]	Coordinates[b]	Probable/known function	gi #	Notes/sequence references[c]
IS2	17662 .. 18992	Insertion sequence	N.A.	(Hu, Ohtsubo, and Davidson, 1975)
YbhB	17673 .. 18578	IS2 hypothetical 34.4kD protein	8918843	
YbiA	18536 .. 18946	IS2 hypothetical 13kD protein	8918844	
YbiB	18858 .. 19146	IS2 putative protein	8918845	
YbjA	19245 .. 19554	Unknown	8918846	
YcaA	20096 .. 20791	Unknown	8918847	
YcbA	21222 .. 21455	Unknown	8918848	
YcbB	21541 .. 26817	AidA-I adhesin/ autotransporter	8918849	(Benz and Schmidt, 1989)
repeat H	24085 .. 25818	repeat region	N.A.	Five copies in region
YcgA	26990 .. 27259	Unknown	8918850	
YchA	27486 .. 31601	AidA-I adhesin/ autotransporter	8918851	(Benz and Schmidt, 1989)
OmpP	32322 .. 33269	Outer membrane protein/protease	8918852	Similar to OmpP of E. coli K-12 (Kaufmann, Stierhof, and Henning, 1994; Matsuo et al., 1999)
YddA	33362 .. 34861	Unknown	8918853	

YdfA	35217 . . 35504	Unknown	8918854	
Int	35618 . . 36358	Integrase or site-specific recombinase, similar to F ResD	8918855	(Saul et al., 1989)
RepFIB	35982 . . 36642	RepFIB repeat sequence	N.A.	Seven repeats in region (Bergquist, Saadi, and Maas, 1986; Lane, 1981; Lane and Gardner, 1979; Willetts and Skurray, 1987)
RepFIB	37816 . . 37878	RepFIB repeat sequence	N.A.	Three repeats in region
RepFIB	38218 . . 38238	RepFIB repeat sequence	N.A.	One repeat in region
RepB	36643 . . 37620	RepFIB replication protein	8918856	(Bergquist, Saadi, and Maas, 1986; Willetts and Skurray, 1987)
YdhA	37942 . . 38403	Unknown	8918857	
YdiA	38460 . . 39101	Unknown	8918858	
YdjA	39088 . . 39318	Unknown	8918859	
YeaA	40634 . . 40870	Unknown	8918860	
YebA	41216 . . 41506	Unknown	8918861	
YebB	41496 . . 42395	Unknown	8918862	
PifA	42445 . . 44670	Exclusion of phage T7	8918863	(Miller and Malamy, 1984)
PifB	42206 . . 42405	Exclusion of phage T7 (predicted)	N.A.	Original map position 43.3–43.7 (Miller and Malamy, 1984; Rotman, Cooney, and Malamy, 1983)

(cont.)

Table 5.1 (*cont.*)

Feature Name[a]	Coordinates[b]	Probable/known function	gi #	Notes/sequence references[c]
RepC (PifC)	44672 . . 45760	Essential for replication from *oriV*	8918864	Transcriptional repressor (Miller and Malamy, 1983, 1984; Santini and Stanisich, 1998)
oriV	45934 . . 46135	Origin of vegetative replication	N.A.	Bidirectional (Eichenlaub, Figurski, and Helinski, 1977; Lane, 1981)
rfsF site	45922 . . 45949	RecA-independent formation of cointegrates between 2 *oriV*-carrying plasmids.	N.A.	Possible *oriV*-mediated integration of F into chromosome (Lane et al., 1986; O'Connor and Malamy, 1984; Willetts and Skurray, 1987)
CcdA	46340 . . 46558	Suppression of CcdB	8918865	(Dao-Thi et al., 2002; Miki, Chang, and Horiuchi, 1984)
CcdB	46560 . . 46865	Coupled cell division/post-segregational killing	8918866	(Miki, Chang, and Horiuchi, 1984)
ResD	46866 . . 47672	Resolvase for cointegrate resolution, similar to F Int	8918867	Formerly protein D; acts at *rfsF* site (Lane et al., 1986; Makino et al., 1998)
oriS (ori-2)	48136 . . 48352	RepFIB replication origin	N.A.	Unidirectional (Eichenlaub, Wehlmann, and Ebbers, 1981; Murotsu, Tsutsui, and Matsubara, 1984; Willetts and Skurray, 1987)
IncB	48136 . . 48349	Direct repeats associated with *oriS*	N.A.	RepE binding sites; four repeats in region; (Rokeach, Sogard-Andersen, and Molin, 1985)

Gene	Coordinates	Function	Accession	Reference
RepE	48446 .. 49201	Replication initiation protein	8918868	(Komori et al., 1999; Masson and Ray, 1986; Uga, Matsunaga, and Wada, 1999)
IncC region	49204 .. 49451	RepE binding sites	N.A.	Five direct repeats in region (Tolun and Helinski, 1981; Uga, Matsunaga, and Wada, 1999)
SopA	49780 .. 50955	essential for RepFIA partitioning	8918869	(Libante, Thion, and Lane, 2001; Mori et al., 1986; Ogura and Hiraga, 1983)
SopB	50956 .. 51926	Essential for RepFIA partitioning	8918870	(Libante, Thion, and Lane, 2001; Mori et al., 1986; Ogura and Hiraga, 1983)
SopC	51999 .. 52472	Partitioning	N.A.	11 direct repeats in region (Libante, Thion, and Lane, 2001; Mori et al., 1986; Ogura and Hiraga, 1983)
YfcA	52569 .. 52841	Unknown	8918871	
YfcB	52855 .. 53361	Unknown	8918872	Orf168 (Manwaring, Skurray, and Firth, 1999)
YfdA	53343 .. 53780	Unknown	8918873	Orf145 (Manwaring, Skurray, and Firth, 1999)
YfdB	53784 .. 54089	Unknown	8918874	Orf101 (Manwaring, Skurray, and Firth, 1999)
YfeA	54166 .. 54849	Putative methyltransferase	8918875	Orf227 (Manwaring, Skurray, and Firth, 1999; Mise and Nakajima, 1984)
YfeB	54850 .. 55071	Unknown	8918876	Orf73 (Manwaring, Skurray, and Firth, 1999)
YffA	55085 .. 55519	Unknown	8918877	Orf144 (Manwaring, Skurray, and Firth, 1999)

(cont.)

Table 5.1 (cont.)

Feature Name[a]	Coordinates[b]	Probable/known function	gi #	Notes/sequence references[c]
YffB	55565 .. 56341	Unknown	8918878	Orf258 (Manwaring, Skurray, and Firth, 1999)
YfgA	56429 .. 56740	Unknown	8918879	
YfgB	56755 .. 57180	Predicted antirestriction protein	8918880	Orf141 (Manwaring, Skurray, and Firth, 1999); similar to KlcA of pO157 (Makino et al., 1998)
YfhA	57227 .. 57649	Unknown	8918881	Orf140 (Manwaring, Skurray, and Firth, 1999)
YfhB	57956 .. 58354	Unknown	8918882	
Ssb	58643 .. 59182	Single-stranded DNA binding protein	8918883	(Chase, Merrill, and Williams, 1983; Manwaring, Skurray, and Firth, 1999)
YfjA	59233 .. 59472	Unknown	8918884	Orf77 (Manwaring, Skurray, and Firth, 1999)
YfjB	59538 .. 61496	Possible function in plasmid partitioning	8918885	Orf652 (Lin and Grossman, 1998; Manwaring, Skurray, and Firth, 1999)
PsiB	61551 .. 61985	Plasmid SOS inhibition	8918886	(Bagdasarian et al., 1992; Bailone et al., 1988)
PsiA	61982 .. 62701	Plasmid SOS inhibition	8918887	(Loh et al., 1990)
flmB	62827 .. 62927	Antisense RNA of flmC mRNA	N.A.	(Loh, Cram, and Skurray, 1988)
FlmC	62918 .. 63130	F leading region maintenance	8918888	(Loh, Cram, and Skurray, 1988)

Name	Location	Function	Accession number	References
FlmA	62976 . . 63134	F leading region maintenance	8918889	(Loh, Cram, and Skurray, 1988)
F*rpo* region	63655 . . 64003	Single-stranded DNA promoter	N.A.	Primer RNA synthesis for DNA replication (Masai and Arai, 1997)
YgdA	63372 . . 63689	Unknown	8918890	
YgeA	64032 . . 64319	Unknown	8918891	Orf95 (Manwaring, Skurray, and Firth, 1999)
YgeB	64438 . . 65259	Unknown	8918892	Orf273 (Manwaring, Skurray, and Firth, 1999)
YgfA	65556 . . 66065	Putative transglycosylase	8918893	Orf169 (Bayer et al., 1995; Loh, Cram, and Skurray, 1989; Manwaring, Skurray, and Firth, 1999)
oriT	66118 . . 66407	Origin of transfer region		(Everett and Willetts, 1980; Frost, Ippen-Ihler, and Skurray, 1994)
nic site	66157^66158	Site of nicking transferred strand		Target of TraI nicking at *oriT* (Thompson, Centola, and Deonier, 1989)
sbmC	66250 . . 66267	Inverted repeat		TraM binding site (Fekete and Frost, 2002)
sbmA/B	66340 . . 66407	Two direct repeats		TraM binding site (Fekete and Frost, 2002)
TraM	66479 . . 66862	DNA binding	8918894	Involved in mating signal and relaxosome formation (; Frost, Ippen-Ihler, and Skurray, 1994)
unnamed	66944 . . 66960	Possible CRP binding site		(Starcic et al., 2003)

(cont.)

Table 5.1 (*cont.*)

Feature Name[a]	Coordinates[b]	Probable/known function	gi #	Notes/sequence references[c]
finP	66977 . . 67055	Antisense RNA; complementary to portion of *traJ* mRNA		Antisense RNA in FinOP fertility inhibition (Jerome, van Biesen, and Frost, 1999)
TraJ	67049 . . 67738	P$_Y$ positive regulator	8918895	(Frost, Ippen-Ihler, and Skurray, 1994)
TraY	67837 . . 68232	*oriT* nicking; *traM, traY-I* regulator	8918896	(Howard, Nelson, and Matson, 1995; Lum, Rodgers, and Schildbach, 2002)
TraA	68265 . . 68630	F/T4SS pilin subunit	8918897	(Frost, Ippen-Ihler, and Skurray, 1994; Frost, Paranchych, and Willetts, 1984)
TraL	68645 . . 68956	F/T4SS pilus assembly/pore formation	8918898	(Frost, Ippen-Ihler, and Skurray, 1994; Lawley et al., 2003)
TraE	68978 . . 69544	F/T4SS pilus assembly/pore formation	8918899	(Frost, Ippen-Ihler, and Skurray, 1994; Lawley et al., 2003)
TraK	69531 . . 70259	F/T4SS pilus assembly/pore formation	8918900	(Frost, Ippen-Ihler, and Skurray, 1994; Lawley et al., 2003)
TraB	70259 . . 71686	F/T4SS pilus assembly/pore formation	8918901	(Frost, Ippen-Ihler, and Skurray, 1994; Lawley et al., 2003)
TraP	71676 . . 72266	F pilus stability	8918902	(Frost, Ippen-Ihler, and Skurray, 1994; Lawley et al., 2003)

Gene	Coordinates	Function	Accession	Reference
TrbD	72253 .. 72450	Unknown	8918903	(Frost, Ippen-Ihler, and Skurray, 1994)
TrbG	72462 .. 72713	Unknown	8918904	(Frost, Ippen-Ihler, and Skurray, 1994)
TraV	72710 .. 73225	F/T4SS pilus assembly; lipoprotein	8918905	(Frost, Ippen-Ihler, and Skurray, 1994; Lawley et al., 2003)
TraR	73360 .. 73581	Unknown	8918906	(Doran et al., 1994; Frost, Ippen-Ihler, and Skurray, 1994)
TraC	73741 .. 76368	F/T4SS pilus assembly/ secretion; ATPase	8918907	(Doran et al., 1994; Frost, Ippen-Ihler, and Skurray, 1994; Lawley et al., 2003)
TrbI	76365 .. 76751	F pilus retraction	8918908	(Frost, Ippen-Ihler, and Skurray, 1994; Lawley et al., 2003)
TraW	76748 .. 77380	F pilus assembly/pore formation	8918909	(Frost, Ippen-Ihler, and Skurray, 1994; Lawley et al., 2003)
TraU	77377 .. 78369	F DNA transfer	8918910	(Frost, Ippen-Ihler, and Skurray, 1994; Lawley et al., 2003)
TrbC	78378 .. 79016	F pilus assembly	8918911	(Frost, Ippen-Ihler, and Skurray, 1994; Lawley et al., 2003)
TraN	79013 .. 80821	Mating pair formation; adhesin	8918912	(Frost, Ippen-Ihler, and Skurray, 1994; Lawley et al., 2003)
TrbE	80845 .. 81105	Unknown	8918913	(Frost, Ippen-Ihler, and Skurray, 1994)
TraF	81068 .. 81841	F pilus assembly; thioredoxin homologue	8918914	(Frost, Ippen-Ihler, and Skurray, 1994; Lawley et al., 2003)

(cont.)

Table 5.1 (*cont.*)

Feature Name[a]	Coordinates[b]	Probable/known function	gi #	Notes/sequence references[c]
TrbA	81857 .. 82204	Unknown	8918915	(Frost, Ippen-Ihler, and Skurray, 1994; Kathir and Ippen-Ihler, 1991)
artA	82206 .. 82520	Unknown	8918916	(Frost, Ippen-Ihler, and Skurray, 1994; Kathir and Ippen-Ihler, 1991)
TraQ	82601 .. 82885	pilin synthesis	8918917	(Frost, Ippen-Ihler, and Skurray, 1994; Kathir and Ippen-Ihler, 1991)
TrbB	82872 .. 83417	Thioredoxin/TraF homolog	8918918	(Lawley, Klimke, Gubbins, and Frost, 2003)
TrbJ	83347 .. 83688	Unknown	8918919	(Frost, Ippen-Ihler, and Skurray, 1994)
TrbF	83675 .. 84055	Unknown	8918920	(Frost, Ippen-Ihler, and Skurray, 1994)
TraH	84052 .. 85428	F pore formation/pilus assembly	8918921	(Frost, Ippen-Ihler, and Skurray, 1994; Lawley et al., 2003)
TraG	85425 .. 88241	T4SS pore formation/mating pair stabilization	8918922	(Frost, Ippen-Ihler, and Skurray, 1994; Lawley et al., 2003)
TraS	88274 .. 88795	Entry exclusion	8918923	(Jalajakumari et al., 1987)
TraT	88817 .. 89551	Surface exclusion	8918924	(Jalajakumari et al., 1987)
TraD	89804 .. 91957	DNA transport; ATPase	8918925	Coupling protein (Cabezon, Sastre, and de la Cruz, 1997)

TrbH	91957 .. 92676	Unknown	8918926	(Frost, Ippen-Ihler, and Skurray, 1994)
TraI	92673 .. 97943	*oriT* nicking and DNA unwinding; relaxase; ATPase	8918927	Helicase I (Frost, Ippen-Ihler, and Skurray, 1994; Stern and Schildbach, 2001; Traxler and Minkley, 1988)
TraX	97963 .. 98709	F pilin acetylation	8918928	(Frost, Ippen-Ihler, and Skurray, 1994; Moore et al., 1993)
FinO	98764 .. 99159	FinP RNA binding/chaperone	N.A.	Interrupted by IS3 in F (Cheah and Skurray, 1986; Frost, Ippen-Ihler, and Skurray, 1994)

[a] Some genes and features previously listed in F are not included (e.g., see Willetts and Skurray, 1987). Only those features contained within the current GenBank entry for F are included in this analysis, for the sake of simplicity. The reader is directed to such references for discussions of putative genes and features that were previously described in F but that are no longer listed in the current F sequence.

[b] Coordinates are from the sequence deposited in GenBank by Shimizu et al. (2000), accession number AP001918.

[c] The reader is directed to numerous papers and reviews on the F transfer region (Frost et al., 1994, accession number U01159; Lawley et al., 2003; Wilkins and Frost, 2001) and leading region (Manwaring et al., 1999; accession number AF106329) for further details regarding these well-characterized regions of F.

involved in conjugation and the establishment of the plasmid in the recipient cell, respectively (Frost, Ippen-Ihler, and Skurray, 1994; Manwaring, Skurray, and Firth, 1999; Willetts and Skurray, 1987). The second region (coordinates 42.6F to 53.3F; GenBank accession number AP001918) contains genes involved in replication, partitioning, and maintenance of the plasmid, including the RepFIA replicon, a homolog of the single-stranded DNA-binding protein encoded by *ssb* or *ssf*, the *sopABC* partitioning genes, the *ccdAB* toxin/antitoxin system, and the *pif* genes, associated with phage inhibition, and the *oriV* replication origin. Further upstream is a second replicon, RepFIB, as well as an integrase-like gene, Int. A third region contains genes of unknown function, mutations in which have no measurable effect of F survival or dissemination.

The leading region, which is the first portion of the plasmid to be transferred into a recipient cell, contains genes that are conserved in many F-like plasmids. Such conserved genes include *orf169*, a homolog of lytic transglycosylases possibly involved in constructing the transfer apparatus spanning the cell envelope; *psiAB*, a system to prevent induction of the SOS response; and *ssb* or *ssf*, which is a single-stranded DNA-binding protein that might be involved in promoting DNA replication in the recipient. Located within the leading region is F*rpo*, which is a single-stranded promoter that provides for the transcription of a primer for synthesis of the complementary F DNA strand in a new host cell (Manwaring, Skurray, and Firth, 1999; Masai and Arai, 1997). The origin of transfer (*oriT*) lies between the first gene in the leading region, *orf169*, and the first gene of the transfer operon, *traM*. Within *oriT* is the *nic* site, the point at which nicking of F and transfer of the leading strand is initiated (Frost, Ippen-Ihler, and Skurray, 1994).

The F plasmid contains three separate replicons: RepFIA (44.6–53.3F), RepFIB (36.0–40.0F), and the nonfunctional RepFIC (3.0–4.2F and 9.9–10F), which is interrupted by a Tn*1000* element. Although both RepFIA and RepFIB are complete, functional replicons, RepFIA is the primary replicon controlling replication of the F plasmid, containing both bidirectional and unidirectional origins of replication, *oriV* and *oriS*, respectively (Lane, 1981). Mini-F plasmids containing the RepFIB replicon are unstable, particularly at high growth rates (Lane and Gardner, 1979).

The F plasmid contains four significant insertion sequences. In addition to Tn*1000* (4.2–9.9F) and IS*3a* (100/0–1.3F) inserted in RepFIC and *finO*, respectively, there are also IS*2* (17.6–18.9F) and IS*3b* (15.0–16.3F) (Cheah and Skurray, 1986; Guyer, 1978; Hu, Ohtsubo, and Davidson, 1975; Hu et al., 1975a, 1975b; Saadi et al., 1987) that appear to be outside any genes encoding proteins of known function. These insertion sequences are believed to have a significant impact on the cell by promoting integration and, subsequently,

MICHAEL J. GUBBINS, WILLIAM R. WILL, AND LAURA S. FROST

imprecise excision in the host chromosome to form Hfr and F' strains (see "The Role of F as a Sex Factor" section). The insertion of IS3a in the *finO* gene at the distal end of the *tra* operon also has a profound impact on F transfer. This interruption leads directly to constitutive expression, or derepression, of the *tra* genes and F transfer by functionally inactivating the FinO protein (see "The Fertility Inhibition (FinOP) System" section; Cheah and Skurray, 1986; Yoshioka, Ohtsubo, and Ohtsubo, 1987).

The "dark side" of F (100/0–38F) contains few genes with known or predicted functions (Sharp, Cohen, and Davidson, 1973; Shimizu et al., 2000). Two putative genes, *ycbB* (21.5–26.8F) and *ychA* (27.4–31.6F), encode proteins that bear partial homology to the *E. coli* AIDA-I adhesin precursor. Another gene, *ybaA*, encodes a homolog of a selenite dissimilatory reduction protein, whereas Int, a phagelike integrase, is adjacent to the RepFIB replicon. Another functional gene, *ompP*, initially believed to be located on the chromosome, encodes a protein that is approximately 70% homologous with the chromosomally encoded OmpT, a protease associated with the outer membrane. The many other "*yxx*" genes of unknown function might reflect the evolutionary history of F as it integrated into and excised from the chromosome or other elements. That these genes are stably maintained argues that they might have a function that has escaped detection. The finding that several of these putative genes share homology at both the (predicted) amino acid and nucleotide levels with other F-like plasmids strengthens this argument. The best characterized feature of F is the transfer (*tra*) region. This region is comprised of two monocistronic operons, *traM* and *traJ*, and the multicistronic 33.3 kb *tra* operon. Most of the genes required for elaboration of the pilus and subsequent conjugative DNA transfer are encoded within the *tra* region. Due to the relatively limited scope of this chapter, only selected features of this region are discussed in detail. The reader is directed to several excellent reviews in which features of the transfer region are discussed in depth (Firth, Ippen-Ihler, and Skurray, 1996; Frost, Ippen-Ihler, and Skurray, 1994; Lawley, Wilkins, and Frost, 2004; Lawley et al., 2003; Wilkins and Frost, 2001; Zechner et al., 2000) .

THE ROLE OF F AS A SEX FACTOR

The value of sex lies in its ability to promote evolution by allowing beneficial traits to be disseminated to other members of a population. This allows for the selection of characteristics that either enhance a lifestyle or allow adaptation to a changing environment. Bacteria that carry these traits quickly outgrow their competitors and the composition of a bacterial population can undergo rapid conversion in the presence of selective pressure. The F plasmid has evolved into a tool used by its preferred host, *E. coli*, to promote the sharing of genetic information by conjugation, a process that confers many of the benefits of sex. It provides opportunities for genetic exchange via homologous recombination by transporting chromosomal DNA either in small sections on F' episomes or large ones via the Hfr (high frequency of recombination) mode. Conjugation also affords the opportunity for cells to acquire a "cargo" of new genes residing on the incoming plasmid; the most notable example being genes encoding antibiotic resistance. F was isolated in its derepressed form, which has proved to be unusual in F-like plasmids. Many researchers have commented on the energy costs of maintaining a derepressed transfer system and the dangers of an increased susceptibility to F-pilus-specific phages. One is left wondering why F is stably maintained and why the transfer region is not further inactivated to prevent "wasteful" F *tra* region expression. In examining the lifestyle of *E. coli*, the variation in its habitats is striking. *E. coli* goes from feast to famine as it is shed from the body and enters either into the water or soil, undergoes wide variations in temperature and osmolarity, and lives both as a free-living organism and in microbial communities or biofilms. The conjugative potential of the F plasmid is sensitive to these variations and expresses the genes for transfer only when conditions are optimal. It monitors the welfare of its host and modulates its gene expression so as not to burden its host with unwelcome energy outlays. A comparison of *E. coli* MG1655 with and without F revealed that the gene expression profile, as monitored by Affymetrix *E. coli* chips, varied very little. The presence of F upregulated a variety of ABC transporters involved in amino acid uptake and downregulated a few amino acid tRNA synthetases (Lau et al. 2005). That F appeared to alter amino acid metabolism might be related to the high levels of F-pilin in the cell (estimated to be about 100,000 copies).

Thus, F transfers efficiently in media replete with excellent carbon sources such as glucose, and at a fairly narrow range of temperatures with 37°C to 42°C being preferred. It transfers equally well in liquid or on solid media and becomes "phenotypically F^-" (i.e., does not express the transfer genes) in stationary phase (reviewed in Hayes, 1964). F appears to have replaced repression by the fertility inhibition (FinOP) system of other F-like plasmids with a complex regulatory program that shuts down transfer gene expression as the host cell enters stationary phase. Considering that stationary phase, brought on by starvation, is probably the normal state of affairs for *E. coli*, the cell is expending little in maintaining F. If conditions improve, the transfer apparatus can be reestablished within a matter of minutes, suggesting that the components of the apparatus are disassembled during stationary phase but are not necessarily degraded. This can be illustrated by the observation that the addition of small amounts of glucose to stationary phase cells can restore full mating ability within 30 minutes at 37°C (Frost and Manchak, 1998).

Transfer of F itself can occur from a donor to a recipient cell during the early exponential phase of growth and also between donor cells, albeit at a much lower rate due to entry and surface exclusion, properties of $TraS_F$ and $TraT_F$, respectively (Achtman et al., 1980). F can also transfer into stationary F^+ cells (F^- phenocopies) because of the downregulation of the exclusion genes. Thus, *E. coli* is constantly casting about for recipient cells, including F^+ stationary recipients and will engage in conjugative DNA transfer, providing conditions for *tra* gene expression are optimal.

What are the benefits of this propensity for conjugation? F has three basic identities: as F, as an F′ episome carrying host genes, and as an Hfr strain whereby F is incorporated into the chromosome. As F, it is able to acquire genes carried on transposons, insertion sequences, integrative phages, and integrons and move them between bacteria. It is also able to encounter other conjugative elements and exchange genes; the mosaicism in conjugative transfer systems is clearly evident, and many examples of gene loss and acquisition have been cataloged (Boyd et al., 1996). T4SS exemplify this with respect to the frequent loss of self-transmissibility, gene shuffling, and interruption of *tra* operons by antibiotic resistance cassettes, among others (Lawley et al., 2004).

As F′ episomes (e.g., F′*lac*, F′*his*, F′*pro*) that can stably maintain sizeable portions of the chromosome, F acts as a "scratchpad" for evolution. In this merodiploid state, there are two copies of each gene on the F′ episome and chromosome, one of which can afford to undergo mutation and contribute to the evolution of the species. An interesting phenomenon of more recent notoriety is the concept of adaptive mutation, in which an F′ promotes point mutations that escape repair during selection for traits that combat starvation or death (see "F DNA Metabolism and Adaptive Mutation" section).

F also has the property of integrating into the chromosome at many different sites at a frequency of about 0.01%. These integrated F plasmids are stably maintained and have a correspondingly low rate of excision, with imprecise excision leading to F′ episome formation. Integration occurs through sequences such as Tn*1000*, IS*2*, or IS*3* elements via RecA-dependent homologous recombination or via transpositional events. Hfr strains have the ability to transfer the entire chromosome, which takes approximately 90 minutes. This is the basis for the techniques used to map the *E. coli* chromosome;

in fact, minutes are still used as a unit of measurement for chromosomal distance in *E. coli*. F rarely, if ever, transfers the whole chromosome and genes distal to F in an Hfr strain are transferred with ever-decreasing frequency, a phenomenon known as the *gradient of transmission*. This property allows genes to be mapped with respect to their distance from the site of F insertion. The reasons for the gradient of transmission are speculative and include a break in the ssDNA being transferred, a break in the conjugation bridge, or the presence of sequences in the DNA that mimic the sequences in F near *oriT* that signal termination. Interestingly, the transfer region is the last sequence to be transferred in Hfr-mediated conjugation, meaning that the recipient is never converted to F$^+$ status. Again, this suggests that F is a tool of *E. coli* rather than acting as "selfish DNA," a property of many mobile elements. F is unusual in that its replicons are quiescent when it is in the Hfr state; they are capable of expressing incompatibility, but they do not initiate replication unless the chromosomal origin is inactivated (Hohn and Korn, 1969; Scaife and Gross, 1962).

Thus, *E. coli* has developed a symbiotic relationship with its sex factor in which it trades maintenance of F for the benefits of genetic exchange. F transfer gene expression only occurs when the conditions are optimal for its host. Unlike more promiscuous plasmids, such as those in the IncP incompatibility group, F and related plasmids have developed many control circuits that involve host regulatory elements, allowing it to monitor the metabolic state of its host. The remainder of this chapter discusses the control elements provided by F and *E. coli* to regulate F transfer, as well as a brief description of the conjugative apparatus.

F TRANSFER REGION IS A TYPE IV SECRETION SYSTEM

Type IV secretion systems (T4SSs) are involved in the transport of macromolecules either into or out of the bacterial cell. The classic T4SS is conjugative DNA transfer; however, T4SSs have been shown to be involved in tumorigenesis in plants by the Ti plasmid of *Agrobacterium tumefaciens*, pertussis toxin excretion in *Bordetella pertussis*, DNA excretion by *Neisseria gonorrhoeae*, Cag protein translocation into target cells, and DNA uptake by *Helicobacter pylori*, among many others (Lawley et al., 2003; Nagai and Roy, 2003) .

The F plasmid represents one of three known conjugative T4SSs: the F, P, and I groups. The P group, represented by the IncP plasmids RP4 and R751, R388 (IncW), and the VirB region of the Ti plasmid, is associated with the transfer of protein or nucleoprotein complexes. The I group appears to be distantly related to the P group but uses a second pilus system to promote mating pair formation. This group is represented by IncI plasmids, such as R64 and ColIb-P9, and the Dot/Icm genomic island of *Legionella pneumophila*. The F group is represented by F and is related to conjugative genomic elements, such as the IncJ integrative phage/plasmids (Boltner et al., 2002), the SXT element of *Vibrio cholerae* (Beaber, Hochhut, and Waldor, 2002), and the DNA excretion system of *N. gonorrhoeae* (Dillard and Seifert, 2001). The F systems seem to be involved in the transport of naked DNA, although a pilot protein might be bound to the 5' end of the transferred strand. F is also involved in the uptake of RNA during infection by class I RNA phages such as MS2 and R17. Again, there is evidence that a pilot protein, the phage attachment protein, might be important in this process.

The F transfer region encodes 23 proteins involved in bacterial conjugation. These include proteins involved in conjugative DNA metabolism, pilus assembly, mating pair stabilization, surface and entry exclusion, and regulation (Fig. 5.2). Conjugative type II pilus assembly requires the core T4SS genes [in F, *traA* (pilin), *traB,-C,-E, -G* (N-terminal region), *-K, -L*, and *-V*], as well as *traF, -H, -W*, and *trbC*, with pilus assembly being energized by $TraC_F$ in an NTP-dependent manner. P and I systems have the $TrbB_P/VirB11_{Ti}/TraJ_I$ homologs that use ATP, form ring structures, and are essential for pilus assembly (Savvides et al., 2003). F T4SS lack this homolog but instead encode *traG* (C-terminal domain), *traN*, and *-U*, which are essential for conjugative DNA transfer. The T4SS proteins form the "transferosome" that interacts with the "coupling protein" (in F, $TraD_F$; Gomis-Ruth, de la Cruz, and Coll, 2002) via $TraB_F$, an inner membrane protein that is predicted to extend into the periplasm and multimerize via coil–coil interactions as demonstrated in the homologous R27 (IncHI1) transfer system (Gilmour et al., 2003). $TraB_F$, a distant TonB homolog that might transduce a signal between the inner and outer membranes, is part of a transenvelope structure that interacts with the secretin-like protein $TraK_F$ that, in turn, interacts with a lipoprotein tethered to the outer membrane, $TraV_F$ (Harris, Hombs, and Silverman, 2001). The coupling protein binds single-stranded DNA and is believed to pump the DNA through the transferosome, using ATP as an energy source (Cascales and Christie, 2004). The role of ATP in this process is complex; intuitively, ATP hydrolysis should be important for DNA transport; however, genetic evidence suggests that it might also play a role in signaling (Schroder and Lanka, 2003). The structure of the cytoplasmic domain of a coupling protein, TrwB from plasmid R388, has been solved revealing strong similarity to ring helicases and F1-ATPases (Gomis-Ruth et al., 2001). It contains a central channel of 2.2 nm in its cytoplasmic domain through which the DNA might pass. In conjugative systems, there is a need for a relaxase (in F, $TraI_F$) that

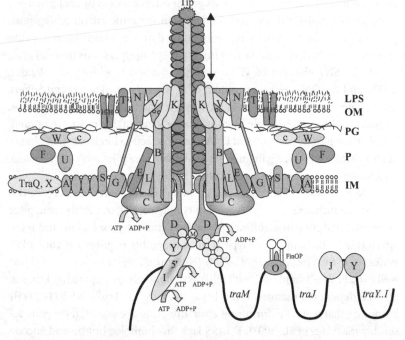

Figure 5.2. The F conjugative apparatus (transferosome) is depicted, based on available data (see text for details). The nature of the pilus tip is unknown. The two-way arrow represents pilus outgrowth and retraction. The relaxosome is depicted with TraY and TraM bound to sites on the 3' end of the cleavage site at *nic*. TraI is bound covalently to the 5' end of the DNA, which will be transferred first into the recipient cell. Only the cleaved strand of the DNA is depicted. The *traM*, *traJ* cistrons and the head of the *traY-I* operon are indicated with the FinOP complex shown bound to the *traJ* mRNA as it emerges from the transcription complex. The position of host proteins that bind the DNA are not indicated for simplicity's sake.

prepares the DNA for transfer (see "F Plasmid DNA Metabolism" section). The relaxase bound to *oriT* is part of the "relaxosome" composed of $TraY_F$, $TraM_F$, and the host-encoded protein IHF (integration host factor). $TraY_F$ is the key regulator of conjugative ability in F both at the level of transcriptional control and regulation of nicking by TraI. TraM binds to three sites in the *oriT* region (sbmA,B,C, Fig. 5.1) and forms a nucleosome-like structure that is believed to control local superhelical density, thus affecting DNA melting and TraI nicking and helicase activity. TraM is positively regulated by TraY and is negatively autoregulated, forming a regulatory loop that is sensitive to the growth cycle of the cell (see "F Plasmid DNA Metabolism" section).

The F-pilus has been implicated in recipient cell recognition, signaling that DNA synthesis and transport should initiate, and formation of the conjugative pore (Manchak, Anthony, and Frost, 2002). The F-pilus is a helical array of subunits arranged as a five-start helix with a periodicity of 25 subunits in two turns of the helix with a rise of 32 nm. The rise per subunit is 1.28 nm and the diameter of the filament is 8 nm with an inner lumen of 2 nm, the same diameter as the channel in the coupling protein. The F-pilus is flexible and can be very long, up to 20 μm, if grown without agitation, suggesting that there is no mechanism to limit its length. The F-pilin subunit is relatively small, 7.2 kDa, with a high alpha-helical content (70%). It is acetylated at its N-terminus after insertion into the inner membrane via $TraQ_F$, a chaperone-like protein, and processing by the host leader peptidase LepB. The pilus tip is formed at the cell surface and requires $TraA_F$, $-C_F$, $-E_F$, $-G_F$ (N-terminal region), $-K_F$, and $-L_F$. Pilus extension requires $TraB_F$, $-F_F$, $-H_F$, $-V_F$, $-W_F$, and $TrbC_F$, whereas pilus retraction or disassembly also requires $TrbI_F$ (Anthony et al., 1999; Firth, Ippen-Ihler, and Skurray, 1996). The nature of the pilus tip remains unknown, although it appears to be composed of pilin subunits in a specific conformation such that subunit–subunit junctions form the sites for recipient cell recognition and filamentous phage attachment. The F-pilus recognizes the phosphorylated sugar residues in the inner core of the lipopolysaccharide of the recipient cell (Anthony et al., 1994). Attachment is believed to trigger pilus retraction and close cell-to-cell contact known as mating pair stabilization (Mps). This process is mimicked by filamentous phage, which attach to the pilus tip via the attachment protein pIII and gain access to their primary receptor, TolA, in the cell envelope via pilus retraction. Purified pIII phage protein binds to the pilus tip and triggers loss of pili from the cell surface (Elton et al., 2003), suggesting that it is attachment and not subsequent steps in phage infection that cause pilus loss. Whether pilus outgrowth and retraction are ongoing processes during cell growth or are triggered by attachment of a receptor at the tip is unknown. The attachment of large objects to the pilus tip might trigger the retraction process by increasing the drag at the end of the pilus making the pilus a primitive touch mechanism.

$TraG_F$ (C-terminal region), $TraN_F$, and $-U_F$ are essential for conjugation and have been implicated in mating pair stabilization involving formation of a more robust conjugative pore. $TraN_F$ (22 cysteines) and $TraU_F$ (10 cysteines) are believed to be very rigid structures with numerous inter- and intramolecular disulfide bonds (Klimke et al., 2004). The host protein DsbA, which catalyzes disulfide bond formation in the periplasm, is essential for F-pilus formation. TraF$_F$ and TrbB$_F$ are homologous to each other and contain a predicted thioredoxin fold, suggesting they play a role in disulfide bond formation. Interestingly, homologs of DsbC, a thiol isomerase, have been identified in many F-like T4SS, which are also homologs of TrbB$_F$ and TraF$_F$. More recent evidence suggests that TrbB$_F$ is an isomerase because it can complement the *dsbA* mutation when overexpressed as demonstrated for DsbC (Elton et al., 2004).

The organization of the genes in the F transfer region (Fig. 5.1) suggests that a T4SS was inserted in the midst of the genes for conjugative DNA replication such that *oriT* and TraM$_F$, -Y$_F$ were separated from TraD$_F$ and -I$_F$. Along with the loss of the TrbB$_P$/VirB11$_{Ti}$ homolog, F underwent a second insertion involving the *traF* to *traH* gene cluster that modified the basic T4SS and gave it the properties of pilus retraction and mating pair stabilization, allowing it to mate very efficiently in any media. The regulatory system FinOP that controls TraJ$_F$ translation, appears to have been inserted between *traM* and *traY* with the *finO* gene displaced to the distal end of the transfer region and inactivated by IS3a insertion.

F PLASMID DNA METABOLISM

The F plasmid origin of transfer (*oriT*) is approximately 450 bp in length (66.7F; Frost, Ippen-Ihler, and Skurray, 1994). Within this region is the *nic* site, necessary for the sequence- and strand-specific, single-stranded nicking of the F plasmid. This nicking reaction is accomplished by the plasmid-encoded relaxase, TraI, which, by way of a transesterification reaction, is covalently linked via a Tyr residue near the N-terminus to the 5' end of the nicked strand of DNA (Byrd et al., 2002; Lanka and Wilkins, 1995; Zechner et al., 2000). The 3' end is buried in a noncovalent manner; its release from this complex allows it to prime continuous replacement strand synthesis. This release of the 3' end has been tied to melting of the DNA and entry of the helicase domain of TraI$_F$ in preparation for unwinding and transport of the DNA (Csitkovits and Zechner, 2003). However, it is not known how contact with a recipient cell triggers this series of events.

The relaxase is also involved in re-ligating the 5' and 3' ends during the termination reaction after one round of transfer. The relaxase can stand alone (P, I systems) or be fused to domains with helicase (F-type systems) or primase (IncQ plasmids) activity. Circumstantial evidence suggests that the relaxase is translocated to the recipient cell at the start of conjugation, although this remains unproved. P-like relaxases have a single tyrosine at their active sites and form dimers that are required for the termination reaction, whereas F-like relaxases have a double pair of tyrosines that are involved in DNA nicking and re-ligation (Grandoso et al., 2000). TraI$_F$ is recruited to oriT by TraY$_F$, which together with TraM$_F$ and IHF form a nucleoprotein complex referred to as the relaxosome (Howard, Nelson, and Matson, 1995; Nelson et al., 1995). IHF induces extensive bending in the DNA and alters local topology, which presumably modulates TraI$_F$ binding and activity (Howard, Nelson, and Matson 1995; Nelson et al., 1995). This architectural role for IHF is supported by the observation that it induces a significant bend of approximately 150 degrees upon binding DNA (Di Laurenzio et al., 1995; Rice et al., 1996), and that IHF also influences local DNA supercoiling (Teter, Goodman, and Galas, 2000).

TraM$_F$, although not necessary for nicking at oriT, is essential for transfer (Fekete and Frost, 2000). It binds to three sites within the oriT region (sbmA, -B, and -C), and interacts with TraD$_F$, linking the relaxosome to the transferosome and possibly providing the signal for transfer to begin (Disque-Kochem and Dreiseikelmann, 1997; Willetts and Wilkins, 1984).

The F TraI$_F$ protein, unlike many other plasmid-encoded relaxases (e.g., RP4 TraI$_P$), contains both transesterase and helicase activities in nonoverlapping domains (Byrd et al., 2002; Pansegrau, Schroder, and Lanka, 1993; Street et al., 2003). The DNA is in an equilibrium between the nicked and ligated states during growth of the cell. It is only upon contact with a recipient cell that the plasmid DNA is unwound by the helicase activity of TraI$_F$ in a 5′ to 3′ direction in an ATP-dependent manner. This unwinding may provide some of the energy necessary to drive strand transfer because TraI helicase mutants are defective for conjugation. The transesterase activity also allows the religation of the nicked strand after unwinding and transport of one copy of the single-stranded DNA to the recipient cell (Luo, Gao, and Deonier, 1994). The transferred strand is spooled into the recipient cell by means of a mechanism similar to rolling circle replication and phage head packaging in spherical ssDNA phages. In the recipient cell, however, complementary strand synthesis uses one or more single-stranded promoters (Frpo) in the leading region to initiate RNA primer synthesis (Masai and Arai, 1997). Because Frpo is present only on the transferred strand and is active only in single-stranded DNA, it could drive the zygotic induction of genes in the leading region that help establish the plasmid in the recipient cell, as has been shown for the IncI plasmid ColIb-P9 (Bates et al., 1999). Other Frpo-like sites, termed ssi, have been marked at intervals around the F plasmid suggesting that a number of primers are formed that initiate DNA synthesis via a discontinuous mode (Nomura et al., 1991).

F DNA METABOLISM AND ADAPTIVE MUTATION

The DNA metabolism of F has been implicated both in recombination and adaptive mutation. In a series of experiments in the mid-1990s, researchers demonstrated that starved *E. coli* could induce frameshift mutations in a mutated *lacZ* reporter gene, thereby restoring the wild-type *lacZ* allele and accessing lactose in the medium, resulting in colony outgrowth (reviewed in Foster, 2000). That the *lacZ* allele was carried on an F' plasmid appeared to be important for this phenomenon. The presence of nicked DNA at *oriT* was invoked as being important in generating a template for RecBCD-based recombination, which, in conjunction with transient downregulation of the mismatch repair system, leads to a hypermutable state. The mechanism for adaptive mutation is still very controversial (Roth et al., 2003). Interestingly, no one has repeated the Foster and Cairns (1992) experiment using a F plasmid deleted for *nic* or *traI*.

Increased recombination in the presence of F was first observed in F42*lac* × λp*lac*5 transductional crosses, where the level of recombination was 20- to 50-fold higher than in λp*lac*5 × chromosomal *lac* crosses (Porter, 1981; Porter, Low, and McLaughlin, 1978). Expression of F conjugative functions induces increased recombination rates, as measured by Lac$^+$ reversion of *E. coli* strains carrying F' *pro*$^+$*lacI33ΩlacZ*, although conjugative transfer is not required (Foster and Trimarchi, 1995a, 1995b). TraY$_F$ and TraI$_F$, which bind and subsequently nick at *oriT*, are required to increase the recombination potential of F (Carter and Porter, 1991). This activity is independent of the TraJ$_F$-mediated activation of expression from P$_Y$, and relies solely on the provision of TraY$_F$ and TraI$_F$ *in trans*, as well as the presence of *oriT* *in cis* on an F' *lac* plasmid. Inhibition of *tra* expression, mediated by the FinOP fertility inhibition system, leads to reduced levels of apparent RecA-dependent adaptive mutation, as measured by Lac$^+$ reversion rates in cultures of *E. coli* carrying a F' *pro*$^+$ *lacI33ΩlacZ* episome (Galitski and Roth, 1995). A decrease in accumulation of TraY$_F$ and TraI$_F$ mediated by the FinOP system, rather than a reduction in conjugation events, likely causes this phenomenon. These authors postulate that under conditions of selective growth, extensive replication of the F plasmid from *oriT*, at a rate higher than that of chromosome replication, contributes to the observed increase in apparent adaptive mutation. Increased replication could lead to an accumulation of all the *tra* proteins, with TraY$_F$ and TraI$_F$ being essential for increased recombination and/or mutation rates mediated by the presence of the F′ episome.

The proposed mechanism for F-mediated increase in recombination rates centers on the activity of the *E. coli* double-stranded break repair pathway. Indeed, elevated rates of adaptive mutation in *E. coli* carrying F′ *pro*$^+$ *lacI33ΩlacZ* rely on functional RecA (Galitski and Roth, 1995). It is believed that TraI$_F$-mediated nicking at *oriT* is necessary to provide an opening for the RecBCD enzyme to enter the DNA duplex (Seifert and Porter, 1984). However, a single-stranded nick is generally not sufficient for RecBCD entry. It has been suggested that the nicked plasmid is more prone to breakage, providing a double-stranded break that is a suitable substrate for RecBCD; however, this has been shown not to be the case (Carter, Patel, and Porter, 1992). In addition, the C-terminal helicase domain of TraI$_F$ appears to be largely dispensable for heightened recombination; only the relaxase activity is essential (Carter and Porter, 1991). Thus, the possibility that the helicase activity is responsible for heightened recombination by providing recombinogenic single-stranded DNA is eliminated. RecBCD itself could act as a helicase, providing a single-stranded substrate for RecA. Alternatively, χ sites within F might be required for RecBCD activity, with nicking modulating enzyme activity via χ. However, it is worth noting that there do not appear to be any exact matches to the χ consensus sequence within the F plasmid sequence, casting some doubt on such a model. Nicking alone will only enhance recombination within the donor cell. The transferred strand exists both in a recombinogenic single-stranded form and in a transient, gapped duplex intermediate during complementary strand synthesis. Both of these forms might be more suitable substrates for RecBCD, promoting recombination in the recipient cell as well.

The answers to questions regarding the role of F in recombination and adaptive mutation are still being sought. Evidence suggests that unselected genes both on the F′ episome and the *E. coli* chromosome can exhibit heightened mutability in conjunction with Lac$^+$ reversion. A mutagenic mechanism might also exist in a subpopulation of cells exposed to selection, with the phenomenon of F′ episome-mediated Lac$^+$ reversion involving factors other than simply the requirement of F *tra* operon expression and a functional RecBCD pathway (Torkelson et al., 1997). In the case of plasmid F′$_{128}$ (*proABlac*), a separate mechanism might exist for inducing increased mutation events in *E. coli*. DinB, an error-prone DNA polymerase that is induced by the SOS-response (Kim et al., 1997), lies within 16.5 kb of *lac* on this plasmid. It has been suggested that amplification of *dinB* expression, coupled to increased *lac* expression under selective conditions for Lac$^+$ reversion, leads to increased general mutagenesis (Kofoid et al., 2003). Clearly, the connection between F conjugation and adaptive mutation exists; however, it appears to be more complicated than originally anticipated.

REGULATION OF F TRANSFER REGION EXPRESSION

The F transfer region is known to be regulated by fertility inhibition imposed by the presence of compatible FinO+ plasmids, the physiological state of the cell, and in response to stress exemplified by the Cpx pathway. Each condition has been tied to the needs of the host and the plasmid, depending on the circumstances. Fertility inhibition, which has been lost by F but is present in most other F-like plasmids, allows cells to modulate transfer gene expression during vegetative growth. If a suitable recipient is found, fertility inhibition is limited for several generations in a process termed *high frequency of transfer* (HFT) or epidemic spread, to allow all available F– cells to be converted to F+ status. F responds to stationary phase and also probably to changes in temperature, pH, osmolarity, and many other physiological parameters by combining the regulatory power of a number of host-encoded factors with F-derived regulatory elements to downregulate transfer region expression to undetectable levels. F also has the ability to respond rapidly to the imposition of sudden stress through systems such as the Cpx pathway, which involves transcriptional and posttranslational control mechanisms. Here we discuss these three known mechanisms in detail with the idea that there might be many other control mechanisms that have yet to be discovered.

The fertility inhibition (FinOP) system

The primary mechanism for inhibiting expression of the *tra* operon and F plasmid transfer is the FinOP fertility inhibition system, which inhibits expression of the main activator of P$_Y$, the TraJ regulatory protein. Using this method, *tra* operon expression can be rapidly and efficiently repressed using only a small number of plasmid-encoded components (Willetts, 1977). FinOP is a two-component system, consisting of a regulatory antisense RNA, FinP, and an RNA chaperone, the FinO protein (Jerome, van Biesen, and Frost, 1999).

The F *traJ* mRNA contains a 105 nucleotide 5′ untranslated region (UTR), which folds into a complex secondary structure consisting of three stem loops (SLs) and an extended single-stranded region in its 5′ proximal end (van Biesen et al., 1993). The three SL regions are designated SL-Ic, SL-IIc, and

SL-III (Fig. 5.2). The ribosome-binding site (RBS) and the AUG start codon for TraJ are within SL-Ic. FinP is transcribed from the strand complementary to the 5′ UTR of *traJ* mRNA from a constitutive promoter, resulting in a 79 nucleotide antisense RNA molecule with complete complementarity to a portion of the *traJ* transcript. FinP folds into two SL domains, designated SL-I and SL-II, separated by a four-nucleotide single-stranded spacer with four and six pair 5′ and 3′ tails.

The loop region of FinP SL-I contains an anti-RBS sequence, which can interact with a portion of the RBS of *traJ* mRNA. This sequence contains the common motif, 5′-YUNR-3′ (Y = C or U; N = any base; R = A or G), which is important in promoting loop–loop interactions in a variety of sense:antisense pairing reactions (Franch et al., 1999). The formation of a duplex between FinP and *traJ* mRNA, mediated by an initial loop–loop kissing interaction, is believed to sequester the RBS of *traJ* mRNA, preventing translation of the message. Formation of a FinP/*traJ* mRNA duplex has been demonstrated *in vitro* (Ghetu et al., 2000; Sandercock and Frost, 1998; van Biesen et al., 1993) and *in vivo*, creating a substrate that is a target for degradation by RNase E (Jerome, van Biesen, and Frost, 1999). Efficient and rapid formation of FinP/*traJ* mRNA duplexes both *in vitro* and *in vivo* requires the FinO protein. FinP-mediated reduction of TraJ accumulation is, therefore, probably controlled by two separate and distinct mechanisms: prevention of translation by sequestration of the RBS, and creation of a double-stranded RNA substrate that is degraded by RNase III.

Eight different alleles of FinP have been described among F-like plasmids. All display a high degree of conservation in the stem and single-stranded tail regions, but less conservation in the loop sequences (Finlay et al., 1986), thereby providing allelic specificity (Koraimann et al., 1991, 1996). Mutational analysis of the loops of SL-I and SL-II of FinP encoded by the F-like plasmid R1 determined that complementarity between the loops of FinP and *traJ* mRNA is critical for the ability of FinP to repress both *traJ* expression and conjugative transfer of the plasmid. Mutation of the stem regions of F and R1 FinP has no effect on the ability of FinP to repress plasmid transfer or form a duplex with *traJ* mRNA *in vitro*. Thus, the initial loop–loop interactions might be sufficient to repress translation of *traJ* mRNA and TraJ accumulation (Gubbins et al., 2003; Koraimann et al., 1996; van Biesen and Frost, 1994).

The intracellular concentration of FinP is a key factor in determining the ability of the FinOP system to control TraJ expression. Previous work has demonstrated that in the presence of FinO the steady-state concentration of FinP increases substantially and the half-life of the RNA can be significantly extended. In the absence of FinO, FinP is a target for degradation by RNase E, which cleaves the RNA within the single-stranded spacer region located between SL-I and SL-II (Jerome, van Biesen, and Frost, 1999). Using *traJ-lacZ* reporter constructs and mating inhibition assays to monitor FinP function, FinP provided *in trans* at an elevated copy number was able to negatively regulate *traJ* expression in the absence of FinO. Thus, the regulatory effect of FinP is highly gene dosage and concentration dependent (Dempsey, 1994; Koraimann et al., 1996).

FinO is absolutely required for the repression of F and F-like plasmid transfer by the FinOP system. F is derepressed for transfer due to an IS3 insertion in *finO* (Cheah and Skurray, 1986; Yoshioka, Ohtsubo, and Ohtsubo, 1987). It is not known whether FinO is expressed from its own promoter, or as part of a multicistronic transcript expressed from P_Y. Two alleles of *finO* have been identified based on their level of repression of transfer (Willetts and Maule, 1986). Type I alleles (R100, R6–5) are characterized by their ability to repress F transfer by approximately 100- to 1000-fold, and by the presence of an open reading frame, *orf286*, immediately upstream of the *finO* coding region. Type II alleles (ColB2) repress mating by 20- to 50-fold and lack *orf286*. The presence of *orf286* appears to stabilize the *finO* transcript, leading to a higher intracellular concentration of FinO, and thus greater repression of plasmid transfer (van Biesen and Frost, 1992). Whereas FinP exhibits allelic specificity among the F-like plasmids, FinO from one plasmid can function to repress transfer among a wide variety of different F-like plasmids (Finnegan and Willetts, 1973). This observation suggests that FinO functions in a manner that is completely independent of the sequence of both FinP antisense RNA and *traJ* mRNA.

FinO is a 21.2-kDa cytoplasmic RNA-binding protein, consisting of 186 amino acids (Ghetu et al., 2000). The high-resolution three-dimensional structure of FinO has recently been solved by X-ray diffraction studies, revealing a unique RNA-binding motif. The N-terminal portion of FinO forms a long solvent-exposed alpha-helix, whereas the central domain of FinO is composed of four short regions of β-sheet structure, interspersed with four short alpha-helices. These regions form a solvent-exposed positively charged face; the C-terminal portion of FinO contains a shorter alpha-helix with its distal portion packing against the base of the long N-terminal alpha-helix.

FinO has been shown to increase the steady-state intracellular concentration of FinP antisense RNA rather than act on the *finP* or *traJ* promoters, with the acidic C-terminal region of FinO required to mediate protection of FinP from RNase E degradation (Sandercock and Frost, 1998). The single-stranded spacer located between SL-I and SL-II could be cradled within the FinO structure, sterically inhibiting RNase E-mediated cleavage of the RNA.

A detailed analysis of the structural features of FinP and *traJ* mRNA recognized by FinO also provided insight into how this RNA–protein interaction occurs. Using *in vitro* transcribed RNAs and purified GST:FinO protein, the minimal RNA-binding target for FinO was determined to be FinP SL-II (van Biesen and Frost, 1994). Binding of FinO to this target was enhanced by the presence of single-stranded tails on either side of the stem loop. Furthermore, the length, but not the sequence, of these single-stranded tails had a major influence on the affinity of FinO for the RNA. Although FinP SL-II provides a minimal target for FinO binding, SL-I is also bound by FinO *in vitro*, albeit with lower affinity. FinO also binds to the analogous stem-loop structure of *traJ* mRNA, leading to the conclusion that FinO binds its RNA targets in a structure-dependent manner (Jerome and Frost, 1999), which explains why FinO exhibits no plasmid specificity.

The observation that FinO increased the intracellular concentration of FinP by approximately two-fold and increased the half-life of the RNA from 2 minutes to more than 40 minutes could not entirely account for the previously observed 10- to 1000-fold repression of F-like plasmid transfer (Koraimann et al., 1996; Lee, Frost, and Paranchych, 1992). FinO also increases the rate of FinP/*traJ* mRNA duplex formation by promoting loop–loop interactions between SL-I and SL-Ic, and by the limited unwinding of SL-II and SL-IIc containing stable stem structures that resist duplexing (Ghetu et al., 2002; Gubbins et al., 2003). Duplex formation occurs from the loops toward the stems with the 5′ and 3′ tails playing a role in FinO binding. Once duplexing begins, FinO appears to dissociate from the complex because FinO-bound duplexing intermediates have not been found.

The phenomenon of fertility inhibition is also tied to that of epidemic spread. In cells containing FinO+ F-like plasmids, conjugative frequency remains unusually high for several generations until all available recipient cells have been converted to donor status. This state is known as *high frequency of transfer* (HFT) and represents a period of time in the new transconjugant before fertility inhibition sets in. The simplest explanation for HFT is the need for the intracellular levels of FinO to rise and stabilize FinP, allowing for repression of *traJ* mRNA translation. However, the presence of FinO in the recipient cell does not impede this process, suggesting a more subtle mechanism. The two promoters for *traM* are extremely strong, and it requires a high concentration of TraM to repress them. In a new transconjugant, the *traM* promoters are unregulated, and there is considerable read-through into *traJ*. This increase in *traJ* mRNA could counteract the presence of FinOP and extends the period of time that the *tra* genes are expressed. Once the levels of TraM increase, the *traM* promoters are repressed, the levels of read-through into *traJ* decrease, and FinOP repression of *traJ* begins. The situation is more complex than mere autoregulation by TraM, however, because overexpressing TraM *in trans* in the donor or recipient cell promotes even higher levels of HFT for a few generations, but eventually represses the system sooner than in cells lacking extra TraM (Lu, Forsyth, and Frost, 2004 [kal2]). The control of *traJ* by *traM* in new transconjugants completes the regulatory loop in F. A similar loop has been proposed for the R1 plasmid (Polzleitner et al., 1997) and for R100 in which a "latch relay" system operates (Dempsey, 1989). Unlike R1 or R100, which are FinO$^+$ and probably require some low level of read-through into *traJ* by *traM* to release the system from complete repression, F lacks FinO; consequently, read-through from *traM* into *traJ* need only occur during epidemic spread because the *traM* terminator effectively blocks read-through when the *traM* promoters are repressed (Lu, Forsyth and Frost, 2004).

The role of host factors in *tra* gene regulation

F plasmid transfer is a highly energy-dependent process. Piliation, conjugative pore formation, strand transfer, and complementary strand synthesis all carry significant energetic costs for the donor cell, not to mention the costs incurred by the recipient cell upon transfer during plasmid establishment and zygotic induction. Uncontrolled, the potential runaway synthesis of the various plasmid-encoded components for these processes could have a highly deleterious effect on the host. To avoid this, the donor cell appears to use several mechanisms and factors for controlling both transfer gene expression and transfer itself (Fig. 5.3).

The effect of host-mediated control is evident in the growth phase–dependent patterns of F transfer. Transfer efficiency has been shown to rapidly diminish as the host cell progresses through exponential phase and into stationary phase (Frost and Manchak, 1998). TraJ, believed to be the primary plasmid-encoded activator of transfer gene expression, appears to be present long after a donor cell population enters stationary phase. Nevertheless, in spite of the presence of TraJ, mating efficiency drops rapidly, presumably due to the presence of a number of host factors that make plasmid gene expression and transfer more sensitive to the physiological state of the host cell. One such factor is the degree of supercoiling, which is directly linked to the state of free energy within the cell, due to modulation of DNA gyrase activity by the ATP/ADP ratio (Jensen et al., 1995; van Workum

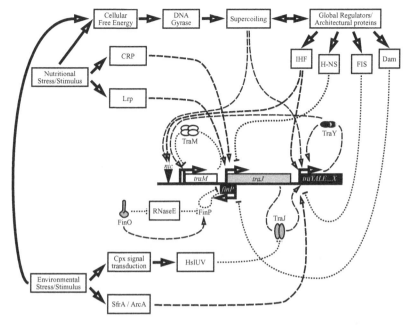

Figure 5.3. The regulatory circuit of the F plasmid. The host and plasmid encoded regulatory factors believed to influence plasmid transfer gene expression, based on work in F and related plasmids. Host factors are enclosed within white boxes. Factors exerting positive control upon transfer genes are indicated by long dashed lines ending in arrowheads, whereas negative gene control is indicated by short dashed lines ending in bars. Further details are discussed within the text.

et al., 1996; Westerhoff et al., 1988). This allows gyrase activity to fluctuate in response to multiple stimuli, including pH (Karem and Foster, 1993), anaerobiosis (Dorman et al., 1988), osmolarity (Higgins et al., 1988), temperature (Goldstein and Drlica, 1984), and nutrient shift (Balke and Gralla, 1987). Supercoiling is required both for transcription of the *tra* operon from P_Y (Gaudin and Silverman, 1993) and for efficient nicking at *oriT* by $TraI_F$ (Matson and Morton, 1991), which may serve to help link transfer efficiency to the physiological state of the cell.

Histone-like proteins, including IHF, Fis, and H-NS, have been observed to provide physiologically dependent control in numerous systems, including rRNA synthesis and chromosomal replication. However, of particular interest is the apparent role of these proteins as physiological coupling factors for genetic "free agents," such as phage, transposons, and plasmids. These proteins sensitize gene regulation and nucleic acid transactions, such as recombination and replication, to the state of the host, and such a role is also evident in F plasmid biology. Because of the highly dynamic expression patterns of these histone-like proteins throughout the growth phase (Ali Azam et al., 1999), they are key elements in the control of F transfer region gene expression. IHF is one such factor known to provide multilevel control of F plasmid transfer. Studies have shown that IHF is necessary for piliation and mating pair formation (Gamas et al., 1987). IHF has also been shown to enhance transfer operon expression somewhat in exponential phase (Silverman, Wickersham, and Harris, 1991). Also, as described previously, IHF is an essential component of the relaxosome and is believed to act in conjunction with TraY$_F$ to alter $oriT$ structure in such a way as to facilitate optimal activity on the part of TraI$_F$ during the nicking and unwinding of the plasmid (Howard, Nelson, and Matson, 1995; Nelson et al., 1995). This is somewhat paradoxical given the growth-phase dependent levels of IHF. IHF levels are known to increase throughout exponential phase, reaching maximal levels in early and midstationary phase (Ali Azam et al., 1999), which does not correspond to its proposed role as an enhancer or activator of plasmid transfer and transfer gene expression. It suggests the presence of additional factors limiting transfer in a physiologically dependent manner. More recent experimental evidence suggests that F transfer gene expression might also be influenced by H-NS and other factors (Will et al., 2004). These factors may act as components of a putative regulatory nucleoprotein complex upstream of the transfer operon promoter or at other transfer gene promoters on the plasmid (Gaudin and Silverman, 1993). H-NS appears to silence the entire tra region by binding to the promoters for $traM$, $traJ$, and $traY$-X. Thus, TraJ$_F$ and possibly TraY$_F$ and TraM$_F$ have additional roles in relieving H-NS repression as F+ cells enter exponential phase. Indeed, TraJ$_F$ appears to be an antagonist of H-NS repression rather than an activator of transcription (Will et al., 2004). In addition, many F-like plasmids, including R100 and pRK100, also carry genes encoding Hha/YmoA family proteins (Nieto et al., 2002). Hha family proteins have been shown to interact directly with H-NS by way of a similar oligomerization domain and are believed to modulate the activity of H-NS through these interactions. RmoA, an Hha/YmoA family protein expressed by R100, has been shown to positively regulate plasmid transfer (Nieto et al., 1998). The possibility that F-like plasmids encode their own modulator protein to interact directly with host H-NS and counteract host repression is particularly intriguing, representing a novel system of transfer gene control.

F plasmid transfer and transfer gene expression is also influenced by many more specific host factors, taking cues from various environmental

MICHAEL J. GUBBINS, WILLIAM R. WILL, AND LAURA S. FROST

and physiological stimuli. One example is ArcA (SfrA), a response regulator of a two-component signal transduction system that senses and responds to the redox state of the cell that is required for maximal transcription from P_Y (Lynch and Lin, 1996; Silverman, Wickersham, and Harris, 1991; Silverman et al., 1991). Direct activation of transcription from P_Y of the F-like plasmid R1 by ArcA has been demonstrated, and the presence of $TraJ_F$ is required for this activity (Strohmaier et al., 1998). The active, phosphorylated form of ArcA was shown to bind to a region directly upstream of P_Y, suggesting that changes in the DNA topology of this promoter on ArcA binding may lead to increased accessibility of RNA polymerase to this region. The precise mechanism of the ArcA/TraJ-mediated control of P_Y remains to be elucidated, but likely involves the potential formation of a nucleoprotein complex involving TraJ, TraY, ArcA, and the P_Y promoter mentioned previously that antagonizes H-NS silencing (Gaudin and Silverman, 1993; Strohmaier et al., 1998).

Host-encoded pathways that sense and respond to nutrient availability also affect *tra* operon expression. The presence of a putative consensus-binding sequence for CRP (cAMP repressor protein) overlapping the initiation site for *traJ* transcription has led to the theory that cAMP-CRP directly influences *tra* operon expression (Kumar and Srivastava, 1983; Paranchych, Finlay, and Frost, 1986). Indeed, a lack of cyclic AMP has been shown to influence piliation of various F-like plasmids (Harwood and Meynell, 1975). More recent analysis has determined that CRP does indeed bind to a specific site in the *traJ* promoter region, directly affecting transcription of *traJ* and plasmid transfer efficiency in the F-like plasmid pRK100 (Starcic et al., 2003).

More recently, the global regulatory protein Lrp leucine-responsive regulatory protein (Lrp) has been shown to directly upregulate expression of P_{traJ} of the F-like plasmid pSLT of *Salmonella typhimurium* (Camacho and Casadesus, 2002), as well as pRK100 from uropathogenic *E. coli* (Starcic-Erjavec et al., 2003). Lrp influences expression of a variety of operons involved in responding to changes in nutrient availability (D'Ari, Lin, and Newman, 1993; Tani et al., 2002). Lrp has been shown to bind *in vitro* to a region upstream of the *traJ* promoter containing three putative Lrp-binding sites (Camacho and Casadesus, 2002; Starcic-Erjavec et al., 2003). However, the observed effect on plasmid transfer is different in both systems. pSLT transfer is reduced by approximately 50-fold in a *lrp* host, whereas pRK100 transfer is reduced only 4-fold (Camacho and Casadesus, 2002; Starcic-Erjavec et al., 2003). Although the reason for the differing degree of activation is unclear, similarities between pSLT, pRK100, and other F-like plasmids has led the authors of both studies to propose that Lrp expression likely influences *tra* expression in multiple F-like plasmids via direct upregulation of P_{traJ}. Given that each putative Lrp binding site at least partially overlaps known H-NS binding sites at P_{traJ}, it seems likely that Lrp, like other transfer regulators, antagonizes H-NS silencing (Will, Lu and Frost, 2004).

Although FinP is transcribed from its own constitutive promoter, its expression is also influenced by cellular factors. The presence of a GATC site in the promoter region of FinP prompted an investigation into the influence of Dam methylation on FinP expression and F-like plasmid transfer. Dam methylation is a well-known mechanism implicated in global regulatory control in *E. coli* (Oshima et al., 2002). Using *finP-lacZ* reporter constructs, Torreblanca, Marques, and Casadesus (1999) demonstrated that FinP promoter activity was significantly reduced in a *dam⁻ E. coli* background, resulting in elevated F plasmid transfer. These authors suggest that the methylation state of the plasmid may be a mechanism used to couple plasmid replication with FinP expression and control of *tra* operon expression. Immediately after plasmid replication, the hemimethylated state of the plasmid DNA might allow for a temporary reduction in FinP expression, thus allowing a short period of *traJ* expression, *tra* operon derepression, and plasmid transfer immediately after the completion of replication. However, because F transfer is derepressed, the benefit of such a mechanism should be of no consequence in F, although it may be important in other F-like plasmids. To date, an extensive analysis of the regulation of a naturally repressed plasmid has not been undertaken because gene expression levels are too low to monitor accurately. Regardless of the potential purpose of controlling FinP expression via *dam* methylation, the evidence suggests that the concentration of FinP antisense RNA is a critical factor in determining the efficacy of repression of *tra* operon expression and that cellular factors influence FinOP-mediated plasmid transfer inhibition.

The response to extracytoplasmic stress

The Cpx regulon was first identified by isolating mutations in the *E. coli* chromosome that affected F conjugative plasmid expression and caused an inability to assemble F-pili on the surface of the cell (McEwen and Silverman, 1980). Silverman et al. (1993) observed that *cpxA* deletion mutants exhibited quasi-wild-type levels of F*lac* transfer. However, F*lac* transfer was decreased in some *cpxA* gain-of-function mutants, and the levels of the F-encoded regulatory protein, TraJ, and expression from a P_{traY}-*lacZ* fusion construct were reduced, suggesting that activation of the Cpx system led to a lowered level of

TraJ. Decreased levels of TraJ, therefore, most likely impaired the conjugative transfer ability of F in *cpxA* gain-of-function mutants.

Extensive analysis has revealed that the *cpx* regulon is controlled by a two-component signal transduction system, which senses and responds to cell envelope stress in *E. coli* (reviewed in Raivio and Silhavy, 2001). This system consists of the inner membrane-bound sensor kinase, CpxA, and its cognate cytoplasmic response regulator, CpxR (Weber and Silverman, 1988). CpxR is a transcriptional activator (Dong et al., 1993), which, when phosphorylated by CpxA, binds target promoters at a consensus sequence (De Wulf, Kwon, and Lin, 1999; Pogliano et al., 1997). The stress-inducing signal is transferred from CpxA to CpxR through a highly conserved phosphotransferase reaction (Hoch and Silhavy, 1995). A variety of signals can activate the Cpx stress response; however, an increased level of misfolded proteins in the cell envelope is believed to be the main activating signal (Raivio et al., 2000; Snyder et al., 1995). Active, phosphorylated CpxR upregulates the transcription of several genes that are involved in protein folding and degradation in the bacterial envelope (Danese et al., 1995; Jones et al., 1997; Dartigalongue and Raina, 1998)

*cpxA** gain-of-function mutants were characterized by their ability to suppress the toxic effects of mislocalized and misfolded proteins in the cell envelope (Cosma et al., 1995). One class of *cpxA** mutants contains point mutations in the cytoplasmic domain of CpxA. An example is *cpxA101**, which contains a single amino acid change located near the putative site for autophosphorylation (Raivio and Silhavy, 1997). This mutant retains its autokinase and kinase functions, but has lost its phosphatase activity. The result is elevated levels of active, phosphorylated CpxR, shifting the Cpx regulon to a constitutively active state. More recent work has determined that reduced TraJ levels in a *cpxA101** mutant occurs at a posttranslational level, and this phenomenon is related to an instability of the TraJ protein in such mutants (Gubbins et al., 2002) with TraJ being a target for the HslVU protease/chaperone pair (Lau et al., 2004).

Expression of the *tra* operon is sensitive to numerous environmental cues, including the metabolic and nutritional state of the host cell, as well as physiological changes induced in the host cell imparted by environmental growth conditions. Global regulatory pathways of *E. coli* are clearly implicated in the control of F conjugation. Host-encoded factors also link the control of expression of the *tra* operon with host cell replication, ensuring concomitant vertical and horizontal transmission of the genetic information encoded within the F plasmid. Furthermore, the F plasmid exhibits multilevel control, exhibiting host sensitivity at multiple promoters as well as at the level of the relaxosome, thus influencing the event of transfer itself. These observations highlight the close ties between F and the genome of its host, and the intricate crossover that exists between host and plasmid gene expression. Similarly, the ability of F to survive and thrive within a population appears to be inextricably linked to the overall health and well-being of that host population. Although this observation seems both obvious and intuitive, only when the extent of linkage between host- and plasmid-encoded control mechanisms is examined does one gain an appreciation of how important the health of the host is to the maintenance of a stable plasmid population.

MICHAEL J. GUBBINS, WILLIAM R. WILL, AND LAURA S. FROST

THE CONNECTION BETWEEN VEGETATIVE AND CONJUGATIVE REPLICATION

The position of the F plasmid during the cell cycle has also been investigated using the elegant GFP:LacI fusion system to monitor the position of a multiple lacO-binding site cassette on the plasmid (Gordon et al., 1997). Previous experiments using mini-F plasmids concluded that F replication initiation was random; however, these plasmids did not contain the partitioning loci needed for synchronized replication (Eliasson, Bernander, and Nordstrom, 1996). Intact F was found to occupy fairly precise locations in the cell, whereby it is replicated at the cell midpoint and then partitioned to the one-fourth and three-fourths positions. This mirrors the situation for the host chromosome and suggests that F and chromosomal replication are in some way coordinated. Because F transfer appears to be possible at any point on the cell (the placement of F-pili is random), there needs to be a mechanism whereby the plasmid senses the position of a competent mating junction within the cell and is translocated to that site in order for transfer to begin. This might be the "signal" envisioned as necessary for transfer initiation by Willetts and Wilkins (1984) with $TraM_F$ being the "signaling" protein. This also predicts that vegetative and conjugative replication might be mutually exclusive. The more recent findings that $TraM_F$ binds to $TraD_F$ (Disque-Kochem and Dreiseikelmann, 1997) and that $TraM_F$ must be at a critical concentration for transfer to occur (Lu, Forsyth and Frost, 2004) are consistent with this model. During cell growth, the intracellular concentration of $TraM_F$ might achieve the level needed to permit transfer only once per generation. As cell growth slows in early stationary phase, the levels of $TraM_F$ could increase beyond a certain threshold, repressing the traM promoters and initiating the shutdown of transfer ability. In light of the ability for certain cell division proteins to oscillate from pole to pole in E. coli (Meinhardt and de Boer, 2001), it would be interesting to know whether $TraM_F$ or indeed any of the other F proteins also oscillate and scan the cell for appropriate times in the cell cycle for vegetative and conjugative replication.

SOLID SURFACE MATING OR MATING IN NATURAL ENVIRONMENTS

Most of the experimental work performed to assess the ability of bacteria to transfer plasmids via conjugation is performed in liquid cultures within the laboratory. Generally, homogeneous cultures of both donors and recipients are employed, resulting in random distributions of each, and promoting an increased probability of donor–recipient contact under essentially ideal conditions (Simonsen, 1990). Under the less ideal conditions of a natural habitat, however, most bacteria are attached to solid surfaces as microcolonies. Bacterial cells are therefore clustered within large groups, exhibiting heterogeneity with respect to growth rate, metabolic state, and the ability to contact potential recipient cells within a given clonal colony.

The yield of transconjugant cells resulting from mating events in liquid medium is largely independent of initial cell density, as a consequence of the random collisions between donors and recipients. However, on more "natural" solid surfaces, the initial density of donor and recipient colonies plays a more important role. Indeed, when colony density is high, the kinetics of plasmid transfer on solid surfaces is virtually indistinguishable from matings performed in liquid. In these situations, the efficiency and kinetics of plasmid transfer approach those values seen in the laboratory under ideal conditions. In fact, high rates of plasmid transfer have been observed in biofilms where cell density is high, suggesting that plasmid transfer on "natural" solid surfaces is very efficient (Christensen et al., 1998; Hausner and Wuertz, 1999; Simonsen, 1990).

The prevalence of F and other plasmids in natural bacterial populations have been previously estimated. Conjugative plasmids have been shown to be present in 2.5% to 67% of natural isolates of *E. coli* and other enteric bacteria (Datta, 1985; Hanzawa et al., 1984; Lederberg, 1951). Clearly, conjugative plasmids are able to spread within natural populations of bacteria with relative ease. More recent studies have shown that the presence of F and F-like conjugative plasmids in *E. coli* contributes greatly to their ability to form biofilms (Ghigo, 2001). In fact, the addition of planktonic F^+ donor cells to preformed biofilms consisting solely of F^- recipient cells resulted in the formation of a biofilm consisting of high numbers of new F^+ donors, illustrating that biofilms are an excellent niche in which plasmid transfer occurs readily (Christensen et al., 1998; Ghigo, 2001; Reisner et al., 2003).

What role might the formation of biofilms play in the evolution of the mechanisms involved in the horizontal transfer of the F plasmid? Considering the propensity for bacteria to colonize solid surfaces (e.g., the gut of humans and animals, sewage treatment facilities), conjugative plasmids appear to have taken advantage of high-efficiency transfer on solid surfaces. Donor cells inherently colonize solid surfaces more efficiently than nondonors, increasing the chance for random encounters with planktonic recipients. Conversely, and more likely, a preexisting recipient population on a solid surface is ideally suited to encounter random plasmid-carrying planktonic donors, allowing the plasmid to spread throughout the recipient population efficiently (Ghigo, 2001). This is accomplished through transient derepression of plasmid transfer in new transconjugant cells, resulting in the phenomenon of epidemic spread (Lundquist and Levin, 1986). Through this mechanism, an entire population of bacteria may acquire a given plasmid efficiently. Thus, it appears as though F and F-like plasmids have developed the ability to spread within populations under various physical and environmental conditions, illustrating the remarkable success these plasmids have achieved in propagating themselves via horizontal transfer.

CONCLUSION

In considering the concept of the "dynamic genome," the role of the F plasmid in the maintenance and evolution of *E. coli* cannot be underestimated. F has rather unique properties in terms of its ability to integrate into the chromosome and to stably maintain large sections of the chromosome in the merodiploid state. Its major replicon, RepFIA, maintains a very low copy number roughly equivalent to that of the host chromosome and is quiescent in the integrated state, properties that permit it to act as an agent for recombination and evolution. The conjugative ability of the F plasmid, in its derepressed state, is central to the ability of *E. coli* to undergo horizontal gene transfer and evolutionary adaptation rapidly. The host maximizes the benefits of carrying F by imposing regulation on F gene expression such that, in Nature, we predict that the conjugative ability of F is present only at optimal times in the growth cycle and life cycle of the cell, and that replication and conjugation are regulated in tandem in a manner that has not yet been defined.

The long history of F as an agent for chromosomal mutation and adaptation, as well as its role in molecular biology and recombinant DNA technology, have given it a unique place in the pantheon of mobile elements. The more recent findings that conjugation is one branch of the T4SS secretion mechanisms adds to the intrigue of F. The mechanism for pilus retraction, the route of DNA transfer, and the ability of F to identify and use a mating bridge within mere minutes of mixing donor and recipient cells together remain the major problems in F biology. With the increased interest in F as a paradigm for T4SS-mediated secretion, these questions will surely be answered in the next few years.

ACKNOWLEDGMENTS

The authors thank Drs. Diane Taylor, Bart Hazes, and Mark Glover for sharing unpublished data and the members of the Frost lab, especially Jan Manchak, Isabella Lau, Trevor Lawley, Jun Lu, and Trevor Elton for sharing unpublished results. We also thank Troy Locke and the Institute for Biomolecular Design for help with the Affymetrix microarray analysis.

REFERENCES

Achtman, M., Manning, P. A., Kusecek, B., Schwuchow, S., and Willetts, N. (1980) A genetic analysis of F sex factor cistrons needed for surface exclusion in *Escherichia coli. J. Mol. Biol* **138**: 779–795.

Ali Azam, T., Iwata, A., Nishimura, A., Ueda, S., and Ishihama, A. (1999) Growth phase-dependent variation in protein composition of the *Escherichia coli* nucleoid. *J. Bacteriol.* **181**: 6361–6370.

Anthony, K. G., Klimke, W. A., Manchak, J., and Frost, L. S. (1999) Comparison of proteins involved in pilus synthesis and mating pair stabilization from the related plasmids F and R100-1: Insights into the mechanism of conjugation. *J. Bacteriol.* **181**: 5149–5159.

Anthony, K. G., Sherburne, C., Sherburne, R., and Frost, L. S. (1994) The role of the pilus in recipient cell recognition during bacterial conjugation mediated by F-like plasmids. *Mol. Microbiol.* **13**: 939–953.

Bagdasarian, M., Bailone, A., Angulo, J. F., Scholz, P., Bagdasarian, M., and Devoret, R. (1992) PsiB, and anti-SOS protein, is transiently expressed by the F sex factor during its transmission to an *Escherichia coli* K-12 recipient. *Mol. Microbiol.* **6**: 885–893.

Bailone, A., Backman, A., Sommer, S., Celerier, J., Bagdasarian, M. M., Bagdasarian, M., and Devoret, R. (1988) PsiB polypeptide prevents activation of RecA protein in *Escherichia coli. Mol. Gen. Genet.* **214**: 389–395.

Balke, V. L., and Gralla, J. D. (1987) Changes in the linking number of supercoiled DNA accompany growth transitions in *Escherichia coli*. *J. Bacteriol.* 169: 4499–4506.

Bates, S., Roscoe, R. A., Althorpe, N.J., Brammar, W. J., and Wilkins, B. M. (1999) Expression of leading region genes on IncI1 plasmid ColIb-P9: Genetic evidence for single-stranded DNA transcription. *Microbiology* 145 (Pt 10): 2655–2662.

Bayer, M., Eferl, R., Zellnig, G., Teferle, K., Dijkstra, A., Koraimann, G., and Hogenauer, G. (1995) Gene 19 of plasmid R1 is required for both efficient conjugative DNA transfer and bacteriophage R17 infection. *J. Bacteriol.* 177: 4279–4288.

Beaber, J. W., Hochhut, B., and Waldor, M. K. (2002) Genomic and functional analyses of SXT, an integrating antibiotic resistance gene transfer element derived from *Vibrio cholerae*. *J. Bacteriol.* 184: 4259–4269.

Benz, I., and Schmidt, M. A. (1989) Cloning and expression of an adhesin (AIDA-I) involved in diffuse adherence of enteropathogenic *Escherichia coli*. *Infect. Immun.* 57: 1506–1511.

Bergquist, P. L., Saadi, S., and Maas, W. K. (1986) Distribution of basic replicons having homology with RepFIA, RepFIB, and RepFIC among IncF group plasmids. *Plasmid* 15: 19–34.

Boltner, D., MacMahon, C., Pembroke, J. T., Strike, P., and Osborn, A. M. (2002) R391: A conjugative integrating mosaic comprised of phage, plasmid, and transposon elements. *J. Bacteriol.* 184: 5158–5169.

Boyd, E. F., Hill, C. W., Rich, S. M., and Hartl, D. L. (1996) Mosaic structure of plasmids from natural populations of *Escherichia coli*. *Genetics* 143: 1091–1100.

Brinton, C. C., Gemski, P., and Carnahan, J. (1964) A new type of bacterial pilus genetically controlled by the fertility factor of *E. coli* K-12 and its role in chromosome transfer. *Proc. Natl. Acad. Sci. U S A* 52: 776–783.

Byrd, D. R., Sampson, J. K., Ragonese, H. M., and Matson, S. W. (2002) Structure-function analysis of *Escherichia coli* DNA helicase I reveals non-overlapping transesterase and helicase domains. *J. Biol. Chem.* 277: 42645–42653.

Cabezon, E., Sastre, J. I., and de la Cruz, F. (1997) Genetic evidence of a coupling role for the TraG protein family in bacterial conjugation. *Mol. Gen. Genet.* 254: 400–406.

Camacho, E. M., and Casadesus, J. (2002) Conjugal transfer of the virulence plasmid of *Salmonella enterica* is regulated by the leucine-responsive regulatory protein and DNA adenine methylation. *Mol. Microbiol.* 44: 1589–1598.

Carter, J. R., Patel, D. R., and Porter, R. D. (1992) The role of *oriT* in *tra*-dependent enhanced recombination between mini-F-*lac*-*oriT* and lambda p*lac*5. *Genet. Res.* 59: 157–165.

Carter, J. R., and Porter, R. D. (1991) *traY* and *traI* are required for *ori*T-dependent enhanced recombination between *lac*-containing plasmids and lambda p*lac*5. *J. Bacteriol.* **173**: 1027–1034.

Cascales, E., and Christie, P. D. (2004) Definition of a bacterial type IV secretion pathway for a DNA substrate. *Science* **304**: 1170–1173.

Chase, J. W., Merrill, B. M., and Williams, K. R. (1983) F sex factor encodes a single-stranded DNA binding protein (SSB) with extensive sequence homology to *Escherichia coli* SSB. *Proc. Natl. Acad. Sci. U S A* **80**: 5480–5484.

Cheah, K. C., and Skurray, R. (1986) The F plasmid carries an IS*3* insertion within *finO*. *J. Gen. Microbiol.* **132** (Pt 12): 3269–3275.

Christensen, B. B., Sternberg, C., Andersen, J. B., Eberl, L., Moller, S., Givskov, M., and Molin, S. (1998) Establishment of new genetic traits in a microbial biofilm community. *Appl. Environ. Microbiol.* **64**: 2247–2255.

Cosma, C. L., Danese, P. N., Carlson, J. H., Silhavy, T. J., and Snyder, W. B. (1995) Mutational activation of the Cpx signal transduction pathway of *Escherichia coli* suppresses the toxicity conferred by certain envelope-associated stresses. *Mol. Microbiol.* **18**: 491–505.

Csitkovits, V. C., and Zechner, E. L. (2003) Extent of single-stranded DNA required for efficient TraI helicase activity *in vitro*. *J. Biol. Chem.* **278**: 48696–48703.

D'Ari, R., Lin, R. T., and Newman, E. B. (1993) The leucine-responsive regulatory protein: More than a regulator? *Trends Biochem. Sci.* **18**: 260–263.

Danese, P. N., Snyder, W. B., Cosma, C. L., Davis, L. J., and Silhavy, T. J. (1995) The Cpx two-component signal transduction pathway of *Escherichia coli* regulates transcription of the gene specifying the stress-inducible periplasmic protease, DegP. *Genes Dev.* **9**: 387–398.

Dao-Thi, M. H., Charlier, D., Loris, R., Maes, D., Messens, J., Wyns, L., and Backmann, J. (2002) Intricate interactions within the *ccd* plasmid addiction system. *J. Biol. Chem.* **277**: 3733–3742.

Dartigalongue, C., and Raina, S. (1998) A new heat-shock gene, *ppiD*, encodes a peptidyl-prolyl isomerase required for folding of outer membrane proteins in *Escherichia coli*. *EMBO J.* **17**: 3968–3980.

Datta, N. (1985) Plasmids as organisms. *Basic Life Sci.* **30**: 3–16.

Davis, B. D. (1950) Non-filtrability of the agents of genetic recombination in *E. coli*. *J. Bacteriol.* **60**: 507–508.

Davison, J. (1999) Genetic exchange between bacteria in the environment. *Plasmid* **42**: 73–91.

De Wulf, P., Kwon, O., and Lin, E. C. (1999) The CpxRA signal transduction system of *Escherichia coli*: Growth-related autoactivation and control of unanticipated target operons. *J. Bacteriol.* **181**: 6772–6778.

Dempsey, W. B. (1989) Sense and antisense transcripts of *traM*, a conjugal transfer gene of the antibiotic resistance plasmid R100. *Mol. Microbiol.* **3**: 561–570.

Dempsey, W. B. (1994) *traJ* sense RNA initiates at two different promoters in R100–1 and forms two stable hybrids with antisense FinP RNA. *Mol. Microbiol.* **13**: 313–326.

Di Laurenzio, L., Scraba, D. G., Paranchych, W., and Frost, L. S. (1995) Studies on the binding of integration host factor (IHF) and TraM to the origin of transfer of the IncFV plasmid pED208. *Mol. Gen. Genet.* **247**: 726–734.

Dillard, J. P., and Seifert, H. S. (2001) A variable genetic island specific for *Neisseria gonorrhoeae* is involved in providing DNA for natural transformation and is found more often in disseminated infection isolates. *Mol. Microbiol.* **41**: 263–277.

Disque-Kochem, C., and Dreiseikelmann, B. (1997) The cytoplasmic DNA-binding protein TraM binds to the inner membrane protein TraD *in vitro*. *J. Bacteriol.* **179**: 6133–6137.

Disque-Kochem, C., Seidel, U., Helsberg, M., and Eichenlaub, R. (1986) The repeated sequences (*incB*) preceding the protein E gene of plasmid mini-F are essential for replication. *Mol. Gen. Genet.* **202**: 132–135.

Dong, J., Iuchi, S., Kwan, H. S., Lu, Z., and Lin, E. C. (1993) The deduced amino-acid sequence of the cloned cpxR gene suggests the protein is the cognate regulator for the membrane sensor, CpxA, in a two-component signal transduction system of *Escherichia coli*. *Gene* **136**: 227–230.

Doran, T. J., Loh, S. M., Firth, N., and Skurray, R. A. (1994) Molecular analysis of the F plasmid *tra*VR region: *tra*V encodes a lipoprotein. *J. Bacteriol.* **176**: 4182–4186.

Dorman, C. J., Barr, G. C., Bhriain, N. N., and Higgins, C. F. (1988) DNA supercoiling and the anaerobic and growth phase regulation of *tonB* gene expression. *J. Bacteriol.* **170**: 2816–2826.

Eichenlaub, R., Figurski, D., and Helinski, D. R. (1977) Bidirection replication from a unique origin in a mini-F plasmid. *Proc. Natl. Acad. Sci. U.S.A* **74**: 1138–1141.

Eichenlaub, R., Wehlmann, H., and Ebbers, J. (1981) Plasmid mini-F encoded functions involved in replication and incompatibility. In Levy, S. B., Clowes, R. C., and Koenig, E. L. (eds). *Molecular biology, pathogenicity, and ecology of bacterial plasmids*. New York: Plenum, pp. 327–336.

Eliasson, A., Bernander, R., and Nordstrom, K. (1996) Random initiation of replication of plasmids P1 and F (oriS) when integrated into the *Escherichia coli* chromosome. *Mol. Microbiol.* **20**: 1025–1032.

Elton, T. C., Ahsan, I. and L. S. Frost. (2003) Induction of F pilus retraction by a filamentous phage pIII protein fragment. Understanding Phage Display 2003, Vancouver, B. C.

Elton, T. C., Frost, L. S., and Hazes, B. (2004) The cryptic gene *trbB* of the F plasmid encodes a protein involved in disulfide bond formation. Submitted.

Everett, R., and Willetts, N. (1980) Characterisation of an *in vivo* system for nicking at the origin of conjugal DNA transfer of the sex factor F. *J. Mol. Biol* **136**: 129–150.

Fekete, R. A., and Frost, L. S. (2000) Mobilization of chimeric *ori*T plasmids by F and R100–1: Role of relaxosome formation in defining plasmid specificity. *J. Bacteriol.* **182**: 4022–4027.

Fekete, R. A., and Frost, L. S. (2002) Characterizing the DNA contacts and cooperative binding of F plasmid TraM to its cognate sites at *ori*T. *J. Biol. Chem.* **277**: 16705–16711.

Finlay, B. B., Frost, L. S., Paranchych, W., and Willetts, N. S. (1986) Nucleotide sequences of five IncF plasmid *fin*P alleles. *J. Bacteriol.* **167**: 754–757.

Finnegan, D., and Willetts, N. (1973) The site of action of the F transfer inhibitor. *Mol. Gen. Genet.* **127**: 307–316.

Firth, N., Ippen-Ihler, K., and Skurray, R. A. (1996) Structure and function of the F factor and mechanism of conjugation. In Neidhardt, F. C., et al., (eds). *Escherichia coli and Salmonella: Cellular and molecular biology.* Washington, DC: ASM Press, pp. 2377–2401.

Foster, P. L. (2000) Adaptive mutation: Implications for evolution. *BioEssays* **22**: 1067–1074.

Foster, P. L., and Cairns, J. (1992) Mechanisms of directed mutation. *Genetics* **131**: 783–789.

Foster, P. L., and Trimarchi, J. M. (1995a) Adaptive reversion of an episomal frameshift mutation in *Escherichia coli* requires conjugal functions but not actual conjugation. *Proc. Natl. Acad. Sci. U S A* **92**: 5487–5490.

Foster, P. L., and Trimarchi, J. M. (1995b) Conjugation is not required for adaptive reversion of an episomal frameshift mutation in *Escherichia coli. J. Bacteriol.* **177**: 6670–6671.

Franch, T., Petersen, M., Wagner, E. G., Jacobsen, J. P., and Gerdes, K. (1999) Antisense RNA regulation in prokaryotes: Rapid RNA/RNA interaction facilitated by a general U-turn loop structure. *J. Mol. Biol.* **294**: 1115–1125.

Frost, L. S., Ippen-Ihler, K., and Skurray, R. A. (1994) Analysis of the sequence and gene products of the transfer region of the F sex factor. *Microbiol. Rev.* **58**: 162–210.

Frost, L. S., and Manchak, J. (1998) F-phenocopies: Characterization of expression of the F transfer region in stationary phase. *Microbiology* **144** (**Pt 9**): 2579–2587.

Frost, L. S., Paranchych, W., and Willetts, N. S. (1984) DNA sequence of the F *traALE* region that includes the gene for F pilin. *J. Bacteriol.* **160**: 395–401.

195

Galitski, T., and Roth, J. R. (1995) Evidence that F plasmid transfer replication underlies apparent adaptive mutation. *Science* **268**: 421–423.

Gamas, P., Caro, L., Galas, D., and Chandler, M. (1987) Expression of F transfer functions depends on the *Escherichia coli* integration host factor. *Mol. Gen. Genet.* **207**: 302–305.

Gaudin, H. M., and Silverman, P. M. (1993) Contributions of promoter context and structure to regulated expression of the F plasmid *traY* promoter in *Escherichia coli* K-12. *Mol. Microbiol.* **8**: 335–342.

Ghetu, A. F., Arthur, D.C., Kerppola, T. K., and Glover, J. N. (2002) Probing FinO-FinP RNA interactions by site-directed protein-RNA crosslinking and gelFRET. *RNA* **8**: 816–823.

Ghetu, A. F., Gubbins, M. J., Frost, L. S., and Glover, J. N. (2000) Crystal structure of the bacterial conjugation repressor FinO. *Nat. Struct. Biol* **7**: 565–569.

Ghigo, J. M. (2001) Natural conjugative plasmids induce bacterial biofilm development. *Nature* **412**: 442–445.

Gilmour, M. W., Gunton, J. E., Lawley, T. D., and Taylor, D. E. (2003) Interaction between the IncHI1 plasmid R27 coupling protein and type IV secretion system: TraG associates with the coiled-coil mating pair formation protein TrhB. *Mol. Microbiol.* **49**: 105–116.

Goldstein, E., and Drlica, K. (1984) Regulation of bacterial DNA supercoiling: Plasmid linking numbers vary with growth temperature. *Proc. Natl. Acad. Sci. U S A* **81**: 4046–4050.

Gomis-Ruth, F. X., de la Cruz, F., and Coll, M. (2002) Structure and role of coupling proteins in conjugal DNA transfer. *Res. Microbiol.* **153**: 199–204.

Gomis-Ruth, F. X., Moncalian, G., Perez-Luque, R., Gonzalez, A., Cabezon, E., de la Cruz, F., and Coll, M. (2001) The bacterial conjugation protein TrwB resembles ring helicases and F1-ATPase. *Nature* **409**: 637–641.

Gordon, G. S., Sitnikov, D., Webb, C. D., Teleman, A., Straight, A., Losick, R., et al. (1997) Chromosome and low copy plasmid segregation in *E. coli*: Visual evidence for distinct mechanisms. *Cell* **90**: 1113–1121.

Grandoso, G., Avila, P., Cayon, A., Hernando, M. A., Llosa, M., and de la Cruz, F. (2000) Two active-site tyrosyl residues of protein TrwC act sequentially at the origin of transfer during plasmid R388 conjugation. *J. Mol. Biol.* **295**: 1163–1172.

Gubbins, M. J., Arthur, D.C., Ghetu, A. F., Glover, J. N., and Frost, L. S. (2003) Characterizing the structural features of RNA/RNA interactions of the F-plasmid FinOP fertility inhibition system. *J. Biol.Chem.* **278**: 27663–27671.

Gubbins, M. J., Lau, I., Will, W. R., Manchak, J. M., Raivio, T. L., and Frost, L. S. (2002) The positive regulator, TraJ, of the *Escherichia coli* F plasmid is unstable in a *cpxA** background. *J. Bacteriol.* **184**: 5781–5788.

Guyer, M. S. (1978) The gamma delta sequence of F is an insertion sequence. *J. Mol. Biol.* **126**: 347–365.

Hanzawa, Y., Oka, C., Ishiguro, N., and Sato, G. (1984) Incompatibility groups of R plasmids in *Escherichia coli* isolated from animal waste. *Nippon Juigaku. Zasshi* **46**: 453–457.

Harris, R. L., Hombs, V., and Silverman, P. M. (2001) Evidence that F-plasmid proteins TraV, TraK and TraB assemble into an envelope-spanning structure in *Escherichia coli*. *Mol. Microbiol.* **42**: 757–766.

Harwood, C. R., and Meynell, E. (1975) Cyclic AMP and the production of sex pili by *E. coli* K-12 carrying derepressed sex factors. *Nature* **254**: 628–660.

Hausner, M., and Wuertz, S. (1999) High rates of conjugation in bacterial biofilms as determined by quantitative *in situ* analysis. *Appl. Environ. Microbiol.* **65**: 3710–3713.

Hayes, W. (1952) Genetic recombination in *Bact. coli* K12: Analysis of the stimulating effect of ultraviolet light. *Nature* **169**: 1017.

Hayes, W. (1953) Observations on a transmissible agent determining sexual differentiation in *Bact. coli*. *J. Gen. Microbiol.* **8**: 72–88.

Hayes, W. (1964) *The genetics of bacteria and their viruses*. New York: John Wiley and Sons, p. 740.

Higgins, C. F., Dorman, C. J., Stirling, D. A., Waddell, L., Booth, I. R., May, G., and Bremer, E. (1988) A physiological role for DNA supercoiling in the osmotic regulation of gene expression in *S. typhimurium* and *E. coli*. *Cell* **52**:569–584.

Hoch, J. A., and Silhavy, T. J. (1995) Two-component signal transduction. Washington, DC: American Society for Microbiology Press.

Hohn, B., and Korn, D. (1969) Cosegregation of a sex factor with the *Escherichia coli* chromosome during curing by acridine orange. *J. Mol. Biol.* **45**: 385–395.

Howard, M. T., Nelson, W. C., and Matson, S. W. (1995) Stepwise assembly of a relaxosome at the F plasmid origin of transfer. *J. Biol. Chem.* **270**: 28381–28386.

Hu, S., Ohtsubo, E., and Davidson, N. (1975) Electron microscopic heteroduplex studies of sequence relations among plasmids of *Escherichia coli*: Structure of F13 and related F-primes. *J. Bacteriol.* **122**: 749–763.

Hu, S., Otsubo, E., Davidson, N., and Saedler, H. (1975a) Electron microscope heteroduplex studies of sequence relations among bacterial plasmids: Identification and mapping of the insertion sequences IS *1* and IS *2* in F and R plasmids. *J. Bacteriol.* **122**: 764–775.

Hu, S., Ptashne, K., Cohen, S. N., and Davidson, N. (1975b) Alphabeta sequence of F is IS *31*. *J. Bacteriol.* **123**: 687–692.

Jalajakumari, M. B., Guidolin, A., Buhk, H. J., Manning, P. A., Ham, L. M., Hodgson, A. L., et al. (1987) Surface exclusion genes *traS* and *traT* of the F sex factor of *Escherichia coli* K-12. Determination of the nucleotide sequence and promoter and terminator activities. *J. Mol. Biol.* **198**: 1–11.

Jensen, P. R., Loman, L., Petra, B., van der Weijden, C., and Westerhoff, H. V. (1995) Energy buffering of DNA structure fails when *Escherichia coli* runs out of substrate. *J. Bacteriol.* **177**: 3420–3426.

Jerome, L. J., and Frost, L. S. (1999) In vitro analysis of the interaction between the FinO protein and FinP antisense RNA of F-like conjugative plasmids. *J. Biol. Chem.* **274**: 10356–10362.

Jerome, L. J., van Biesen, T., and Frost, L. S. (1999) Degradation of FinP antisense RNA from F-like plasmids: The RNA-binding protein, FinO, protects FinP from ribonuclease E. *J. Mol. Biol.* **285**: 1457–1473.

Jones, C. H., Danese, P. N., Pinkner, J. S., Silhavy, T. J., and Hultgren, S. J. (1997) The chaperone-assisted membrane release and folding pathway is sensed by two signal transduction systems. *EMBO J.* **16**: 6394–6406.

Karem, K., and Foster, J. W. (1993) The influence of DNA topology on the environmental regulation of a pH-regulated locus in *Salmonella typhimurium*. *Mol. Microbiol.* **10**: 75–86.

Kathir, P., and Ippen-Ihler, K. (1991) Construction and characterization of derivatives carrying insertion mutations in F plasmid transfer region genes, *trbA*, *artA*, *traQ*, and *trbB*. *Plasmid* **26**: 40–54.

Kaufmann, A., Stierhof, Y. D., and Henning, U. (1994) New outer membrane-associated protease of *Escherichia coli* K-12. *J. Bacteriol.* **176**: 359–367.

Kim, S. R., Maenhaut-Michel, G., Yamada, M., Yamamoto, Y., Matsui, K., Sofuni, T., Nohmi, T., et al. (1997) Multiple pathways for SOS-induced mutagenesis in *Escherichia coli*: An overexpression of *dinB/dinP* results in strongly enhancing mutagenesis in the absence of any exogenous treatment to damage DNA. *Proc. Natl. Acad. Sci. U S A* **94**: 13792–13797.

Kofoid, E., Bergthorsson, U., Slechta, E. S., and Roth, J. R. (2003) Formation of an F′ plasmid by recombination between imperfectly repeated chromosomal Rep sequences: A closer look at an old friend (F′(128) *pro lac*). *J. Bacteriol.* **185**: 660–663.

Komori, H., Matsunaga, F., Higuchi, Y., Ishiai, M., Wada, C., and Miki, K. (1999) Crystal structure of a prokaryotic replication initiator protein bound to DNA at 2.6 Å resolution. *EMBO J.* **18**: 4597–4607.

Koraimann, G., Koraimann, C., Koronakis, V., Schlager, S., and Hogenauer, G. (1991) Repression and derepression of conjugation of plasmid R1 by wild-type and mutated *fin*P antisense RNA. *Mol. Microbiol.* **5**: 77–87.

Koraimann, G., Teferle, K., Markolin, G., Woger, W., and Hogenauer, G. (1996) The FinOP repressor system of plasmid R1: Analysis of the antisense RNA control of *traJ* expression and conjugative DNA transfer. *Mol. Microbiol.* **21**: 811–821.

Kumar, S., and Srivastava, S. (1983) Cyclic AMP and its receptor protein are required for expression of transfer genes of conjugative plasmid F in *Escherichia coli*. *Mol. Gen. Genet.* **190**: 27–34.

Lane, D., de Feyter, R., Kennedy, M., Phua, S. H., and Semon, D. (1986) D protein of miniF plasmid acts as a repressor of transcription and as a site-specific resolvase. *Nucleic Acids Res.* **14**: 9713–9728.

Lane, D., and Gardner, R. C. (1979) Second *Eco*RI fragment of F capable of self-replication. *J. Bacteriol.* **139**: 141–151.

Lane, H. E. (1981) Replication and incompatibility of F and plasmids in the IncFI group. *Plasmid* **5**: 100–126.

Lanka, E., and Wilkins, B. M. (1995) DNA processing reactions in bacterial conjugation. *Annu. Rev. Biochem.* **64**: 141–169.

Lau, I. C. Y., Locke, T., Ellison, M., Raivio, T. L., and Frost, L. S. (2004) Activation of the Cpx envelope stress response destabilizes the F plasmid transfer activator, TraJ, via the HslVU protease in *Escherichia coli*. Submitted.

Lawley, T., Wilkins, B. M., and Frost, L. S. (2003) Conjugation in gram-negative bacteria. In Funnell, B. (ed). *The biology of plasmids.* pp. 203–226, Washington, DC: ASM Press.

Lawley, T. D., Klimke, W. A., Gubbins, M. J., and Frost, L. S. (2003) F factor conjugation is a true type IV secretion system. *FEMS Microbiol. Lett* **269**: 1–15.

Lederberg, J. (1951) Prevalence of *E. coli* strains exhibiting genetic recombination. *Science* **114**: 68.

Lederberg, J., Cavalli, L. L., and Lederberg, E. M. (1952) Sex compatibility in *E. coli. Genetics* **37**: 720–730.

Lederberg, J., and Tatum, E. L. (1946) Gene recombination in *E. coli. Nature* **158**: 558.

Lee, S. H., Frost, L. S., and Paranchych, W. (1992) FinOP repression of the F plasmid involves extension of the half-life of FinP antisense RNA by FinO. *Mol. Gen. Genet.* **235**: 131–139.

Libante, V., Thion, L., and Lane, D. (2001) Role of the ATP-binding site of SopA protein in partition of the F plasmid. *J. Mol. Biol.* **314**: 387–399.

Lin, D.C., and Grossman, A. D. (1998) Identification and characterization of a bacterial chromosome partitioning site. *Cell* **92**: 675–685.

Loh, S., Cram, D., and Skurray, R. (1989) Nucleotide sequence of the leading region adjacent to the origin of transfer on plasmid F and its conservation among conjugative plasmids. *Mol. Gen. Genet.* **219**: 177–186.

Loh, S., Skurray, R., Celerier, J., Bagdasarian, M., Bailone, A., and Devoret, R. (1990) Nucleotide sequence of the *psiA* (plasmid SOS inhibition) gene located on the leading region of plasmids F and R6–5. *Nucleic Acids Res.* 18: 4597.

Loh, S. M., Cram, D. S., and Skurray, R. A. (1988) Nucleotide sequence and transcriptional analysis of a third function (Flm) involved in F-plasmid maintenance. *Gene* 66: 259–268.

Lu, J., Forsyth, H., and Frost, L. S. (2004) The strong, autoregulated *traM* promoters in the F plasmid facilitate conjugative DNA transfer and prevent cell toxicity. In preparation.

Lum, P. L., Rodgers, M. E., and Schildbach, J. F. (2002) TraY DNA recognition of its two F factor binding sites. *J. Mol. Biol.* 321: 563–578.

Lundquist, P. D., and Levin, B. R. (1986) Transitory derepression and the maintenance of conjugative plasmids. *Genetics* 113: 483–497.

Luo, Y., Gao, Q., and Deonier, R. C. (1994) Mutational and physical analysis of F plasmid *traY* protein binding to *oriT*. *Mol. Microbiol.* 11: 459–469.

Lynch, A. S., and Lin, E. C. (1996) Transcriptional control mediated by the ArcA two-component response regulator protein of *Escherichia coli*: Characterization of DNA binding at target promoters. *J. Bacteriol.* 178: 6238–6249.

Makino, K., Ishii, K., Yasunaga, T., Hattori, M., Yokoyama, K., Yutsudo, C. H., Kubota, Y., et al. (1998) Complete nucleotide sequences of 93-kb and 3.3-kb plasmids of an enterohemorrhagic *Escherichia coli* O157:H7 derived from Sakai outbreak. *DNA Res.* 5: 1–9.

Manchak, J., Anthony, K. G., and Frost, L. S. (2002) Mutational analysis of F-pilin reveals domains for pilus assembly, phage infection and DNA transfer. *Mol. Microbiol.* 43: 195–205.

Manwaring, N. P., Skurray, R. A., and Firth, N. (1999) Nucleotide sequence of the F plasmid leading region. *Plasmid* 41: 219–225.

Marmur, J., Rownd, R., Falkow, S., Baron, L. S., Shildkraut, C., and Doty, P. (1961) The nature of intergeneric episomal infection. *Proc. Natl. Acad. Sci. U S A* 47: 972–979.

Masai, H., and Arai, K. (1997) Frpo: A novel single-stranded DNA promoter for transcription and for primer RNA synthesis of DNA replication. *Cell* 89: 897–907.

Masson, L., and Ray, D. S. (1986) Mechanism of autonomous control of the *Escherichia coli* F plasmid: Different complexes of the initiator/repressor protein are bound to its operator and to an F plasmid replication origin. *Nucleic Acids Res.* 14: 5693–5711.

Matson, S. W., and Morton, B. S. (1991) *Escherichia coli* DNA helicase I catalyzes a site- and strand-specific nicking reaction at the F plasmid *oriT*. *J. Biol. Chem.* 266: 16232–16237.

Matsuo, E., Sampei, G., Mizobuchi, K., and Ito, K. (1999) The plasmid F OmpP protease, a homologue of OmpT, as a potential obstacle to *E. coli*-based protein production. *FEBS Lett.* **461**: 6–8.

Mazodier, P., and Davies, J. (1991) Gene transfer between distantly related bacteria. *Annu. Rev. Genet.* **25**: 147–171.

McEwen, J., and Silverman, P. (1980) Chromosomal mutations of *Escherichia coli* that alter expression of conjugative plasmid functions. *Proc. Natl. Acad. Sci. U S A* **77**: 513–517.

Meinhardt, H., and de Boer, P. A. (2001) Pattern formation in *Escherichia coli*: A model for the pole-to-pole oscillations of Min proteins and the localization of the division site. *Proc. Natl. Acad. Sci. U S A* **98**: 14202–14207.

Miki, T., Chang, Z. T., and Horiuchi, T. (1984) Control of cell division by sex factor F in *Escherichia coli*. II. Identification of genes for inhibitor protein and trigger protein on the 42.84–43.6 F segment. *J. Mol. Biol.* **174**: 627–646.

Miller, J. F., and Malamy, M. H. (1983) Identification of the *pifC* gene and its role in negative control of F factor *pif* gene expression. *J. Bacteriol.* **156**: 338–347.

Miller, J. F., and Malamy, M. H. (1984) Regulation of the F-factor *pif* operon: *pifO*, a site required in cis for autoregulation, titrates the *pifC* product in *trans*. *J. Bacteriol.* **160**: 192–198.

Mise, K., and Nakajima, K. (1984) Isolation of restriction enzyme *Eco*VIII, an isoschizomer of *HindIII*, produced by *Escherichia coli* E1585–68. *Gene* **30**: 79–85.

Moore, D., Hamilton, C. M., Maneewannakul, K., Mintz, Y., Frost, L. S., and Ippen-Ihler, K. (1993) The *Escherichia coli* K-12 F plasmid gene *traX* is required for acetylation of F pilin. *J. Bacteriol.* **175**: 1375–1383.

Mori, H., Kondo, A., Ohshima, A., Ogura, T., and Hiraga, S. (1986) Structure and function of the F plasmid genes essential for partitioning. *J. Mol. Biol.* **192**: 1–15.

Murotsu, T., Tsutsui, H., and Matsubara, K. (1984) Identification of the minimal essential region for the replication origin of miniF plasmid. *Mol. Gen. Genet.* **196**: 373–378.

Nagai, H., and Roy, C. R. (2003) Show me the substrates: Modulation of host cell function by type IV secretion systems. *Cell Microbiol.* **5**: 373–383.

Nelson, W. C., Howard, M. T., Sherman, J. A., and Matson, S. W. (1995) The *traY* gene product and integration host factor stimulate *Escherichia coli* DNA helicase I-catalyzed nicking at the F plasmid *oriT*. *J. Biol. Chem.* **270**: 28374–28380.

Nielsen, A. K., Thorsted, P., Thisted, T., Wagner, E. G., and Gerdes, K. (1991) The rifampicin-inducible genes *srnB* from F and *pnd* from R483 are regulated by antisense RNAs and mediate plasmid maintenance by killing of plasmid-free segregants. *Mol. Microbiol.* **5**: 1961–1973.

Nieto, J. M., Prenafeta, A., Miquelay, E., Torrades, S., and Juarez, A. (1998) Sequence, identification and effect on conjugation of the *rmoA* gene of plasmid R100–1. *FEMS Microbiol. Lett.* **169**: 59–66.

Nieto, J. M., Madrid, C., Miquelay, E., Parra, J. L., Rodriguez, S., and Juarez, A. (2002) Evidence for direct protein–protein interaction between members of the enterobacterial Hha/YmoA and H-NS families of proteins. *J. Bacteriol.* **184**: 629–635.

Nomura, N., Masai, H., Inuzuka, M., Miyazaki, C., Ohtsubo, E., Itoh, T., et al. (1991) Identification of eleven single-strand initiation sequences (*ssi*) for priming of DNA replication in the F, R6K, R100 and ColE2 plasmids. *Gene* **108**: 15–22.

O'Connor, M. B., and Malamy, M. H. (1984) Role of the F factor *oriV*1 region in *recA*-independent illegitimate recombination. Stable replicon fusions of the F derivative pOX38 and pBR322-related plasmids. *J. Mol. Biol.* **175**: 263–284.

Ogura, T., and Hiraga, S. (1983) Partition mechanism of F plasmid: Two plasmid gene-encoded products and a *cis*-acting region are involved in partition. *Cell* **32**: 351–360.

Oshima, T., Wada, C., Kawagoe, Y., Ara, T., Maeda, M., Masuda, Y., et al. (2002) Genome-wide analysis of deoxyadenosine methyltransferase-mediated control of gene expression in *Escherichia coli*. *Mol. Microbiol.* **45**: 673–695.

Pansegrau, W., Schroder, W., and Lanka, E. (1993) Relaxase (TraI) of IncP alpha plasmid RP4 catalyzes a site-specific cleaving-joining reaction of single-stranded DNA. *Proc. Natl. Acad. Sci. U S A* **90**: 2925–2929.

Paranchych, W., Finlay, B. B., and Frost, L. S. (1986) Studies on the regulation of IncF plasmid transfer operon expression. *Banbury Rep.* **24**: 117–129.

Pogliano, J., Lynch, A. S., Belin, D., Lin, E. C., and Beckwith, J. (1997) Regulation of *Escherichia coli* cell envelope proteins involved in protein folding and degradation by the Cpx two-component system. *Genes Dev.* **11**: 1169–1182.

Polzleitner, E., Zechner, E. L., Renner, W., Fratte, R., Jauk, B., Hogenauer, G., and Koraimann, G. (1997) TraM of plasmid R1 controls transfer gene expression as an integrated control element in a complex regulatory network. *Mol. Microbiol.* **25**: 495–507.

Porter, R. D. (1981) Enhanced recombination between F42*lac* and lambda p*lac*5: Dependence on F42*lac* fertility functions. *Mol. Gen. Genet.* **184**: 355–358.

Porter, R. D., Low, B., and McLaughlin, T. (1978) Transduction versus' "conjuduction": Evidence for multiple roles for exonuclease V in genetic recombination in *Escherichia coli. Cold Spring Harb. Symp. Quant. Biol.* **43**: 1043–1047.

Raivio, T. L., Laird, M. W., Joly, J. C., and Silhavy, T. J. (2000) Tethering of CpxP to the inner membrane prevents spheroplast induction of the Cpx envelope stress response. *Mol. Microbiol.* **37**: 1186–1197.

Raivio, T. L., and Silhavy, T. J. (1997) Transduction of envelope stress in *Escherichia coli* by the Cpx two-component system. *J. Bacteriol.* **179**: 7724–7733.

Raivio, T. L., and Silhavy, T. J. (2001) Periplasmic stress and ECF sigma factors. *Annu. Rev. Microbiol.* **55**: 591–624.

Reisner, A., Haagensen, J. A., Schembri, M. A., Zechner, E. L., and Molin, S. (2003) Development and maturation of *Escherichia coli* K-12 biofilms. *Mol. Microbiol.* **48**: 933–946.

Rice, P. A., Yang, S., Mizuuchi, K., and Nash, H. A. (1996) Crystal structure of an IHF-DNA complex: A protein-induced DNA U-turn. *Cell* **87**: 1295–1306.

Rokeach, L. A., Sogaard-Andersen, L., and Molin, S. (1985) Two functions of the E protein are key elements in the plasmid F replication control system. *J. Bacteriol.* **164**: 1262–1270.

Roth, J. R., Kofoid, E., Roth, F. P., Berg, O. G., Seger, J., and Andersson, D. I. (2003) Regulating general mutation rates. Examination of the hypermutable state model for cairnsian adaptive mutation. *Genetics* **163**: 1483–1496.

Rotman, G. S., Cooney, R., and Malamy, M. H. (1983) Cloning of the *pif* region of the F sex factor and identification of a *pif* protein product. *J. Bacteriol.* **155**: 254–264.

Saadi, S., Maas, W. K., Hill, D. F., and Bergquist, P. L. (1987) Nucleotide sequence analysis of RepFIC, a basic replicon present in IncFI plasmids P307 and F, and its relation to the RepA replicon of IncFII plasmids. *J. Bacteriol.* **169**: 1836–1846.

Sandercock, J. R., and Frost, L. S. (1998) Analysis of the major domains of the F fertility inhibition protein, FinO. *Mol. Gen. Genet.* **259**: 622–629.

Santini, J. M., and Stanisich, V. A. (1998) Both the *fipA* gene of pKM101 and the *pifC* gene of F inhibit conjugal transfer of RP1 by an effect on *traG*. *J. Bacteriol.* **180**: 4093–4101.

Saul, D., Spiers, A. J., McAnulty, J., Gibbs, M. G., Bergquist, P. L., and Hill, D. F. (1989) Nucleotide sequence and replication characteristics of RepFIB, a basic replicon of IncF plasmids. *J. Bacteriol.* **171**: 2697–2707.

Savvides, S. N., Yeo, H. J., Beck, M. R., Blaesing, F., Lurz, R., Lanka, E., Buhrdorf, R., et al. (2003) VirB11 ATPases are dynamic hexameric assemblies: New insights into bacterial type IV secretion. *EMBO J.* **22**: 1969–1980.

Scaife, J., and Gross, J. D. (1962) Inhibition of multiplication of an F*lac* factor in Hfr cells in *Escherichia coli* K-12. *Biochem. Biophys. Res. Commun.* **7**: 403–407.

Schroder, G., and Lanka, E. (2003) TraG-like proteins of type IV secretion systems: Functional dissection of the multiple activities of TraG (RP4) and TrwB (R388). *J. Bacteriol.* **185**: 4371–4381.

Seifert, H. S., and Porter, R. D. (1984) Enhanced recombination between lambda p*lac*5 and F42*lac*: Identification of *cis*- and *trans*-acting factors. *Proc. Natl. Acad. Sci. U S A* **81**: 7500–7504.

Sharp, P. A., Cohen, S. N., and Davidson, N. (1973) Electron microscope heteroduplex studies of sequence relations among plasmids of *Escherichia coli*. II. Structure of drug resistance (R) factors and F factors. *J. Mol. Biol.* **75**: 235–255.

Shimizu, H., Saitoh, Y., Suda, Y., Uehara, K., Sampei, G., and Mizobuchi, K. (2000) Complete nucleotide sequence of the F plasmid: Its implications for organization and diversification of plasmid genomes. GenBank accession number AP001918.

Silverman, P. M., Tran, L., Harris, R., and Gaudin, H. M. (1993) Accumulation of the F plasmid TraJ protein in *cpx* mutants of *Escherichia coli*. *J. Bacteriol.* **175**: 921–925.

Silverman, P. M., Wickersham, E., and Harris, R. (1991) Regulation of the F plasmid *tra*Y promoter in *Escherichia coli* by host and plasmid factors. *J. Mol. Biol.* **218**: 119–128.

Silverman, P. M., Wickersham, E., Rainwater, S., and Harris, R. (1991) Regulation of the F plasmid *tra*Y promoter in *Escherichia coli* K12 as a function of sequence context. *J. Mol. Biol.* **220**: 271–279.

Simonsen, L. (1990) Dynamics of plasmid transfer on surfaces. *J. Gen. Microbiol.* **136 (Pt 6)**: 1001–1007.

Snyder, W. B., Davis, L. J., Danese, P. N., Cosma, C. L., and Silhavy, T. J. (1995) Overproduction of NlpE, a new outer membrane lipoprotein, suppresses the toxicity of periplasmic LacZ by activation of the Cpx signal transduction pathway. *J. Bacteriol.* **177**: 4216–4223.

Starcic, M., Zgur-Bertok, D., Jordi, B. J., Wosten, M. M., Gaastra, W., and van Putten, J. P. (2003) The cyclic AMP–cyclic AMP receptor protein complex regulates activity of the *tra*J promoter of the *Escherichia coli* conjugative plasmid pRK100. *J. Bacteriol.* **185**: 1616–1623.

Starcic-Erjavec, M., van Putten, J. P., Gaastra, W., Jordi, B. J., Grabnar, M., and Zgur-Bertok, D. (2003) H-NS and Lrp serve as positive modulators of *tra*J expression from the *Escherichia coli* plasmid pRK100. *Mol. Genet. Genomics* **270**: 94–102.

Stern, J. C., and Schildbach, J. F. (2001) DNA recognition by F factor TraI36: Highly sequence-specific binding of single-stranded DNA. *Biochemistry* **40**: 11586–11595.

Street, L. M., Harley, M. J., Stern, J. C., Larkin, C., Williams, S. L., Miller, D. L., et al. (2003) Subdomain organization and catalytic residues of the F factor TraI relaxase domain. *Biochim. Biophys. Acta* **1646**: 86–99.

Strohmaier, H., Noiges, R., Kotschan, S., Sawers, G., Hogenauer, G., Zechner, E. L., and Koraimann, G. (1998) Signal transduction and bacterial conjugation: Characterization of the role of ArcA in regulating conjugative transfer of the resistance plasmid R1. *J. Mol. Biol.* **277**: 309–316.

Tani, T. H., Khodursky, A., Blumenthal, R. M., Brown, P. O., and Matthews, R. G. (2002) Adaptation to famine: A family of stationary-phase genes revealed by microarray analysis. *Proc. Natl. Acad. Sci. U S A* **99**: 13471–13476.

Teter, B., Goodman, S. D., and Galas, D. J. (2000) DNA bending and twisting properties of integration host factor determined by DNA cyclization. *Plasmid* **43**: 73–84.

Thisted, T., Nielsen, A. K., and Gerdes, K. (1994) Mechanism of post-segregational killing: Translation of Hok, SrnB and Pnd mRNAs of plasmids R1, F and R483 is activated by 3'-end processing. *EMBO J.* **13**: 1950–1959.

Thompson, T. L., Centola, M. B., and Deonier, R. C. (1989) Location of the nick at *oriT* of the F plasmid. *J. Mol. Biol.* **207**: 505–512.

Tolun, A., and Helinski, D. R. (1981) Direct repeats of the F plasmid *incC* region express F incompatibility. *Cell* **24**: 687–694.

Torkelson, J., Harris, R. S., Lombardo, M. J., Nagendran, J., Thulin, C., and Rosenberg, S. M. (1997) Genome-wide hypermutation in a subpopulation of stationary-phase cells underlies recombination-dependent adaptive mutation. *EMBO J.* **16**: 3303–3311.

Torreblanca, J., Marques, S., and Casadesus, J. (1999) Synthesis of FinP RNA by plasmids F and pSLT is regulated by DNA adenine methylation. *Genetics* **152**: 31–45.

Traxler, B. A., and Minkley, E. G., Jr. (1988) Evidence that DNA helicase I and *oriT* site-specific nicking are both functions of the F TraI protein. *J. Mol. Biol.* **204**: 205–209.

Uga, H., Matsunaga, F., and Wada, C. (1999) Regulation of DNA replication by iterons: An interaction between the *ori2* and *incC* regions mediated by RepE-bound iterons inhibits DNA replication of mini-F plasmid in *Escherichia coli*. *EMBO J.* **18**: 3856–3867.

van Biesen, T., and Frost, L. S. (1992) Differential levels of fertility inhibition among F-like plasmids are related to the cellular concentration of *fin*O mRNA. *Mol. Microbiol.* **6**: 771–780.

van Biesen, T., and Frost, L. S. (1994) The FinO protein of IncF plasmids binds FinP antisense RNA and its target, *traJ* mRNA, and promotes duplex formation. *Mol. Microbiol.* **14**: 427–436.

van Biesen, T., Soderbom, F., Wagner, E. G., and Frost, L. S. (1993) Structural and functional analyses of the FinP antisense RNA regulatory system of the F conjugative plasmid. *Mol. Microbiol.* **10**: 35–43.

van Workum, M., van Dooren, S. J., Oldenburg, N., Molenaar, D., Jensen, P. R., Snoep, J. L., and Westerhoff, H. V. (1996) DNA supercoiling depends on the phosphorylation potential in *Escherichia coli*. *Mol. Microbiol.* **20**: 351–360.

Watanabe, T. (1963) Infective heredity of multiple drug resistance in bacteria. *Bacteriol. Rev.* **27**: 87–115.

Watanabe, T. (1966) Infectious drug resistance in enteric bacteria. *N. Engl. J. Med.* **275**: 888.

Weber, R. F., and Silverman, P. M. (1988) The Cpx proteins of *Escherichia coli* K12. Structure of the CpxA polypeptide as an inner membrane component. *J. Mol. Biol.* **203**: 467–478.

Westerhoff, H. V., O'Dea, M. H., Maxwell, A., and Gellert, M. (1988) DNA super-coiling by DNA gyrase. A static head analysis. *Cell Biophys.* **12**: 157–181.

Wilkins, B. M., and Frost, L. S. (2001) Mechanisms of gene exchange between bacteria. In Sussman, M. (ed). *Molecular medical microbiology*. London: Academic Press, pp. 355–400.

Will, W. R., Lu, J. and Frost, L. S. The role of H-NS in silencing F transfer gene expression during entry into stationary phase. *Mol. Microbiol.* **54**: 769–782.

Willetts, N. (1977) The transcriptional control of fertility in F-like plasmids. *J. Mol. Biol.* **112**: 141–148.

Willetts, N., and Maule, J. (1986) Specificities of IncF plasmid conjugation genes. *Genet. Res.* **47**: 1–11.

Willetts, N., and Skurray, R. (1987) Structure and function of the F factor and mechanism of conjugation. In Neidhardt, F. C., et al., (eds). *Escherichia coli and Salmonella typhimurium: Cellular and molecular biology*. Washington, DC: American Society for Microbiology, pp. 1110–1133.

Willetts, N., and Wilkins, B. (1984) Processing of plasmid DNA during bacterial conjugation. *Microbiol. Rev.* **48**: 24–41.

Yoshioka, Y., Ohtsubo, H., and Ohtsubo, E. (1987) Repressor gene *fin*O in plasmids R100 and F: Constitutive transfer of plasmid F is caused by insertion of IS*3* into F *fin*O. *J. Bacteriol.* **169**: 619–623.

Zechner, E. L., de la Cruz, F., Eisenbrandt, R., Grahn, A. M., Koraimann, G., Lanka, E., et al. (2000) Conjugative-DNA transfer processes. In Thomas, C. M. (ed). *The horizontal gene pool: Bacterial plasmids and gene spread*. Amsterdam: Harwood Academic Publishers, pp. 87–174.

CHAPTER 6

The conjugative transposons: Integrative gene transfer elements

Adam P. Roberts and Peter Mullany

(207)

Until the late 1970s, it was believed that the majority of gene transfer events in bacteria were mediated by plasmids. However, at that time, Don Clewell and co-workers isolated a strain of *Enterococcus faecalis* that could transfer tetracycline resistance in the absence of plasmid DNA (Tomich et al., 1979). The element responsible was an 18-kb segment of DNA that was integrated into the bacterial chromosome. As well as being capable of conjugative transfer to a new host, this element was also capable of intercellular transposition, so the term *conjugative transposon* was coined and this particular element was designated Tn916 (Franke and Clewell, 1981). Subsequently, it has become apparent that conjugative transposons are probably ubiquitous and are highly heterogeneous. This heterogeneity in form and function has led to some confusion and controversy about what a conjugative transposon actually is and how these elements should be named. This issue has more recently been addressed in a number of review articles (Burrus et al., 2002; Mullany et al., 2002; Osborn and Boltner, 2002). Therefore, for the purposes of this chapter, we define conjugative transposons in the loosest possible terms, as discrete DNA elements, usually integrated into the bacterial genome, which can transfer from a donor to a recipient cell by conjugation.

Because conjugative transposons have such a broad host range, they are very important in bacterial evolution (i.e., in disseminating genes between distantly related organisms, induction of deletion, rearrangements) and as a substrate for recombination events (Beaber et al., 2002; Mahairas and Minion, 1989; O'Keeffe et al., 1999; Swartley et al., 1993). They are also clinically important because they have been shown to mediate the transfer of antibiotic resistance between unrelated bacteria (Bahl et al., 2004; Bertram et al., 1991; Roberts and Kenny, 1987; Roberts et al., 2001b; Sen and Oriel, 1990; Woolley et al., 1989). Finally, the broad host range of these elements has been exploited

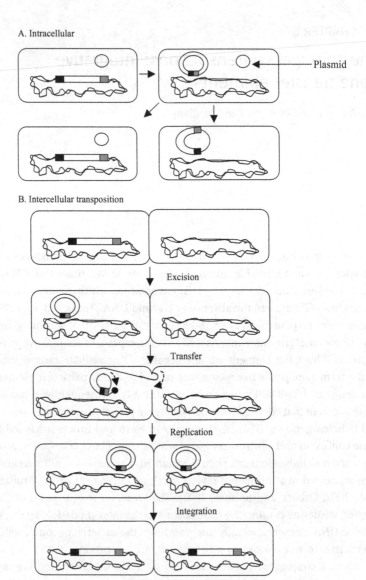

ADAM P. ROBERTS AND PETER MULLANY

A. Intracellular

Plasmid

B. Intercellular transposition

Excision

Transfer

Replication

Integration

Figure 6.1. Generalized model for the translocation of conjugative transposons. **(A)** The integrated form of the conjugative transposon excises from the chromosome to form a covalently closed circular, double-stranded, nonreplicative molecule that is the transposition intermediate. Filled boxes indicate the two ends of the conjugative transposon. Integration can occur into a plasmid or the chromosome. **(B)** The integrated form excises to form the circular intermediate. The circular form is nicked at the *oriT* (*filled circle*) and a single strand is transferred in a similar fashion to plasmid transfer. Regeneration of double-stranded molecules (*dashed arrows*) in both the donor and recipient cells is followed by integration (Salyers et al., 1995).

extensively by researchers to investigate the molecular genetics of a diverse range of bacteria that were previously intractable to genetic analysis (Geist et al., 1993; Mahairas and Minion, 1989; Manganelli et al., 1998; Natarajan and Oriel, 1991; Norgren et al., 1989; Roberts et al., 2003).

This chapter reviews our current understanding of the mechanisms of movement of conjugative transposons at the molecular level. For more information regarding the epidemiology and evolution of these elements, the reader is referred to the following reviews (Pembroke et al., 2002; Toussaint and Merlin, 2002).

MECHANISMS OF CONJUGATIVE TRANSPOSITION

The movement of all conjugative transposons investigated so far appears to proceed via excision of the mobile element from the host DNA molecule (chromosome or plasmid), followed by formation of a circular molecule, which is the conjugative transposition intermediate. This molecule is capable of transposition to a new target site within the host's genome (or transposition into the same target site) or alternatively, a single strand is transferred to a new host. The circular form in both donor and recipient can integrate into the host's genome (Fig. 6.1). The molecular mechanisms underlying these processes, however, can vary dramatically between different conjugative transposons.

Conjugative transposons have a modular organization with regions required for conjugation, regulation, antibiotic resistance, and recombination (Fig. 6.2). This chapter reviews what is known of these processes in different conjugative transposons.

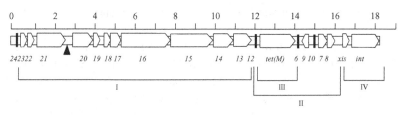

Figure 6.2. The top line shows the scale in kb. The second line represents the open reading frames with the direction of the arrows representing the orientation of the gene. The small orfs, represented by thick vertical lines, are all orientated from left to right. The designated names are underneath. The filled triangle represents the position of the *oriT*. The final line represents the functional grouping of the orfs: I, conjugation functions; II, regulatory functions; III, antibiotic resistance functions; and IV, excision and insertion (adapted from Flannagan et al. 1994).

MECHANISM OF INSERTION AND EXCISION

In all conjugative transposons investigated so far, insertion and excision is mediated by a site-specific recombinase. There are two general families – the tyrosine recombinase and the serine recombinase – both of which are discussed in detail in chapters 1.2 and 1.3.

INSERTION AND EXCISION OF Tn916

Tn916 (Fig. 6.2) requires the integrase Int (a tyrosine recombinase) and the excisionase Xis (a small basic protein) for integration and excision. Int is absolutely required for both processes and Xis greatly stimulates excision, and in some hosts, is absolutely required. It has also been observed that at high concentration Xis inhibits integration (Hinerfeld and Churchward, 2001b). The initial step in excision of Tn916 is the generation, by Int, of 5′ protruding staggered endonucleolytic cuts at each end of the element with one strand of the DNA being cut five bases from the end of the transposon and the other immediately adjacent to the other end. The overhangs that are generated at each end of the transposon are not identical in nucleotide sequence and are called the *coupling sequences*. These are ligated together to form a circular molecule containing a short heteroduplex region at its joint (formed by the 5′ overhangs at each end of the element). The chromosomal DNA is also ligated together (Scott et al., 1988; Storrs et al., 1991; Fig. 6.3). It is believed that, prior to transfer to a recipient cell, this molecule is nicked at an *oriT* site (Fig. 6.2) and a single strand is transferred (Scott et al., 1994; Fig. 6.1). Second-strand synthesis then takes place in both cells thus repairing the heteroduplex region located at the joint of the circular molecule. The heteroduplex in the genomic target site is repaired by replication of the DNA molecule or by the mismatch repair system. Excision and circularization is followed by insertion in which Int-mediated staggered cuts are made at the joint of the circular molecule and at the target site, again as the 5′ overhanging regions are noncomplementary and heteroduplexes will occur after strand exchange. Once more these heteroduplexes are resolved by replication or by mismatch repair of the DNA molecule (Fig. 6.3). Insertion of Tn916 is not random because it shows a preference for targets that have a T-rich region separated from an A-rich region usually by 6 bp (Lu and Churchward, 1995; Mullany et al., 2002). However, Tn916 does not behave the same in every host; in some hosts, it appears to have one specific insertion site, but in others it can insert at multiple sites. Indeed, in certain strains of *Clostridium difficile*, Tn916 has been shown to be extremely site specific (Mullany et al., 1991;

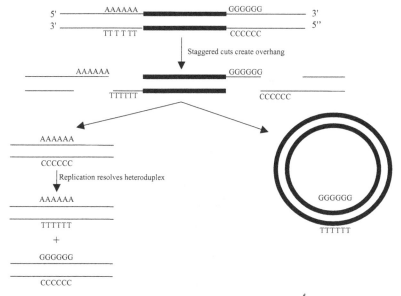

Figure 6.3. (A) Model for the excision of Tn916 from the donor DNA molecule. The coupling sequences flanking the element are represented by the base pairings A-T and G-C. The conjugative transposon is shown by the thick line; the thin line represents the host DNA. The 5'-3' orientation of the DNA strands is shown in the top of the figure. Prior to excision, a staggered cut is made at each end of the conjugative transposon, and on excision, 5' single-stranded overhangs are created. These are ligated to form a heteroduplex molecule (G-T) in the circular intermediate molecule. Heteroduplexes are also formed in the chromosome (A-C); these are repaired by the next round of replication. (B) Insertion of Tn916. G-C represents the complementary base pairings of the target. Other base pairings are as in panel A. Insertion occurs by staggered cleavage of the new target site and of the circular intermediate; ligation of these DNA molecules results in heteroduplex DNA. This is repaired by replication or DNA mismatch repair. (Adapted from Salyers et al., 1995, and Scott, 1992.)

Wang et al., 2000b), whereas in other strains of the same species, it appears to enter the genome in a more random fashion (Roberts et al., 2003; Hussain et al., 2005). The reasons for this are not completely understood, but it is plausible that Tn916 has a preferred attachment (att) site; however, in some hosts it has the ability to recognize degenerate or alternative att sites, or the preferred att sites may not be present in all hosts so that other sites can be used.

As Tn916 Int is related to phage λ Int, similar terminology is used to describe the recombination regions in Tn916 as in λ. Therefore, the region at

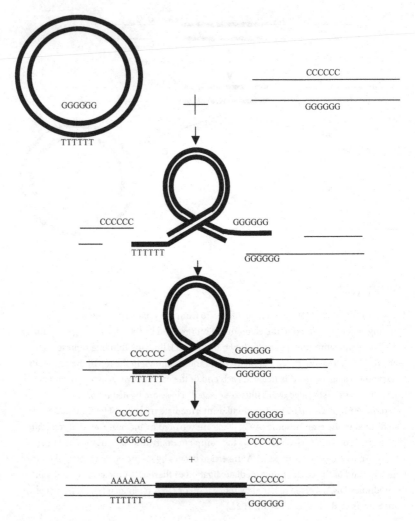

Figure 6.3. (continued)

which recombination takes place is the *core site* and the ends of the transposon adjacent to the core are termed the *arm sites*.

FUNCTIONAL ANALYSIS OF THE GENE PRODUCTS OF *XIS* AND *INT*

Int contains two DNA binding domains (Lu and Churchward, 1994). The C-terminal domain binds to both ends of the transposon and protects a 40-bp region of transposon and flanking DNA in a DNase protection assay (Lu and Churchward, 1994). A 40-bp region is also protected in the target, centered on the insertion site. Int mediates specific DNA cleavage at both the left and right ends of Tn*916* (Taylor and Churchward, 1997). Double-stranded staggered cuts are made, which leave 5′ single-stranded protruding ends (Fig. 6.3A). It is these ends that form the joint in the circular molecule.

Dnase protection assays with Xis showed that it bound specifically to sites close to the ends of Tn*916* (Rudy et al., 1997). The binding site contains the 11-bp direct repeat sequence DR2 and is located in the same relative position to the binding sites for the N-terminal domain of Int (Scott and Churchward, 1995). The results suggest that Xis is involved in the formation of nucleoprotein structures at the ends of the element that help to align the ends so excision and circularization can take place.

The DNA protection pattern exceeded the region that Xis was expected to protect (based on the size of the protein), suggesting that a structure may be formed where the DNA is wrapped around Xis. A bent region of DNA will be protected over a larger region than a comparable straight stretch. Another possible explanation is that Xis may bind to the DNA in the crook of the static bend. Circular permutation assays showed a static bend in the target sequence. Therefore, Xis may facilitate the binding of Int to its specific target site by bending the DNA or by the formation of protein–protein or protein–DNA complexes, or a combination of both. Wojciak et al. (1999) demonstrated the nuclear magnetic resonance structure of the Tn*916* integrase–DNA complex and showed that the DNA was bent approximately 35 degrees. These experiments were performed in the absence of Xis, but the DNA substrate for these experiments was a 13-mer, which included the binding site for Int. Therefore, it is still possible that Xis could facilitate the bending of the DNA and that Xis is required for the correct alignment of the DNA strands and subsequent binding to the Int molecules.

Marra and Scott (1999) reported that Xis is needed for excision; in the absence of Xis, excision of Tn*916* from a chloramphenicol resistance gene was observed at less than 3.1×10^{-10} events per cell (beyond the limits of detection in this experiment). When Xis was provided *in trans*, the normal levels of excision were restored. When *xis* was overexpressed with normal expression of *int*, there was no increase in the excision frequency. This was also true when *int* was overexpressed and *xis* was expressed normally. When both *xis* and *int* were overexpressed in the same cell, however, increased levels of excision were detected. This strongly supports the view that both *xis* and *int* are needed for excision, and the relative concentration of each is the rate-limiting step in the excision reaction that precedes all natural transposition and conjugation events of this element.

The excision and circularization of the element and closure or repair of the donor molecule occurs at identical frequencies, strongly suggesting that a single reaction causes both events (Marra and Scott, 1999). Overexpression of both *xis* and *int* caused the circular molecule to be present in high enough quantities to be detected by a plasmid preparation. The excised molecules were, however, lost from the cells, suggesting that the molecules were not reinserting into the host genome. One factor that might prevent insertion into the host DNA is an excess of Xis, as is the case in the λ system. For λ, insertion requires the presence of Int λ alone and is inhibited by the overexpression of Xisλ (Leffers and Gottesman, 1998). The Xisλ protein prevents the binding of Intλ to a site required for integration (Moitoso and Landy, 1991). In *Esherichia coli* (*E. coli*), only Int is required for integration of Tn916 (Lu and Churchward, 1994). Using primer extensions and reverse-transcriptase polymerase chain reaction the presence of two transcripts containing *int* have been shown from the RNA of *Bacillus subtilis* (*B. subtilis*). One of these transcripts was initiated at a promoter located between *int* and *xis*; the other transcript contained *xis* initiating presumably at a promoter upstream of *xis*. This suggests that there are steps in conjugative transposition that require Int but not Xis (Marra and Scott, 1999).

The various protection experiments can, together with a comparison of the bacteriophage λ system, allow a model to be proposed for the relative positions of the binding interactions between Int, Xis, and the arms of the transposon during excision of Tn916 (Fig. 6.4). The model, proposed by Scott and Churchward (1995), assumes that four Int molecules align the two DNA strands in an antiparallel arrangement; this is presumably catalyzed by the presence of Xis. One Int molecule bound to the right end of Tn916 at the T' site by its C-terminus and to N'1 by its N-terminus forms an intrastrand loop. A second Int molecule binds the left end of the transposon at the T site by its C-terminus and the N'2 site of the right end of the transposon by its N-terminus, forming a bridge between the left and right end of the transposon. The third Int molecule binds to the B site of the right end and the N2 site at the left end, forming another bridge between the two transposon ends. The fourth Int molecule binds to the B' and N1 sites of the left end forming another intrastrand loop. It may, however, be a truncated Int molecule possessing only the C-terminus; sites N1 and N2 have only a single corresponding site in phage λ. Therefore, in the diagram, the fourth Int molecule is shown bound only by its C-terminus. The *orf* encoding Int has other plausible start codons with ribosome-binding sites in the correct positions, and these could

A.

B.

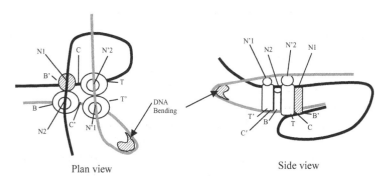

Plan view   Side view

Figure 6.4. **(A)** Position of Xis- and Int-binding sites within Tn916. The thick black line represents Tn916, and the thin black line represents flanking bacterial DNA. L and R represent the left and right end of the element, respectively. The open triangles represent the regions protected by Xis. The filled triangles labeled N1, N2, N3, N1′, and N2′ represent the regions bound by the N-terminal domain of Int. The filled diamonds labeled B and T′ and labeled B′ and T represent the regions bound by the C-terminus of Int. The area labeled C and C′ represent the coupling sequences at each end of the element. **(B)** Model for the alignment of the left and right ends of Tn916 during excision. The right and left parts of the figure represent two different views of the same structure. The grey line represents the right end of Tn916, and the black line represents the left end. The positions of the Int-binding sites are indicated, as is the relative positions of the coupling sequences present at either end of the transposon. In the left part of **B**, the large circles represent the C-terminal domain of Int and the small circles represent the N-terminal domain. On the right, the large cylinders represent the C-terminal domain and the small circles represent the N-terminal domain. The hatched circle on the left and the hatched cylinder on the right represent an Int molecule, which may have two DNA-binding domains or only a C-terminal domain. The hatched shape labeled DNA bending represents a hypothetical host factor that binds and bends the DNA between T′ and N′1. (Adapted from Rudy et al., 1997, and Scott and Churchward, 1995.)

provide the truncated form of Int that may function in this model (Scott and Churchward, 1995; Fig. 6.4B)

Investigation of the role of other factors in Tn916 insertion and excision in E. coli demonstrated that the histone-like factor HU greatly enhances excision (Connolly et al., 2002). This is not surprising because Tn916 is adapted to survive in a huge diversity of bacteria, and it would make use of proteins such as HU, which are well conserved in prokaryotes.

REGULATION OF INSERTION AND EXCISION IN Tn916

Uncontrolled movement of Tn916 would be deleterious for both the element and the host; therefore, it is subject to regulation. A detailed transcriptional analysis of Tn916 has shown that the key components of the regulatory system are orf7, orf8, orf9, and orf12 (Figs. 6.2 and 6.5). It has been postulated that the orf12 region is involved in regulating the system by detecting the presence of tetracycline via a transcription attenuation-type mechanism (Su et al., 1992).

A plausible model for the regulation of transcription of genes in Tn916 has been proposed (Celli and Trieu-Cuot, 1998). It states that in the absence of tetracycline the majority of transcripts initiating at Ptet terminate at the palindromic sequences within orf12. In this situation, orf9, which is believed to be a repressor of transcription at Porf7, is transcribed, resulting in a basal level of transcription from this promoter, which presumably extends into the downstream genes. If the element is in its circular form, transcription will run into the tra region of the element. If it is integrated, it will run into flanking sequences. It is also believed that there is a basal level of transcription from Pxis and Pint into all downstream genes, although this is not proved.

When tetracycline is present, a transcriptional attenuation-type mechanism allows transcription through the terminator upstream of orf12, resulting in increased transcription of tet(M) and downstream genes. This will have the effect of reducing the amount of Orf9 (possibly by an antisense RNA mechanism), resulting in increased expression from Porf7 (and possibly Pxis and Pint) into the downstream genes. This should result in the excision of the element and read-through of the joint in the circular form into the conjugation region (Figs. 6.2 and 6.5). No promoters have been detected in this region, so it is possible that the promoters in the regulatory region are all that is required for transcription of the genes in the conjugation region and that circularization is absolutely required for this process.

The directionality (i.e., integration versus excision) could be, in part, down to the relative concentrations of Xis and Int. Xis has an inhibitory effect on insertion and a stimulatory effect on excision. The ratio of these two

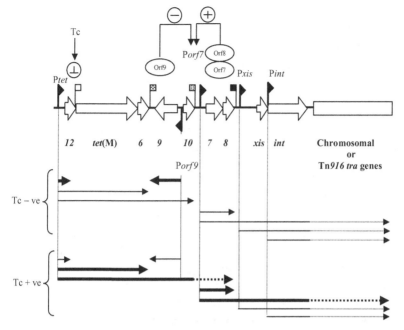

Figure 6.5. Regulation of expression of the transfer genes within Tn916. Symbols: promoter; ⌐, pal_{orf12}; ⌐, pal_{orf9}; ⌐, pal_{orf10}; ⌐, pal_{orf8}, ⌐, antitermination; ⊕, transcriptional repression; ⊖, transcriptional activation; ⊕. pal; palindromic sequences representing putative terminators. The thick arrows underneath the figure represent the majority of the transcripts; the thin lines represent lower levels of transcription. The dotted lines represent possible basal level, read-through transcripts. In the absence of tetracycline (Tc), most of the transcripts initiated at P*tet* terminate at pal_{orf12}, P*orf9* transcribes *orf9* efficiently, whereas P*orf7* directs a low level of transcription through *orf7* and *orf8*. In these conditions, P*orf7* and P*xis* direct a low level of transcription through the transposition-associated and downstream genes. In the presence of Tc, the transcripts initiated at P*tet* read through pal_{orf12}, pal_{orf9}, and pal_{orf10}, which leads to a decreased transcription of *orf9* and an increased transcription of *orf7* and *orf8*. The resulting overexpression of *orf7* and *orf8* stimulates the activity of P*orf7*, which leads to an increased transcription of the downstream genes. *orf9* could repress the activity of P*orf7* (Celli and Trieu-Cuot, 1998).

proteins could be affected by the promoter P*int*, which directs transcription of *int* but not *xis*. However, more work is required to understand how this promoter is regulated.

A further potential role for Int has been proposed because this protein, as well as being able to bind to the ends of the element as discussed previously, can also bind to the *oriT* sequence (see the following and Fig. 6.2). It has been proposed that binding to *oriT* could prevent transcriptional read-through from the promoters P*int*, P*xis*, and P*orf7* (Fig. 6.5) into the transfer genes. Because these promoters are believed to be constitutively active, even if at a low level, they have the capacity to produce Int in all cells, not just that minority that has the capacity to act as conjugal donors (Hinerfeld and Churchward, 2001a). The interaction of *oriT* with Int may provide a further level of control within the element, possible by preventing premature transfer of the element and also preventing excision occurring before until the appropriate signals have been received (e.g., the presence of a suitable recipient).

CONJUGATION IN Tn*916*

The role of the genes in the putative conjugation region of Tn*916* is probably the least well understood part of this genetic element. The genes involved in conjugative transfer have been mapped by mutation studies to the left side of the conjugative transposon (Fig. 6.2). A functional *oriT* has been identified between *orf20* and *orf21* (Fig. 6.2). Limited functional studies have been carried out; Orf21 has been partially purified and its C-terminus exhibits nonspecific DNA nicking activity, which could be involved in nicking the transposon at its *oriT* prior to transfer (Jaworski and Clewell, 1995).

Homologs of some of the orfs in the conjugation module also provide indications as to the function of the various proteins produced. Orf15 and Orf16 have homologs in plasmids pAD1 and pAM373, and are both required for conjugation, although their role in this process is not understood. Orf18 has some homology to antirestriction proteins, and this could be involved in protecting the transposon from the host's restriction system, thus contributing to its extremely wide distribution. Due to the location of *orf18* relative to the *oriT* it is believed that this gene is one of the first to be transferred to a new host. Consequent expression of *orf18* would lead to the protection of the incoming DNA. Despite this information, the process by which conjugative transposons form mating aggregates is unclear. However, there is some evidence that the transposon Tn*925* (this element is very closely related to Tn*916*) is capable of mobilizing chromosomal genes. However, it is not clear if this is by formation of a zygotic cell by cell fusion or by mobilization of linked genes in a Hfr-like manor (Showsh and Andrews, 1996). Clearly, there is scope for much more study on the mechanism of conjugation.

Tn*5397* FROM *CLOSTRIDIUM DIFFICILE*

Tn*5397* is a conjugative transposon that was originally isolated from *Clostridium difficile* (Mullany et al., 1990). The element can transfer between *C. difficile*

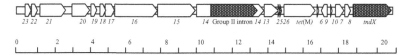

0　　　2　　　4　　　6　　　8　　　10　　　12　　　14　　　16　　　18　　　20

Figure 6.6. Schematic representation of Tn5397 from *Clostridium difficile*. The arrows represent the orfs pointing in the probable direction of transcription. The orfs that are shaded are unique to Tn5397, all other orfs have homologs in Tn916. The bottom line represents that scale in kb (Roberts et al., 2001a).

strains and to and from *B. subtilis* and from *B. subtilis* to streptococci and enterococci (Hachler et al., 1987; Mullany et al., 1990). Tn5397 has been completely sequenced (Fig. 6.6) and is very closely related to Tn916; however, there are some important differences. It contains a group II intron, and instead of the Tn916 genes *int* and *xis*, Tn5397 contains the gene *tndX*, which encodes a protein of the family of serine site-specific recombinases (Mullany et al., 1996; Roberts et al., 2001b; Wang et al., 2000a). The serine recombinases form a diverse group (the mechanisms of action of these are discussed in more detail in chapter 1.3 that has been classified into five groups based on evolutionary mechanistic and structural properties (Smith and Thorpe, 2002). TndX belongs to class V, which contains the most diverse members of this group. Members of class V have also been called the large resolvases because they are related in the N-terminus to the "classical" resolvases, but compared with the latter they have an extended and divergent C-terminal region. Members of this family promote insertion and excision of phage genomes, excision of DNA from genes involved in bacterial developmental processes, insertion and excision of transposons that can be mobilized transposons as well as insertion and excision of Tn5397 (Smith and Thorpe, 2002).

INSERTION AND EXCISION OF Tn5397

Inserted copies of Tn5397 are flanked by GA dinucleotides (Wang et al., 2000a). Furthermore, the joint between the left and right ends in the circular form of the transposon is separated by a GA dinucleotide, as is the target site of the element in the host genome (Wang et al., 2000a). After excision of Tn5397 from the host genome, the target site is completely regenerated (Fig. 6.7).

Mutational analysis has confirmed that TndX is required for insertion and excision of Tn5397 (Wang et al., 2000a). Further work undertaken in *E. coli* using a minitransposon that just contains the ends of Tn5397 with TndX provided *in trans* demonstrated that TndX was required and sufficient for insertion and excision of the minielement. Furthermore, the transposition

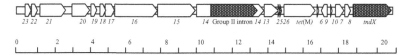

219

THE CONJUGATIVE TRANSPOSONS: INTEGRATIVE GENE TRANSFER ELEMENTS

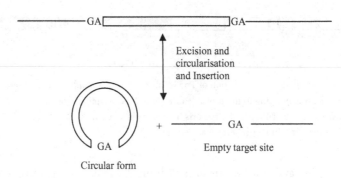

Figure 6.7. Tn*5397* is represented by the unfilled box and the host genomic DNA is represented by the thin line. When Tn*5397* is present in the host genome, it is flanked by a GA dinucleotide. This GA dinucleotide is also present at the joint of the circular form that results from excision of the molecule and at the perfectly reformed target site.

ADAM P. ROBERTS AND PETER MULLANY

products are the same as those observed in *C. difficile* proving that TndX is the only Tn*5397*-encoded protein required for transposition. The following model has been proposed for TndX-mediated insertion and excision of Tn*5397*: TndX mediates excision of Tn*5397* by introducing 2-bp staggered cuts at the 3′ ends of directly repeated GA dinucleotides at each end of the transposon. This is followed by strand exchange, resulting in excision of the circular form of the transposon and regeneration of the original target site. Transfer to a recipient cell can occur, probably by a similar mechanism to that described for Tn*916* previously (i.e., by transfer of a single DNA strand). In the recipient and donor, a new second DNA strand is synthesized. The element can now insert into the genome; insertion is believed to be the reverse of excision.

The TndX system of Tn*5397* appears to require minimal host factors because we have shown that it can mediate excision of miniTn*5397* *in vitro* in a solution that just contains DNA substrates TndX and buffer. This lack of a requirement for host factors could help contribute to the ability of this element to exist in a range of unrelated bacteria.

REGULATION OF TRANSPOSITION IN Tn*5397*

Although Tn*5397* contains genes very closely related to those proposed to be involved in regulation of Tn*916* (i.e., *orf7*, *orf8*, and *orf9*), the putative regulatory region upstream of *tet*(M) is different and there are no obvious promoters upstream of *tndX* equivalent to P*xis* or P*int*. Therefore, the regulatory circuits in Tn*5397* are likely to be appreciably different from those in Tn*916*.

Preliminary experimental results have shown that both the growth of cells containing Tn5397 and transcription of the tet(M) gene are inducible by the addition of tetracycline to the growth media (Roberts et al., 2002, unpublished data). Perusal of the DNA sequence upstream of the tet(M) gene of Tn5397 shows that there are two open reading frames. These orfs are predicted to encode RNA that has the ability to adopt different secondary structures, depending on the presence or absence of tetracycline.

Some of the secondary structures also have the ability to block strong ribosome-binding sites. This mechanism is probably a variation on a theme of transcriptional attenuation that is predicted to occur in Tn916, but in the case of Tn5397, there is a further level of control in that occlusion of the ribosome-binding site is also likely to occur.

THE CONJUGATIVE TRANSPOSONS FROM THE *BACTEROIDES*

The *Bacteroides* family of conjugative transposons forms a distinct group of conjugative elements. Like the conjugative transposons described previously, the first step of conjugative transposition is excision of the element to form a circular intermediate molecule, which has the ability to transfer to a suitable recipient cell. These elements can also mediate the transfer of other elements that can be mobilized, such as nonreplicating *Bacteroides* units (NBUs), which are small, nonconjugative transposons that can be mobilized and that encode their own mob protein (Li et al., 1995). Mobilization of NBUs can be both *in cis* and *in trans*. The Tn916 family has not been observed to be able to mobilize other elements *in cis*.

The best understood of the *Bacteroides* conjugative transposons are CTnERL and CTnDOT (Bonheyo et al., 2001; Cheng et al., 2000, 2001; Gupta et al., 2003; Whittle et al., 2001, 2002a; Fig. 6.8). These elements differ only in the presence of a 13-kb region encoding macrolide resistance on CTnDOT. This element is part of a family of related conjugative transposons that share extensive DNA sequence homology (Whittle et al., 2002b). The elements contain an integrase gene *int* located at one end that encodes a tyrosine recombinase that is required for integration and excision of the element. Unlike the Tn916 system, CTnDOT has a small number of highly preferred integration sites that may be due to a 10-bp sequence in the transposon, which is also adjacent to the insertion site. Excision of the element is completely different to the Tn916 system. Essentially, an unknown mechanism senses the presence of tetracycline upstream of the tetracycline resistance gene *tet(Q)*. This leads to the increased transcription of *rteB* and increased levels of RteB protein. RteB acts as a positive regulator on the promoter controlling *rteC*.

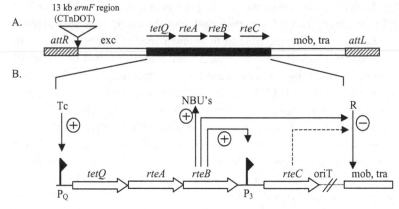

Figure 6.8. Schematic representation of CTnDOT and CTnERL and the proposed model for the regulation of expression of the transfer and mobilization genes. **(A)** CTnDOT/ERL family contain homologous excision (exc), mobilization (mob), and transfer (tra) genes and homologous right (*attR*) and left (*attL*) end sequences. CTnDOT contains a 13-kb ermF region encoding resistance to macrolides. The regulatory region extends from *tetQ* to *rteC* and is expanded in **B**. The promoter P_Q senses the presence of tetracycline by an unknown mechanism and undergoes 20-fold enhanced transcription. Increased levels of *rteB* transcript gives rise to increased levels of RteB protein. RteB activates transcription of *rteC* by acting on the promoter P_3. RteB also activates the mobility of NBUs by an as yet undetermined mechanism. In addition, RteB, and to a lesser extent RteC, act as an antirepressor of R (an unknown repressor) and lead to increased levels of transcription of the downstream mob and tra genes (Salyers et al., 1995; Whittle et al., 2002b).

In addition, experimental evidence (Li et al., 1995) suggests that RteB and, to a lesser extent RteC, act as an antirepressor of an as yet unidentified repressor controlling the expression of the downstream *mob* and *tra* genes (Fig. 6.8B). Increased levels of the proteins encoded by these are believed to lead to elevated levels of mobility. Furthermore, it has been shown that RteB can mobilize NBUs in trans (Salyers et al., 1995).

Four *Bacteroides*, conjugative transposons that do not belong to the CTnDOT family have been discovered – CTn*12256*, CTn*7853*, CTnGERM1, and CTnBst. CTn*12256* is a large composite element that contains a copy of CTnDOT integrated into a large (150-kb) conjugative transposon (Whittle et al., 2002b). Transfer of this CTn*12256*, also known as Tn*5030* (Halula and Macrina, 1990), is not dependent on the internal CTnDOT element because the transfer of the erythromycin resistance present on CTn*12256* is not stimulated by tetracycline; instead, it is observed for the internal CTnDOT element. The detailed structure of the other elements has not yet been elucidated. However, they do appear to have an even larger host range than CTnDOT, but are not stimulated by tetracycline; rather, they transfer at a frequency of around 10^{-6} transconjugants per recipient cfu. Interestingly, the linkage of *ermF* from the larger element in CTn*12256* and *tet*(Q) from the internal CTnDOT element, plus other genes such as *rteA* and *rteB*, has been demonstrated for a variety of bacteria, including *Bacteroides forsythus, B. fragilis, Prevotella bivia, Clostridium butyricum,* and *Veillonella parvula,* all of which could transfer these genes to *E. faecalis* (Chung et al., 1999). A more recent review covers detailed aspects of these other *Bacteroides* elements (Whittle et al., 2002b).

CONJUGATIVE TRANSPOSONS IN THE *ENTEROBACTERIACEAE*

Conjugative transposons have only more recently been discovered in this group of organisms, initially, in *Proteus rettgeri* (Coetzee et al., 1972). These elements encode a plasmid-like conjugation system and a phage-like integration system. Some of the *Enterobacteriaceae* conjugative transposons exhibit a degree of incompatibility and have been classified as belonging to the IncJ incompatibility group. One of these elements, CTnR391, has been completely sequenced (Boltner et al., 2002). The element contains 96 *orfs* and is made up of several modules containing a phage λ-like integration/excision system (like that of the *Salmonella* plasmid R27), resistance determinants, and genes encoding DNA repair enzymes (Boltner et al., 2002).

When two IncJ elements are present in the same cell circular forms of the element are observed (Pembroke and Murphy, 2000). It is not clear if these are replicative or if they result from increased amounts of integrase being present, leading to more circular transposon molecules being produced by excision. It has also been reported that related IncJ elements also undergo recombination with each other when present in a recombinase-proficient host, leading to the generation of new genetic elements (Beaber et al., 2002); however, no quantification as to the frequency of this recombination was provided. Despite these means of generating diversity, these elements also have some properties that are likely to limit their spread. They have a highly preferred insertion site within *prfC*, the gene encoding peptide chain release factor 3 (Hochhut et al., 2001). These elements also have a low frequency of transfer (e.g., 10^{-5}–10^{-7} transconjugants per donor cfu for CTnR391; Pembroke et al., 2002). Finally, the sequence analysis has shown that the transfer region of these elements is related to that of enterobacterial plasmids possibly limiting the spread of these elements to the Gram-negative *Enterobacteriaceae*.

The 100-kb CTnscr has more recently been identified in *Salmonella senftberg*; this element can transfer to *E. coli*, and it contains a plasmid-type transfer region, where it integrates in a *recA*-independent manor, probably via its λ-type integrase, into the phenylalanine tRNA *pheV* gene. The element contains the 3′ portion of the *pheV* gene on one end, so if it inserts in the correct orientation, the gene is reconstituted. It confers a sucrose-metabolizing ability on the host (Hochhut et al., 1997).

The SXT element was discovered in *Vibrio cholerae* O139 (Waldor et al., 1996). It is a 62-kb self-transmissible element that encodes resistance to sulphamethoxazole, trimethoprim, chloramphenicol, and streptomycin. It has a very similar integrase to the IncJ elements and also inserts into a specific site in the genome where it regenerates the original target. Despite the relationship between these elements and the IncJ conjugative transposons, the two do contain different genes and there are some differences in the behavior of the two elements.

SXT has a specific insertion site in *V. cholerae* and *E. coli* (Hochhut et al., 2001) and appears to excise in a *recA*-dependent manner (Waldor et al., 1996). SXT has evolved to detect the global stress response of the host due to DNA damage. Activated RecA, when present within the cell, will cleave the SXT repressor leading to the transcriptional activation of *setC* and *setD*. Increased levels of SetC and SetD lead to autoactivation of *setC* and *setD*, and also to the activation of the integrase gene and the transfer genes (Beaber et al., 2004; Fig. 6.9). The presence of a self-transmissible conjugative transposon encoding multiple antibiotic resistance in an organism that has been responsible for a major cholera epidemic in India and Bangladesh (No author, 1993) illustrates the importance of conjugative transposons in the spread of antibiotic resistance.

OTHER CONJUGATIVE TRANSPOSONS

The conjugative transposons described in the previous sections have had some detailed biochemical characterization. There are also a whole range of conjugative transposons that have been identified just by mating experiments and by DNA sequence analysis and they await more detailed biochemical characterization. In this section, we briefly review what is known about the molecular aspects of movement of three of these less well-understood elements.

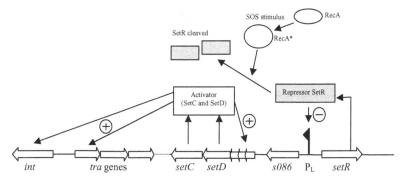

Figure 6.9. Stimulation of transcription of the SXT element by RecA. RecA (*unfilled ovoid*), activated by the cellular stress response (*unfilled circle*), cleaves the repressor protein SetR (*shaded box*). This alleviates the repression at promter P_L and leads to an increase in transcription of the *s086* and the activators *setC* and *setD*. Increased levels of SetC and SetD proteins lead to the autoactivation of *setC* and *setD* genes, and an increase in the transcription of integrase and transfer genes (Beaber et al., 2004).

Tn*5253*

Originally identified in the chromosome of *Streptococcus pneumoniae* (Vijayakumar et al., 1986), this conjugative transposon has subsequently been shown to be a composite structure. Tn*5253* consists of a Tn*916*-like conjugative transposon, designated Tn*5251*, which has integrated into a different, larger conjugative transposon, designated Tn*5252*. The independent conjugative transposition of both of these elements has been demonstrated (Ayoubi et al., 1991). Transfer of Tn*5251* is believed to be identical to Tn*916*; however, the movement of Tn*5252* appears to proceed by a different mechanism. Although this mechanism has not been elucidated experimentally, it has been shown to involve the site-specific nicking of the conjugative transposon by the relaxase Orf4 encoded by Tn*5252*. This strongly suggests that DNA transfer proceeds by a single-stranded mechanism, possibly by pore formation because there are no pili-associated genes reported on the element. The control of transcription of this element also appears to be different, and transfer is greatly reduced in the presence of a repressor.

Tn*1549*

Tn*1549* was originally identified in the chromosome of *E. faecalis*. The element is 34-kb and contains a vancomycin resistance operon (Garnier et al.,

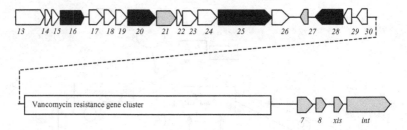

Figure 6.10. Schematic of Tn*1549*. The genes are shown as arrows pointing in the probable direction of transcription. Shaded arrows represent genes with homologs in Tn*916*, and the filled black arrows represent genes with homologs in Staphylococcal conjugative plasmids. The vancomycin resistance gene cluster is represented as an unfilled rectangle. Orf names are underneath their cognate gene (Garnier et al., 2000).

2000). Excision and insertion are believed to proceed in much the same way as in the Tn*916* system with a heteroduplex existing at the joint of an intermediate circular molecule. However, there are some important differences between Tn*916* and Tn*1549*. Tn*1549* contains different genes in the transfer region, which are related to the genes involved in the conjugative transfer of the Staphylococcal plasmids pG01 and pSK41. In addition, there is another homolog of *orf8* within the transfer region (Garnier et al., 2000; Fig. 6.10).

Tn*5276*

Tn*5276* is a 70-kb conjugative transposon isolated from *Lactococcus lactis* NIZO R5 (Rauch and de Vos, 1992a). It encodes the production of and immunity to the antibiotic nisin, as well as a metabolic pathway for the fermentation of sucrose. These traits can be conjugatively transferred to other *L. lactis* strains. Tn*5276* integrates into a preferred target site designated site 1, but has at least four other sites where integration has been observed (Rauch and de Vos, 1994). Tn*5276* contains *xis* and *int* genes that share limited homology to those found in Tn*916*. The integrated element is flanked by a TTTTTG hexanucleotide, which is also present at the junction of the circular form and at target site 1. This suggests that there are some subtle differences in the mechanism of excision when compared with Tn*916* because Tn*916* is not flanked by a directly repeated hexanucleotide sequence. In addition, host factors that are believed to be involved in excision of Tn*916* in *L. lactis* MG1363 (Bringel et al., 1991) have been shown not to be required for the transposition of Tn*5276* in this strain (Rauch and de Vos, 1992a, 1992b, 1994).

CONCLUDING REMARKS

This chapter has demonstrated that conjugative transposons are important mediators of gene transfer in bacteria and that different families of these elements employ a wide range of mechanisms. This chapter is by no means exhaustive because new and novel conjugative transposons are being described.

These elements are modular, and these modules can be exchanged between different mobile elements, which can alter the host range of the elements. The examples discussed also illustrate the varied ways in which this large group of mobile elements has evolved to regulate their mobility, allowing movement at the correct times and, as is the case with the SXT element, to utilize the hosts cells response to detect the optimal time for mobility.

The different modules within conjugative transposons share some relationship with those of plasmids and phage; therefore, these elements should be considered as part of a continuum of genetic flux mediated by mobile genetic elements.

227

ACKNOWLEDGMENTS

The work in our laboratory has been funded by the Wellcome Trust, the BBSRC, and in part by a small research grant from the Eastman Foundation for Oral Research and Training (EFFORT). Thanks go to Dr. Ian Davis for critical reading of the manuscript.

REFERENCES

No author. (1993) Large epidemic of cholera-like disease in Bangladesh caused by *Vibrio cholerae* O139 synonym Bengal. Cholera Working Group, International Centre for Diarrhoeal Diseases Research, Bangladesh *Lancet* **342**: 387–390.

Ayoubi, P., Kilic, A. O., and Vijayakumar, M. N. (1991) Tn5253, the pneumococcal omega (cat tet) BM6001 element, is a composite structure of two conjugative transposons, Tn5251 and Tn5252. *J. Bacteriol.* **173**: 1617–1622.

Bahl, M. I., Sorensen, S. J., Hansen, L. H., and Licht, T. R. (2004) Effect of tetracycline on transfer and establishment of the tetracycline-inducible conjugative transposon Tn916 in the guts of gnotobiotic rats. *Appl. Environ. Microbiol.* **70**: 758–764.

Beaber, J. W., Hochhut, B., and Waldor, M. K. (2002) Genomic and functional analyses of SXT, an integrating antibiotic resistance gene transfer element derived from *Vibrio cholerae. J. Bacteriol.* **184**: 4259–4269.

Beaber, J. W., Hochhut, B., and Waldor, M. K. (2004) SOS response promotes horizontal dissemination of antibiotic resistance genes. *Nature* **427**: 72–74.

Bertram, J., Stratz, M., and Durre, P. (1991) Natural transfer of conjugative transposon Tn916 between gram-positive and gram-negative bacteria. *J. Bacteriol.* **173**: 443–448.

Boltner, D., MacMahon, C., Pembroke, J. T., Strike, P., and Osborn, A. M. (2002) R391: A conjugative integrating mosaic comprised of phage, plasmid, and transposon elements. *J. Bacteriol.* **184**: 5158–5169.

Bonheyo, G., Graham, D., Shoemaker, N. B., and Salyers, A. A. (2001) Transfer region of a *Bacteroides* conjugative transposon, CTnDOT. *Plasmid* **45**: 41–51.

Bringel, F., Van Alstine, G. L., and Scott, J. R. (1991) A host factor absent from *Lactococcus lactis* subspecies lactis MG1363 is required for conjugative transposition. *Mol. Microbiol.* **5**: 2983–2993.

Burrus, V., Pavlovic, G., Decaris, B., and Guedon, G. (2002) Conjugative transposons: The tip of the iceberg. *Mol. Microbiol.* **46**: 601–610.

Celli, J., and Trieu-Cuot, P. (1998) Circularization of Tn916 is required for expression of the transposon-encoded transfer functions: Characterization of long tetracycline-inducible transcripts reading through the attachment site. *Mol. Microbiol.* **28**: 103–117.

Cheng, Q., Paszkiet, B. J., Shoemaker, N. B., Gardner, J. F., and Salyers, A. A. (2000) Integration and excision of a *Bacteroides* conjugative transposon, CTnDOT. *J. Bacteriol.* **182**: 4035–4043.

Cheng, Q., Sutanto, Y., Shoemaker, N. B., Gardner, J. F., and Salyers, A. A. (2001) Identification of genes required for excision of CTnDOT, a *Bacteroides* conjugative transposon. *Mol. Microbiol.* **41**: 625–632.

Chung, W. O., Young, K., Leng, Z., and Roberts, M. C. (1999) Mobile elements carrying ermF and tetQ genes in gram-positive and gram-negative bacteria. *J. Antimicrob. Chemother.* **44**: 329–335.

Coetzee, J. N., Datta, N., and Hedges, R. W. (1972) R factors from Proteus rettgeri. *J. Gen. Microbiol.* **72**: 543–552.

Connolly, K. M., Iwahara, M., and Clubb, R. T. (2002) Xis protein binding to the left arm stimulates excision of conjugative transposon Tn916. *J. Bacteriol.* **184**: 2088–2099.

Franke, A. E., and Clewell, D. B. (1981) Evidence for a chromosome-borne resistance transposon (Tn916) in *Streptococcus faecalis* that is capable of "conjugal" transfer in the absence of a conjugative plasmid. *J. Bacteriol.* **145**: 494–502.

Garnier, F., Taourit, S., Glaser, P., Courvalin, P., and Galimand, M. (2000) Characterization of transposon Tn1549, conferring VanB-type resistance in *Enterococcus* spp. *Microbiology* **146 (Pt 6)**: 1481–1489.

Geist, R. T., Okada, N., and Caparon, M. G. (1993) Analysis of *Streptococcus pyogenes* promoters by using novel Tn916-based shuttle vectors for the construction of transcriptional fusions to chloramphenicol acetyltransferase. *J. Bacteriol.* **175**: 7561–7570.

Gupta, A., Vlamakis, H., Shoemaker, N., and Salyers, A. A. (2003) A new *Bacteroides* conjugative transposon that carries an ermB gene. *Appl. Environ. Microbiol.* **69**: 6455–6463.

Hachler, H., Kayser, F. H., and Berger-Bachi, B. (1987) Homology of a transferable tetracycline resistance determinant of *Clostridium difficile* with *Streptococcus (Enterococcus) faecalis* transposon Tn916. *Antimicrob. Agents Chemother.* **31**: 1033–1038.

Halula, M., and Macrina, F. L. (1990) Tn5030: A conjugative transposon conferring clindamycin resistance in *Bacteroides* species. *Rev. Infect. Dis.* **12** (**Suppl 2**): S235–S242.

Hinerfeld, D., and Churchward, G. (2001a) Specific binding of integrase to the origin of transfer (oriT) of the conjugative transposon Tn916. *J. Bacteriol.* **183**: 2947–2951.

Hinerfeld, D., and Churchward, G. (2001b) Xis protein of the conjugative transposon Tn916 plays dual opposing roles in transposon excision. *Mol. Microbiol.* **41**: 1459–1467.

Hochhut, B., Beaber, J. W., Woodgate, R., and Waldor, M. K. (2001) Formation of chromosomal tandem arrays of the SXT element and R391, two conjugative chromosomally integrating elements that share an attachment site. *J. Bacteriol.* **183**: 1124–1132.

Hochhut, B., Jahreis, K., Lengeler, J. W., and Schmid, K. (1997) CTnscr94, a conjugative transposon found in enterobacteria. *J. Bacteriol.* **179**: 2097–2102.

Hussain, H., Roberts, A. P., and Mullany, (2005). Generation of an erythromycin sensitive derivative of *Clostridium difficile* strain 630 (630Δerm) and demonstration that the conjugative transposon Tn916ΔE enters the genome of this strain at multiple sites. *J. Med. Micro.* 52(pts): 137–141.

Jaworski, D. D., and Clewell, D. B. (1995) A functional origin of transfer (oriT) on the conjugative transposon Tn916. *J. Bacteriol.* **177**: 6644–6651.

Leffers, G. G., Jr., and Gottesman, S. (1998) Lambda Xis degradation in vivo by Lon and FtsH. *J. Bacteriol.* **180**: 1573–1577.

Li, L. Y., Shoemaker, N. B., and Salyers, A. A. (1995) Location and characteristics of the transfer region of a *Bacteroides* conjugative transposon and regulation of transfer genes. *J. Bacteriol.* **177**: 4992–4999.

Lu, F., and Churchward, G. (1994) Conjugative transposition: Tn916 integrase contains two independent DNA binding domains that recognize different DNA sequences *EMBO J.* **13**: 1541–1548.

Lu, F., and Churchward, G. (1995) Tn916 target DNA sequences bind the C-terminal domain of integrase protein with different affinities that correlate with transposon insertion frequency. *J. Bacteriol.* **177**: 1938–1946.

Mahairas, G. G., and Minion, F. C. (1989) Transformation of Mycoplasma pulmonis: Demonstration of homologous recombination, introduction of cloned genes, and preliminary description of an integrating shuttle system. *J. Bacteriol.* **171**: 1775–1780.

Manganelli, R., Provvedi, R., Berneri, C., Oggioni, M. R., and Pozzi, G. (1998) Insertion vectors for construction of recombinant conjugative transposons in *Bacillus subtilis* and *Enterococcus faecalis*. *FEMS Microbiol. Lett.* **168**: 259–268.

Marra, D., and Scott, J. R. (1999) Regulation of excision of the conjugative transposon Tn916. *Mol. Microbiol.* **31**: 609–621.

Moitoso, D., and Landy, A. (1991) A switch in the formation of alternative DNA loops modulates lambda site-specific recombination. *Proc. Natl. Acad. Sci. U S A* **88**: 588–592.

Mullany, P., Pallen, M., Wilks, M., Stephen, J. R., and Tabaqchali, S. (1996) A group II intron in a conjugative transposon from the gram-positive bacterium, *Clostridium difficile*. *Gene* **174**: 145–150.

Mullany, P., Roberts, A. P., and Wang, H. (2002) Mechanism of integration and excision in conjugative transposons. *Cell Mol. Life Sci.* **59**: 2017–2022.

Mullany, P., Wilks, M., Lamb, I., Clayton, C., Wren, B., and Tabaqchali, S. (1990) Genetic analysis of a tetracycline resistance element from *Clostridium difficile* and its conjugal transfer to and from *Bacillus subtilis*. *J. Gen. Microbiol.* **136** (Pt 7): 1343–1349.

Mullany, P., Wilks, M., and Tabaqchali, S. (1991) Transfer of Tn916 and Tn916 delta E into *Clostridium difficile*: Demonstration of a hot-spot for these elements in the *C. difficile* genome. *FEMS Microbiol. Lett.* **63**: 191–194.

Natarajan, M. R., and Oriel, P. (1991) Conjugal transfer of recombinant transposon Tn916 from *Escherichia coli* to *Bacillus stearothermophilus*. *Plasmid* **26**: 67–73.

Norgren, M., Caparon, M. G., and Scott, J. R. (1989) A method for allelic replacement that uses the conjugative transposon Tn916: Deletion of the emm6.1 allele in Streptococcus pyogenes JRS4. *Infect. Immun.* **57**: 3846–3850.

O'Keeffe, T., Hill, C., and Ross, R. P. (1999) In situ inversion of the conjugative transposon Tn916 in *Enterococcus faecium* DPC3675. *FEMS Microbiol. Lett.* **173**: 265–271.

Osborn, A. M., and Boltner, D. (2002) When phage, plasmids, and transposons collide: Genomic islands, and conjugative- and mobilizable-transposons as a mosaic continuum. *Plasmid* **48**: 202–212.

Pembroke, J. T., MacMahon, C., and McGrath, B. (2002) The role of conjugative transposons in the Enterobacteriaceae. *Cell Mol. Life Sci.* **59**: 2055–2064.

Pembroke, J. T., and Murphy, D. B. (2000) Isolation and analysis of a circular form of the IncJ conjugative transposon-like elements, R391 and R997: Implications for IncJ incompatibility. *FEMS Microbiol. Lett.* **187**: 133–138.

Rauch, P. J., and de Vos, W. M. (1992a) Characterization of the novel nisin-sucrose conjugative transposon Tn5276 and its insertion in *Lactococcus lactis*. *J. Bacteriol.* **174**: 1280–1287.

Rauch, P. J., and de Vos, W. M. (1992b) Transcriptional regulation of the Tn5276-located *Lactococcus lactis* sucrose operon and characterization of the sacA gene encoding sucrose-6-phosphate hydrolase. *Gene* **121**: 55–61.

Rauch, P. J., and de Vos, W. M. (1994) Identification and characterization of genes involved in excision of the *Lactococcus lactis* conjugative transposon Tn5276. *J. Bacteriol.* **176**: 2165–2171.

Roberts, A. P., Cheah, G., Ready, D., Pratten, J., Wilson, M., and Mullany, P. (2001a) Transfer of Tn916-like elements in microcosm dental plaques. *Antimicrob. Agents Chemother.* **45**: 2943–2946.

Roberts, A. P., Hennequin, C., Elmore, M., Collignon, A., Karjalainen, T., Minton, N., and Mullany, P. (2003) Development of an integrative vector for the expression of antisense RNA in *Clostridium difficile*. *J. Microbiol. Methods* **55**: 617–624.

Roberts, A. P., Johanesen, P. A., Lyras, D., Mullany, P., and Rood, J. I. (2001b) Comparison of Tn5397 from *Clostridium difficile*, Tn916 from *Enterococcus faecalis* and the CW459tet(M) element from *Clostridium perfringens* shows that they have similar conjugation regions but different insertion and excision modules. *Microbiology* **147**: 1243–1251.

Roberts, A. P., Mullany, P. (2002) Unpublished data.

Roberts, M. C., and Kenny, G. E. (1987) Conjugal transfer of transposon Tn916 from *Streptococcus faecalis* to *Mycoplasma hominis*. *J. Bacteriol.* **169**: 3836–3839.

Rudy, C. K., Scott, J. R., and Churchward, G. (1997) DNA binding by the Xis protein of the conjugative transposon Tn916. *J. Bacteriol.* **179**: 2567–2572.

Salyers, A. A., Shoemaker, N. B., and Li, L. Y. (1995) In the driver's seat: The *Bacteroides* conjugative transposons and the elements they mobilize. *J. Bacteriol.* **177**: 5727–5731.

Scott, J. R. (1992) Sex and the single circle: Conjugative transposition. *J. Bacteriol.* **174**: 6005–6010.

Scott, J. R., Bringel, F., Marra, D., Van Alstine, G., and Rudy, C. K. (1994) Conjugative transposition of Tn916: Preferred targets and evidence for conjugative transfer of a single strand and for a double-stranded circular intermediate. *Mol. Microbiol.* **11**: 1099–1108.

Scott, J. R., and Churchward, G. G. (1995) Conjugative transposition. *Annu. Rev. Microbiol.* **49**: 367–397.

Scott, J. R., Kirchman, P. A., and Caparon, M. G. (1988) An intermediate in transposition of the conjugative transposon Tn916. *Proc. Natl. Acad. Sci. U S A* **85**: 4809–4813.

Sen, S., and Oriel, P. (1990) Transfer of transposon Tn916 from *Bacillus subtilis* to *Thermus aquaticus*. *FEMS Microbiol. Lett.* **55**: 131–134.

Showsh, S. A., and Andrews, R. E., Jr. (1996) Functional comparison of conjugative transposons Tn916 and Tn925. *Plasmid* **35**: 164–173.

Smith, M. C., and Thorpe, H. M. (2002) Diversity in the serine recombinases *Mol. Microbiol.* **44(2)**: 299–307.

Storrs, M. J., Poyart-Salmeron, C., Trieu-Cuot, P., and Courvalin, P. (1991) Conjugative transposition of Tn916 requires the excisive and integrative activities of the transposon-encoded integrase. *J. Bacteriol.* **173**: 4347–4352.

Su, Y. A., He, P., and Clewell, D. B. (1992) Characterization of the tet(M) determinant of Tn916: Evidence for regulation by transcription attenuation. *Antimicrob. Agents Chemother.* **36**: 769–778.

Swartley, J. S., McAllister, C. F., Hajjeh, R. A., Heinrich, D. W., and Stephens, D. S. (1993) Deletions of Tn916-like transposons are implicated in tetM-mediated resistance in pathogenic *Neisseria*. *Mol. Microbiol.* **10**: 299–310.

Taylor, K. L., and Churchward, G. (1997) Specific DNA cleavage mediated by the integrase of conjugative transposon Tn916. *J. Bacteriol.* **179**: 1117–1125.

Tomich, P. K., An, F. Y., Damle, S. P., and Clewell, D. B. (1979) Plasmid-related transmissibility and multiple drug resistance in *Streptococcus faecalis* subsp. *zymogenes* strain DS16. *Antimicrob. Agents Chemother.* **15**: 828–830.

Toussaint, A., and Merlin, C. (2002) Mobile elements as a combination of functional modules. *Plasmid* **47**: 26–35.

Vijayakumar, M. N., Priebe, S. D., Pozzi, G., Hageman, J. M., and Guild, W. R. (1986) Cloning and physical characterization of chromosomal conjugative elements in streptococci. *J. Bacteriol.* **166**: 972–977.

Waldor, M. K., Tschape, H., and Mekalanos, J. J. (1996) A new type of conjugative transposon encodes resistance to sulfamethoxazole, trimethoprim, and streptomycin in *Vibrio cholerae* O139. *J. Bacteriol.* **178**: 4157–4165.

Wang, H., Roberts, A. P., Lyras, D., Rood, J. I., Wilks, M., and Mullany, P. (2000a) Characterization of the ends and target sites of the novel conjugative transposon Tn5397 from *Clostridium difficile*: Excision and circularization is mediated by the large resolvase, TndX. *J. Bacteriol.* **182**: 3775–3783.

Wang, H., Roberts, A. P., and Mullany, P. (2000b) DNA sequence of the insertional hot spot of Tn916 in the *Clostridium difficile* genome and discovery of a Tn916-like element in an environmental isolate integrated in the same hot spot. *FEMS Microbiol. Lett.* **192**: 15–20.

Whittle, G., Hund, B. D., Shoemaker, N. B., and Salyers, A. A. (2001) Characterization of the 13-kilobase ermF region of the *Bacteroides* conjugative transposon CTnDOT. *Appl. Environ. Microbiol.* **67**: 3488–3495.

Whittle, G., Shoemaker, N. B., and Salyers, A. A. (2002a) Characterization of genes involved in modulation of conjugal transfer of the *Bacteroides* conjugative transposon CTnDOT 21. *J. Bacteriol.* **184**: 3839–3847.

Whittle, G., Shoemaker, N. B., and Salyers, A. A. (2002b) The role of *Bacteroides* conjugative transposons in the dissemination of antibiotic resistance genes. *Cell Mol. Life Sci.* **59**: 2044–2054.

Wojciak, J. M., Connolly, K. M., and Clubb, R. T. (1999) NMR structure of the Tn916 integrase–DNA complex. *Nat. Struct. Biol.* **6**: 366–373.

Woolley, R. C., Pennock, A., Ashton, R. J., Davies, A., and Young, M. (1989) Transfer of Tn1545 and Tn916 to *Clostridium acetobutylicum*. *Plasmid* **22**: 169–174.

CHAPTER 7

Competence for genetic transformation

Irena Draskovic and David Dubnau

Competence for genetic transformation is a physiological state that enables the uptake of exogenous DNA. Although competence is widespread in nature (Lorenz and Wackernagel, 1994), it appears to be a variable phenotype because natural isolates of a given species may or may not be transformable, and genome sequencing has revealed the presence of competence genes in isolates that are not known to be transformable.

WHAT USE IS COMPETENCE?

In this chapter, we consider three disputed hypotheses that have been advanced for the biological role of competence: DNA uptake for new genetic information, DNA uptake for repair, and DNA uptake for nutrition.

DNA for genetic diversity

The evolutionary pressure for the maintenance of competence genes may be explained by the advantages gained from the acquisition of fitness-enhancing genes; competence may expand the repertoire of genes available to improve the chances of survival in harsh conditions. Several examples of the acquisition by transformation of new genes that confer a selective advantage have been suggested. For instance, in *Neisseria gonorrhoeae*, in which transformation is the only known mode of DNA transfer, the expression of new allelic variants may facilitate antigenic variation, which presumably plays a role in the evasion of the host immune response. Thus, although new pilin variants may be formed by intracellular recombination between a silent gene segment and an expressed pilin gene, the intercellular exchange of pilin alleles may also occur by transformation (Gibbs et al., 1989; Seifert

et al., 1988). Another example may be provided by heterologous transformation between *Haemophilus influenzae* and *Neisseria meningitidis*, which was
suggested because *Haemophilus*-like DNA uptake sequences (DUSs) were
found in the meningococcal chromosome (Kroll et al., 1998). Three independent neisserial loci were proposed to have been introduced in this way
from *H. influenzae*. Remarkably, *Helicobacter pylori* has been shown to undergo such frequent genetic exchange leading to recombination in the wild
that the organism is essentially at linkage equilibrium and this exchange is
almost certainly transformation mediated (Suerbaum et al., 1998). Although
it is conceivable that this prevalent genetic exchange is a secondary effect of
a transformation mechanism that evolved or has been maintained for some
other function, it is simpler to conclude that the ability to exchange genetic
information itself increases fitness. Other examples may include the horizontal transfer of antibiotic resistance determinants (reviewed in Davies, 1994)
and the transfer of pathogenicity islands.

DNA for repair

It has been reported that transformation in *Bacillus subtilis* (*B. subtilis*),
can serve to repair ultraviolet (UV)-induced DNA lesions, even when the
donor DNA is isolated from a similarly irradiated culture (Hoelzer and Michod, 1991). This supports the hypothesis that transformation may have
evolved to ensure the uptake of template molecules for the repair of damaged
DNA. Experiments with UV-irradiated *H. influenzae* confirmed the higher
survival rate of a transformed culture. Remarkably, this enhanced survival
was seen even when transformation was carried out using a unique cloned
fragment of chromosomal DNA (Mongold, 1992). This result is not consistent with the simple repair hypothesis. The comparable experiment, using a
cloned fragment, has apparently not been carried out in *B. subtilis*. No induction of competence by treatments that damage DNA was observed in either
B. subtilis or *H. influenzae* (Redfield, 1993), and this was offered as evidence
against the repair hypothesis. However, it seems rather teleological to predict
that DNA damage must necessarily induce all possible repair mechanisms.
If competence is selected as a means to repair DNA damage, it may be advantageous to induce it under conditions in which such damage is likely to
occur, possibly as the cells approach stationary phase. In this regard, it is
noteworthy that in *B. subtilis* the *recA* gene is transcriptionally activated by
ComK, the competence regulator, and hence, the SOS pathway is turned on
specifically in competent cells (Haijema et al., 1996; Love et al., 1985). *recA*
transcription is also under competence control in *Streptococcus pneumoniae*

(Martin et al., 1995; Pearce et al., 1995). Although the data are suggestive and the hypothesis is reasonable, no firm conclusion is yet possible concerning the role of transformation in DNA repair.

DNA as food

The idea of transformation as a way to acquire DNA for nutrition is consistent with the fact that in some organisms competence is induced in stationary phase and in response to nutritional signals. In support of this hypothesis, it was demonstrated that extracellular DNA, both homospecific and heterospecific, can serve as the sole source of carbon and energy in *Escherichia coli* (*E. coli*; Finkel and Kolter, 2001). Mutants unable to consume DNA suffered a significant loss of fitness during stationary-phase competition. In *E. coli*, the use of DNA as a nutrient depends on orthologs of proteins involved in natural genetic competence. If these genes do indeed function to import macromolecular DNA to the cytosolic compartment by a competence-like pathway, these interesting experiments will have demonstrated that under certain conditions, "transformation" can provide a source of nutrients. However, many bacteria, including *B. subtilis*, secrete nonspecific nucleases and express uptake systems for purines and pyrimidines, suggesting that selection pressure for the evolution or maintenance of elaborate mechanisms for the uptake of macromolecular DNA as food might not be particularly strong. It is also not straightforward to rationalize the "DNA-for-food" hypothesis with the evolution of DNA uptake specificity in *Neisseria* and *Haemophilus*. Although it appears doubtful that fully functional competence systems have evolved solely for nutrition, it may well be that DNA uptake can provide a source of nutrients under some conditions.

Factors that limit genetic exchange by transformation

It may be supposed that a population of bacteria of a given species residing in a given location will usually consist of a single clone, thereby reducing the introduction of new genetic information. However, Roberts and Cohan (1995) reported that in *B. subtilis* and the closely related *Bacillus mojavensis*, migration rates are high, suggesting that the exchange of genes is not limited by geography, at least in these sporulating bacteria. At least four molecular mechanisms are known that do limit the exchange of DNA between populations. The first mechanism, DNA restriction, may not be important for transformation because DNA is taken up as single strands (Ikawa et al., 1980; Trautner et al., 1974), although this conclusion has been challenged (Cohan et al., 1991). Inhomology certainly limits transformational recombination (Roberts and Cohan, 1993). In certain genera (*Neisseria* and *Haemophilus*), DNA uptake sequences DUSs are required for uptake (Smith et al., 1999). In others (*Bacillus* and *Streptococcus*), quorum-sensing systems induce competence in response to population density of genetically identical or similar bacteria. Any or all these isolating mechanisms may limit the exchange of genes by transformation between distantly related populations. We have more to say about the evolution of competence in the following sections, when we discuss the quorum-sensing systems of the bacilli and streptococci.

DNA UPTAKE MECHANISMS

Known DNA uptake systems can be classified into two groups based on protein homologies: systems related to type II secretion and the assembly of type IV pili, and those related to type IV secretion and conjugation. The competence system of *H. pylori* exemplifies the second, whereas the other competence systems we discuss belong to the first group. We now briefly describe these systems, which have been reviewed more recently in detail (Chen and Dubnau, 2003; Dubnau, 1999).

DNA uptake machinery related to type II secretion and type IV pilus assembly

This appears to be the most common type of transformation machinery, which is present in both G-positive (G+) and G-negative (G−) organisms. The major differences between the transformation systems of G+ and G− bacteria are probably imposed by the presence in the latter group of an outer membrane.

In two well-studied G+ species, *B. subtilis* and *S. pneumoniae*, DNA is bound with no apparent sequence specificity. In *B. subtilis* (Fig. 7.1), DNA binding requires ComEA, an integral membrane protein with a double-strand DNA-binding domain exterior to the membrane (Hahn et al., 1993; Inamine and Dubnau, 1995; Provvedi and Dubnau, 1999). This protein is also required for transformation in *S. pneumoniae* (Campbell et al., 1998; Pestova and Morrison, 1998). The products of the *comG* operon (Albano and Dubnau, 1989; Albano et al., 1989), which resemble type IV pilus system proteins, are needed for the incoming DNA to gain access to ComEA, and may provide a pathway through the cell wall (Provvedi and Dubnau, 1999). Again, ComG-like proteins are essential for competence in the streptococci, although the organization of the operon differs from that in *B. subtilis* (Campbell et al.,1998; Lunsford and Roble, 1997; Pestova and Morrison, 1998). In *B. subtilis*, DNA is

Figure 7.1. DNA uptake in *Bacillus subtilis*. **(a)** Double-stranded DNA binds to the membrane receptor ComEA, which is made accessible to the external medium by the ComG proteins that may form a pilus-like channel through the cell wall. The bound DNA is cleaved by a nuclease, NucA, to create a newly formed DNA terminus, which is presented to the putative channel protein, ComEC **(b)**. Uptake of one DNA strand proceeds linearly from a newly formed end with degradation of the opposite strand. Transport is facilitated by the DNA translocase ComFA.

fragmented by the NucA nuclease (Vosman et al., 1987; Vosman et al., 1988) after binding (Provvedi et al., 2001). Following double-stranded cleavage of DNA bound at the cell surface, uptake of one strand proceeds linearly from a newly formed end with degradation of the opposite strand. At least in *Streptococcus*, uptake proceeds from the 3′-end (Mejean and Claverys, 1988, 1993). Uptake across the membrane requires two proteins, ComEC, a polytopic membrane protein that probably forms an aqueous pore (Dubnau and Lovett Jr., 2001; Hahn et al., 1993; Inamine and Dubnau, 1995), and ComFA, an ATPase located at the inner face of the membrane, which probably serves as a DNA translocase (Londono-Vallejo and Dubnau, 1993, 1994a, 1994b). A *ComFA* ortholog functions as well in streptococcal transformation (Lee and Morrison, 1999). ComEC and ComEA orthologs exist in all the competence systems of this class, whereas ComFA appears to play a role only in the G+ systems (reviewed in Chen and Dubnau, 2003; Dubnau, 1999). In *S. pneumoniae* the membrane-localized *endA* gene product is responsible for degradation of the nontransforming strand (Lacks and Neuberger, 1975; Lacks et al., 1975; Puyet et al., 1990; Rosenthal and Lacks, 1980). No *endA* ortholog exists in *B. subtilis*, and the equivalent gene is unknown.

The DNA uptake pathways of the two best studied G− species, *N. meningitidis* and *H. influenzae*, differ from those in the gram positives in a number of respects. First, DNA uptake in *N. meningitidis* and *H. influenzae* is species specific, although this is apparently not a general rule for G− bacteria (Palmen et al., 1993). Unique motifs, 10 bp in *N. meningitidis* (5′-GCCGTCTGAA; Elkins et al., 1991) and 9 bp in *H. influenzae* (5′-AAGTGCGGT; Danner et al., 1980) are required for efficient uptake (reviewed in Lacks, 1999). The specificity receptors are not known.

Transport across the outer membrane requires a pore-forming secretin protein (Drake and Koomey, 1995; Drake et al., 1997; Tomb et al., 1991). However, the core of the DNA uptake process is probably similar in both G+ and G− bacteria. Both require ComEA-like proteins (Chen and Gotschlich, 2001), orthologs of the *B. subtilis* ComG proteins (Porstendorfer et al., 1997; Seifert et al., 1990; Tønjum et al., 1995; Wolfgang et al., 1998), and a ComEC ortholog (Clifton et al., 1994; Facius and Meyer, 1993). In both groups, single-stranded DNA enters the cytosolic compartment, whereas the other is degraded.

DNA uptake machinery related to type IV secretion and conjugation

Pathogenic type I *H. pylori* strains contain two functionally independent type IV transport systems, one for protein translocation encoded by the *cag* pathogenicity island (Odenbreit et al., 2000) and one for uptake of DNA by natural transformation (Hofreuter et al., 1998; Hofreuter et al., 2000; Hofreuter et al., 2001; Smeets and Kusters, 2002). The latter system bears particular resemblance to the DNA transfer machinery of *Agrobacterium tumefaciens*. The *H. pylori* genes known to be required for this novel system are encoded in the *comB* operon, and at least four proteins with orthologs in the *A. tumefaciens* system are involved in DNA uptake. The individual roles of these proteins are not as yet understood.

REGULATION OF COMPETENCE: WHY REGULATE COMPETENCE?

Not all bacteria regulate competence. Two neisserial species, *N. meningitidis* and *N. gonhorreae*, express competence constitutively. *Neisseria* is an obligate human pathogen, and the lack of regulation may reflect its residence in a relatively constant environment in which competence is always advantageous. Complex signal transduction pathways regulate the other well-studied competence systems. This complexity suggests that competence regulation in these bacteria requires the integration of multiple input signals and that competence is only expressed when precise conditions are fullfilled. *S. pneumoniae* is competent during early log phase. In both *B. subtilis* and *H. influenzae*, competence expression develops late in growth when nutrients are limiting, whereas nutrient limitation during exponential growth induces the earlier onset of competence. In *B. subtilis* and *S. pneumoniae*, competence is regulated by quorum sensing, whereas this does not appear to be the case in *Haemophilus*, although in this organism DNA uptake is restricted by uptake specificity to the acquisition of DNA from similar bacteria, or perhaps importantly, to rare genes from distant species that happen to posses uptake sequences. Little is known about competence regulation in *H. pylori*.

From the discussion just presented, it is clear that most of the transformable bacteria that have been studied take up DNA using similar proteins, indicating that the uptake machinery of this large group has evolved from that of a common ancestor. In contrast, the regulatory pathways of these bacteria differ markedly and respond to distinct signals. This suggests that the regulatory mechanisms have evolved individually and that they have done so in response to selective forces that are species specific.

The remaining part of this chapter deals separately with competence regulation in the three most studied organisms: *S. pneumoniae*, *B. subtilis*, and *H. influenzae*.

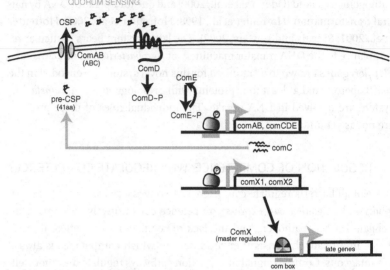

Figure 7.2. Competence regulation in *Streptococcus pneumoniae*. The quorum-sensing pheromone initiates a phosphorylation cascade that results in synthesis of ComX, the competence sigma factor. This in turn results in transcriptional activation of the late competence genes.

COMPETENCE REGULATION IN *S. PNEUMONIAE*

The capacity for DNA uptake develops suddenly and simultaneously in all cells of an exponentially growing culture of *S. pneumoniae*, is maintained for about 15 to 30 minutes, and then decays nearly as rapidly as it arose. The initiation of a cycle of competence induction has been found to be dependent on quorum sensing, mediated by a small peptide, termed *competence-stimulating peptide* (CSP). Quorum sensing apparently provides the input responsible for the abrupt and simultaneous induction of competence in the entire culture. The regulation of competence in *S. pneumoniae* is summarized in Fig. 7.2.

Upon the addition of CSP, the transcription of early competence genes reaches a maximum after 5 minutes and transcription of late genes after 10 minutes (Alloing et al., 1998). This was confirmed in two independent studies using global transcriptional profiling (Peterson et al., 2000; Rimini et al., 2000). DNA microarray analysis, as well as older studies, revealed that there are three regulatory classes of competence genes. The first group contains constitutively expressed competence genes, not affected by the addition of CSP (e.g., *endA*). Group 2 contains quorum-sensing genes with maximal expression at 5 to 10 minutes, whereas group 3 contains late competence genes with maximal expression after 10 minutes. The transcription of the early genes shuts down after 15 minutes, whereas transcription of the late genes ceases by 20 to 30 minutes.

Regulatory proteins (quorum sensing)

A protease-sensitive intercellular signal was shown to coordinate the transformation process (Tomasz and Hotchkiss, 1964). Thirty years later, this signaling molecule (CSP) was identified as a small unmodified peptide of 17 residues (Håvarstein et al., 1995). CSP is synthesized as a 41 amino-acid precursor peptide, encoded by *comC*, containing an N-terminal double-glycine type leader. ComC is processed and exported by an ABC transporter specified by *comA* and *comB* (Hamoen et al., 1995; Hui and Morrison, 1991). Induction of competence by CSP depends on ComD and ComE, members of a two-component regulatory system encoded by genes adjacent to *comC* (Pestova et al., 1996). The histidine kinase ComD is the receptor for CSP binding (Håvarstein et al., 1996). A mutation of *comD* (D299N), located in the cytoplasmic carboxyl terminal domain, activates ComD, making it independent of signals from the transmembrane domain (Lacks and Greenberg, 2001). Activated ComD presumably donates a phosphoryl group to its cognate response regulator, ComE. ComE~P binds to direct repeats adjacent to the promoters of the *comCDE* and *comAB* operons (Ween et al., 1999), creating an autoregulatory loop that augments the synthesis of CSP and the signal transduction proteins. This provides an amplification of the signal response and may be responsible for the dramatic onset of competence in the entire culture (Alloing et al., 1998; Pestova et al., 1996; Tomasz and Mosser, 1966). In addition to pheromone autoinduction, a metabolic control of the *comAB* and *comCDE* operons is apparently mediated directly or indirectly by another response regulator CiaR (Claverys et al., 2000; Guenzi et al., 1994), suggesting that the system is more complex than the account just presented.

ComX: the link between early and late competence genes

Late competence genes (those mediating the binding and uptake of transforming DNA) share an unusual promoter consensus sequence termed either a *com box* or a *cin box*, which is absent from the promoters of *comAB* and *comCDE* (Campbell et al., 1998; Pestova and Morrison, 1998). This suggested that a factor regulating transcription of late competence genes might be an alternative sigma factor. A search for such a global transcriptional regulator identified ComX as a molecule that co-purified with the pneumococcal RNA polymerase holoenzyme (Lee and Morrison, 1999). Expression of late competence genes, but not that of the quorum-sensing operons *comAB* and *comCDE*, depends on *comX*, whereas *comX* expression requires ComE but not ComX itself. Therefore, ComX was proposed to be a competence-specific transcription factor, which links quorum-sensing information to competence. ComX is encoded in two copies (*comX1* and *comX2*), although normal levels of competence can be achieved when only one copy is present (Lee and Morrison, 1999). This model, based on *in vivo* experiments, has received important confirmation from a rigorous demonstration that core RNA polymerase, supplemented with ComX protein, transcribes a number of late competence promoters *in vitro* (Luo and Morrison, 2003). The cause of the shutoff that rapidly reverses the induction of competence genes is not understood, although *comX* has been implicated in this control (Peterson et al., 2000). The marked instability of ComX certainly facilitates this rapid shutoff (Luo and Morrison, 2003). Although many late competence genes have been identified and their roles in DNA uptake surmised or established by experiment, the competence regulon appears to be more extensive than previously suspected because transcriptional profiling has identified about 30 new candidate genes under competence control (Peterson et al., 2000; Rimini et al., 2000).

REGULATION OF COMPETENCE IN *B. SUBTILIS*

As in *S. pneumoniae*, competence in *B. subtilis* is highly regulated, and is precisely controlled with respect to cell density and nutritional status. Only ~10% to 20% of the cells in a given culture reach the competent state. It is hypothesized that this represents a specialization of function, allowing subpopulations to explore different survival strategies in the stationary phase: competence, sporulation, motility, and so on.

It is useful to consider the regulation of competence as consisting of three modules (Fig. 7.3). Module 1 integrates quorum-sensing signals and nutritional status to produce an output molecule, the small protein ComS (46 residues; D'Souza et al., 1994; Hamoen et al., 1995). ComS is the input for module 2, which regulates the stability of the master regulator of competence, ComK. ComK is a transcription factor that is necessary and sufficient for the expression of late competence genes (Hahn et al., 1994, 1996; van Sinderen and Venema, 1994; van Sinderen et al., 1994, 1995). Module 3 consists of proteins that act at the promoter of *comK*. These include ComK itself, which is positively autoregulatory (van Sinderen and Venema, 1994). We describe how these three modules operate to ensure the explosive synthesis of ComK in the subpopulation of cells that become competent.

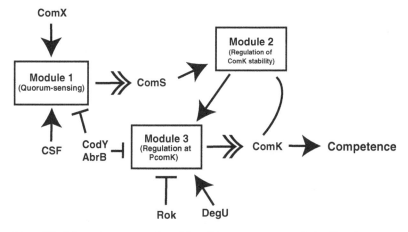

Figure 7.3. Schematic representation of ComK (competence transcription factor) regulation in *Bacillus subtilis*. The three modules described in the text are indicated. The principle positive inputs are indicated by solid arrowheads and the output molecules by double arrowheads. Lines ending in perpendiculars indicate negative inputs.

Module 1: quorum sensing

Module 1 is summarized in Fig. 7.4. *B. subtilis* uses at least two extra-cellular peptides, the ComX pheromone (Magnuson et al., 1994) and the competence and sporulation factor (CSF; Solomon et al., 1996), as inputs to

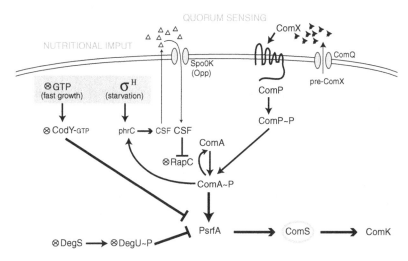

Figure 7.4. Module 1, the quorum-sensing mechanism, is diagrammed. Lines ending with perpendiculars indicate negative effects. Negative regulators are marked (⊗).

signal the onset of competence development. (It is unfortunate that ComX is the name for both the competence sigma factor of *S. pneumoniae* and a *Bacillus* competence pheromone). ComX pheromone and CSF act through convergent signaling pathways to activate the transcription of quorum-responsive genes, whose expression is therefore regulated in a density-dependent manner (Lazazzera et al., 1999b; Solomon et al., 1995). The two signaling pathways converge to increase the concentration of ComA~P. ComA is a response regulator protein that, when phosphorylated, drives expression of *comS*. ComS is encoded in the *srfA* operon, which also specifies proteins required for the synthesis of surfactin (D'Souza et al., 1994; Hamoen et al., 1995). ComA~P binds to ComA boxes upstream from the promoter of *srfA* (Nakano et al., 1991; Roggiani and Dubnau, 1993).

Why should competence in *B. subtilis* require two different extracellular signaling factors (reviewed in Lazazzera, 2000)? In the absence of CSF, ComA-regulated genes are still controlled by cell density through the ComX pheromone, although they are expressed at a lower level (Solomon et al., 1996). In fact, elimination of ComX decreases transformability about 1000-fold, whereas CSF is responsible for only a 3- to 4-fold stimulation. It has been proposed that CSF modulates the quorum response in response to nutrient limitation (Lazazzera and Grossman, 1998; Lazazzera et al., 1999b). Two promoters control *phrC*, the gene encoding CSF (Lazazzera et al., 1999a). P1 precedes the *rapC-phrC* operon and is induced by ComA~P. P2 precedes *phrC* and is controlled by an alternative sigma factor (sigma H), which is maximally active during the transition from exponential to stationary phase. The levels and activity of sigma H are stimulated by starvation (Healy et al., 1991). As a result, genes regulated by ComA are induced at a lower cell density under conditions of starvation. ComX appears to be produced at a constant rate during growth, and its accumulation in the cell medium therefore reflects cell density (Bacon Schneider et al., 2002). It is reasonable then, to conclude that ComX monitors cell density, whereas CSF serves to modulate *srfA* (*comS*) transcription in response to nutritional status.

Competence and sporulation factor

Competence and sporulation factor (CSF) was purified from a cell-free culture supernatant based on its ability to stimulate transcription from *srfA* during early logarithmic growth (Solomon et al., 1996). CSF is an unmodified pentapeptide (ERGMT) that corresponds to the C-terminal 5 amino acids of the 40-amino acid polypeptide encoded by *phrC*. CSF exerts at least three activities on competence and sporulation. By midexponential phase, CSF accumulates to concentrations (1 to 5 nM) that stimulate ComA-dependent gene expression, possibly by inhibiting activity of the RapC phosphatase, which is hypothesized to act on ComA~P (Perego, 1999). However, this has not yet been verified experimentally, and RapC may act in some other way. Upon entry into stationary phase, CSF reaches a concentration of 50 to 100 nM (due to sigma H activation at P2; McQuade et al., 2001). This excess CSF stimulates sporulation and inhibits ComA-dependent gene expression (including that of *comS*, probably by affecting the level of Spo0A~P; Lazazzera et al., 1999b; Solomon et al., 1996). By changing each residue of the CSF pentapeptide to alanine, it was shown that different residues are required for each of these three activities of CSF (Lazazzera et al., 1997). CSF pheromone acts intracellularly and is imported by the oligopeptide permease Spo0K, a member of the ATP-binding cassette (ABC) transporter family (Lazazzera et al., 1997; Perego et al., 1991; Rudner et al., 1991; Solomon et al., 1996).

ComQ and ComX

When produced by the commonly used laboratory strain *B. subtilis* 168, the ComX pheromone is a small peptide (9 or 10 residues) with an iso-prenyl modification on a tryptophan residue (Ansaldi et al., 2002; Magnuson et al., 1994). The peptide backbone corresponds to the C-terminal part of a longer precursor (55 amino acids in *B. subtilis* 168) from which it is cleaved. The pheromone precursor has neither a typical leader sequence for SecA-dependent secretion nor the double glycine motif found in the *S. pneumoniae* pheromone precursors, and the mechanism of its export is unknown. Both the cleavage of the ComX precursor and its isoprenylation are probably accomplished by ComQ or by ComQ together with pre-ComX (Weinrauch et al., 1991), because when expressed in *E. coli*, *comX* and *comQ* are sufficient to produce correctly processed and modified active pheromone (Tortosa et al., 2000). ComQ contains motifs found in isoprenyl transferases and mutations in this domain eliminate production of ComX pheromone (Bacon Schneider et al., 2002). An unmodified synthetic peptide with the sequence of ComX has no activity, nor does it inhibit the activity of the native ComX, proving that the isoprenylation is essential for function (Magnuson et al., 1994). As expected, mutation of the tryptophan residue eliminates function (Bacon Schneider et al., 2002).

The ComX pheromones from a number of natural isolates of *B. subtilis* and the closely related species *B. natto* and *B. mojavensis*, vary with respect to peptide length and amino acid sequence (Ansaldi et al., 2002; Tortosa et al., 2000; Tran et al., 2000). Characterization of the masses of the isoprenyl mod-ification on ComX molecules from different natural isolates has shown that at least three distinct modifications exist. Although these modifications are usually the same within a given phenotype specificity group, and always occur on a conserved tryptophan residue, in at least one case, members of the same group have differing modifications (Ansaldi et al., 2002).

The variations in peptide sequence and in the posttranslational modification determine the specificity of the quorum-sensing responses, although the precise contributions of each factor to phenotype specificity remain to be determined.

The ComP/ComA two-component signal transduction system

ComX activates the histidine kinase ComP (Weinrauch et al., 1990), presumably leading to ComP autophosphorylation and the subsequent transfer of its phosphoryl group to the response regulator ComA. ComP contains six or eight membrane-spanning segments and two large extracytoplasmic loops in its N-terminus (Piazza et al., 1999). Deletion of the second loop conferred a phenotype in which ComP was constitutively active in the absence of ComX (Piazza et al., 1999). It is likely that this loop imposes a conformation that inhibits autophosphorylation and that interaction with ComX reverses this inhibition. It is not known how the ComX interaction signal is transmitted across the membrane.

The variation in *comX* sequence among natural isolates discussed previously is accompanied by an extensive polymorphism extending through *comQ*, *comX*, and the 5′ two-thirds of *comP*. ComX response specificities among these natural isolates are reflected by the differences and similarities among the ComQ, ComX, and ComP sequences, providing strong evidence that the pheromone contacts ComP directly (Ansaldi et al., 2002; Tortosa et al., 2000; Tran et al., 2000). Trees exemplifying the relatedness of the ComX, ComP, and ComQ sequences from the various natural isolates are congruent, demonstrating that the genes encoding these three proteins have co-evolved. The sequences fall into four groups, which correspond to pheromone specificity response groups (phenotype specificities). Among the transformable streptococci, quorum-sensing polymorphisms and phenotype specificity groups have also been described (Håvarstein et al., 1997; Håvarstein and Morrison, 1999; Pozzi et al., 1996). The selective forces that generate these parallel patterns of polymorphism and response specificities in different transformable species are not understood. However, it is remarkable that the evolution of phenotype specificities among the bacilli does not correspond to that of housekeeping genes (*rpoB* and *gyrA*; Ansaldi et al., 2002; Tortosa et al., 2000). These polymorphisms are suggestive of negative, frequency-dependent selection, in which an advantage accrues to a rare genotype. This advantage leads to an increase in the frequency of the rare genotype until it becomes common, at which point a new (rare) genotype may emerge that thereby enjoys increased fitness. This type of situation leads to polymorphism in natural populations. What is unknown in the case of competence is the nature of the advantage conferred by being uncommon. Because a new pheromone specificity will enable a bacterial clone to become competent only in response to its own pheromone, and not that of the most prevalent pheromone, the rare pheromone may offer an opportunity to explore a unique niche with reduced competition.

Regulation on the level of the srfA promoter

ComS, the output of module 1, is subjected to extensive and complex regulation, which is not completely understood (Fig. 7.4). In addition to the quorum-sensing system just described, other inputs serve to regulate *srfA* transcription, and hence, the synthesis of ComS. *srfA* is transcribed by the major sigma factor sigma A, and its transcription is negatively regulated by the phosphorylated response regulator DegU~P, by the transition state regulator AbrB, and by CodY.

CodY binds directly to the *srfA* promoter and inhibits transcription (Serror and Sonenshein, 1996) in response to a high concentration of GTP. Repression by CodY therefore monitors nutritional status, as reflected by the size of the GTP pool.

The second regulator, DegU, is similar in its sequence to ComA, and DegU~P may therefore inhibit *srfA* transcription by competing with ComA~P for binding to *PsrfA* (Hahn and Dubnau, 1991). The factors that control the level of DegU phosphorylation are poorly understood.

SpoOA~P represses the transcription of the third regulator, *abrB* (Perego et al., 1988; Strauch et al., 1990). SpoOA is a master regulator of many stationary-phase phenomena and is itself subject to multiple inputs (reviewed in Burkholder and Grossman, 2000).

SinR (Liu et al., 1996) and polynucleotide phosphorylase (PnpA ComR; Luttinger et al., 1996) act positively on ComS synthesis, and both appear to act posttranscriptionally. Evidently, the complex regulation of ComS synthesis serves as an important point of integration of multiple signals at the output from module 1.

Module 2: regulation of ComK stability

ComS is the input to module 2 (Figs. 7.3 and 7.5), which consists of three additional proteins; MecA, ClpC, and ClpP. *mecA* and *clpC* (originally

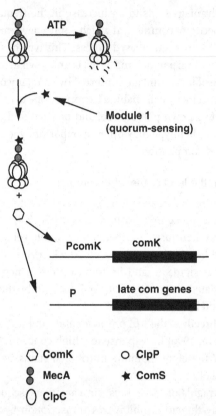

IRENA DRASKOVIC AND DAVID DUBNAU

Figure 7.5. Module 2, the MecA switch, regulates the stability of ComK. ComK is targeted by MecA for degradation by ClpC/ClpP until ComS, the product of module 1, causes its release from binding to MecA. The binding stoichiometries are not known, nor is the exact subunit composition of the ClpC/ClpP complex.

mecB) were discovered in a screen for medium independent expression of competence (*mec* mutants; Dubnau and Roggiani, 1990). Loss of function mutations of either *mecA* or *clpC* result in a dramatic increase in ComK synthesis and bypass the requirements for the module 1 genes (e.g., *comA*, *comP*, and *comQ*), as well as for DegU (Roggiani et al., 1990). However, ComK is still required for transcription from its own promoter in *mecA* or *clpC* backgrounds (Dubnau and Roggiani, 1990; Hahn et al., 1994; Roggiani et al., 1990). It has been shown that MecA negatively regulates competence by sequestering ComK (Kong et al., 1993; Kong and Dubnau, 1994) and by targeting it for proteolysis by ClpC/ClpP (Turgay et al., 1997, 1998) as described in the following section on the switch model. In addition to its role in competence,

Clpc was shown to be essential for growth at high temperature, as expected for a protein belonging to the HSP100 family of heat shock proteins (Msadek et al., 1994) and for sporulation (Pan et al., 2001). These roles of ClpC were partially separated by individual point mutations in the Walker A motifs of the two ATP-binding domains of this protein, suggesting that the role of ClpC in heat shock and spore formation may be mechanistically distinct from its role in regulating ComK stability (Turgay et al., 2001). It is pertinent in this regard to note that ClpC plays a role not only in protein turnover, but also as a chaperone (Schlothauer et al., 2003).

The switch model

Meca is a two-domain protein (Persuh et al., 1999). Its N-terminal domain has affinity for ComK and for ComS, whereas the C-terminal domain binds to ClpC. ClpC is an ATPase and binds to ClpP to form an ATP-activated protease. Like other members of its family of proteins (Neuwald et al., 1999; Schirmer et al., 1996), ClpC forms a multimeric complex (probably a hexamer) in the presence of a nucleotide (Persuh and Dubnau, unpublished data, 2001). *In vivo* experiments, as well as *in vitro* work using purified proteins, have shown that MecA targets ComK for degradation by ClpC/ClpP, in the presence of ATP (Turgay et al., 1997). As is the case with ClpA and ClpX in other systems (Horwich et al., 1999), it is likely that ClpC acts to unfold ComK and then translocates the unfolded protein into the lumen of the ClpC/ClpP protease, where it is degraded. The addition of MecA and ComK to ClpC/ClpP *in vitro* stimulates the ATPase activity of ClpC, suggesting that the unfolding and translocation reactions are probably powered by ATP hydrolysis. The MecA/ComK complex dissociates when ComS is added, liberating ComK to activate its own transcription and that of the downstream late competence genes (Turgay et al., 1997, 1998). As noted previously, ComS is produced in response to the accumulation of CSP and the ComX pheromone, and increased population density therefore results in the stabilization of ComK. The mechanism of ComS-induced dissociation of ComK from the quarternary complex of ComK/MecA/ClpC/ClpP is unclear. MecA binds to ComK in the absence of ClpC and ClpP, and ComS can cause dissociation from this binary complex *in vitro* (Persuh and Dubnau, unpublished data, 2000). It is not known whether ComS causes the dissociation of the complex by competition for a common ComS/ComK-binding site on MecA or by inducing a conformational change in MecA that reduces its affinity for ComK. However, this release is likely to be more complex than a simple competition for an identical common binding site because a mutation (W43A) near the C-terminus of ComS inactivates the *in vivo* activity of ComS, but does not appear to decrease its affinity for MecA binding (Ogura et al., 1999). When ComK is released from the quarternary complex, MecA and ComS are degraded instead (Turgay et al., 1998). MecA degradation is expected to amplify the activity of ComK, both because in its absence ComK will not be targeted for degradation, and because ComK can be sequestered by MecA binding and prevented from interaction with DNA, even in the absence of ClpC and ClpP (Kong and Dubnau, 1994). The degradation of ComS may provide a timing mechanism, facilitating the escape from competence.

Spx, another highly conserved protein among G+ bacteria, that may function together with the adaptor protein MecA, was discovered in a suppressor screen for mutations that bypass the *clpX* requirement for *comK* expression, but still require ComS for this expression (Nakano et al., 2001). Spx, expressed from a gene just upstream of *mecA*, forms a complex with ClpC-MecA-ComK and enhances ComK binding to ClpC-MecA (Nakano et al., 2002a). However, because elimination of *spx* by mutation has little effect on the development of competence (Nakano et al., 2002a) its role in the competence regulatory pathway is unclear, although it most likely does function as part of the heat shock response mechanism.

Because ComK is positively autoregulated, its protection from degradation results in an explosive increase in ComK synthesis. In fact there are about 100,000 ComK monomers per competent cell, a truly astounding concentration for a transcriptional regulator (Turgay et al., 1998). The reason for this high concentration is not known.

The MecA family of adapter molecules

In this scheme, MecA acts as an adapter protein; its N-terminal domain binds ComK and its C-terminal domain binds to ClpC, thereby targeting ComK for degradation by ClpC/ClpP. Does MecA perform similarly with other substrates? Clearly, MecA binds to ComS and targets this small protein for degradation. In addition, it probably performs the same task with Spx (Nakano et al., 2002b). *mecA* loss-of-function mutants form noticeably rough colonies on plates, and this phenotype is not dependent on the overproduction of ComK. Another indication of MecA-related pleiotropy is obtained by transcriptional profiling; many genes are up- or downregulated in the absence of MecA, independently of ComK (Albano and Dubnau, unpublished data, 1999). It remains to be seen whether MecA targets additional regulatory genes for degradation by ClpC/ClpP or whether these pleiotropic effects are due to some other mechanism. In this connection, it is noteworthy that MecA orthologs are present in all the low GC G+ bacteria so far sequenced, even in several that are not known to be competent and that lack *comK* genes (Persuh

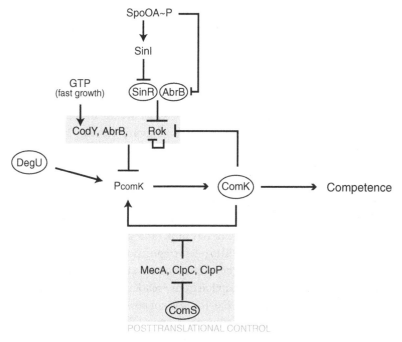

Figure 7.6. Module 3, regulation at the *comK* promoter. The autorepression of Rok and the action of ComK at the *rok* promoter (Hoa et al., 2002) are not discussed in the text. Lines ending in perpendiculars indicate negative inputs. Positive regulators are circled.

et al., 1999). The genome of *B. subtilis* contains a paralog of *mecA*, *ypbH*. Unlike *mecA*, which is constitutively expressed (Kong et al., 1993), ypbH is induced in stationary phase (Persuh et al., 2002). YpbH, like MecA, binds ClpC and its elimination or overproduction has modulating effects on competence and sporulation (Persuh et al., 2002).

Module 3: regulation at PcomK

Module 3 is represented in Fig. 7.6. ComK appears to bind in the minor groove of DNA (Hamoen et al., 1998). The recognition sequences for ComK binding to DNA contain two so-called ComK boxes, each with the consensus AAAANNNNNTTTT. The ComK boxes in various promoters are separated from one another by either one, two, or three turns of the DNA helix. *PcomK* is the only promoter known with a separation of three helical turns, and this may reflect the complexity of regulation at *PcomK* because four proteins, in addition to ComK itself, act at this promoter. The response regulator DegU binds between the ComK boxes and acts positively, apparently by increasing the affinity of ComK for *PcomK* (Hamoen et al., 2000). It was proposed that DegU acts as a priming protein, allowing ComK to act on *PcomK* early in competence development, when the concentration of ComK is low. It has also been proposed that only unphosphorylated DegU has this activity because its cognate histidine kinase (DegS) is not needed for competence (Dahl et al., 1992; Msadek et al., 1995). This has not been tested *in vitro*. In addition to the positively acting ComK and DegU proteins, three repressors (CodY, AbrB, and Rok) act at *PcomK* (Hamoen et al., 2003; Hoa et al., 2002; Serror and Sonenshein, 1996). CodY and AbrB have been footprinted to *PcomK* using DNAse, and like DegU, protect sequences that overlap the ComK-binding site (Hamoen et al., 2003; Serror and Sonenshein, 1996). It is not clear why ComK is needed for transcription from *PcomK*. This promoter is active *in vitro* and also *in vivo* in *E. coli* in the absence of ComK, and it is therefore likely that ComK is needed *in vivo* in *B. subtilis* to relieve repression due to AbrB, Rok, and CodY. Consistent with this is the observation that in a *codY rok* double mutant, the *comK* requirement for transcription from *PcomK* is partially relieved (Hoa and Dubnau, unpublished data, 2001). It was not possible to construct a *codY rok abrB* triple mutant. Strong additional support for this hypothesis comes from the report that ComK also antagonizes LexA repression at the *B. subtilis recA* promoter (Hamoen et al., 2001). Remarkably, LexA and ComK can bind simultaneously to *PrecA*. ComK not only overcomes repression at *PcomK*, but functions as a direct transcriptional activator for genes encoding the DNA uptake machinery (Hamoen et al., 1998; van Sinderen et al., 1995). This has been confirmed *in vitro* using *PcomG* (Hamoen et al., 1998). The regulatory complexity at *PcomK* (module 3) probably serves to integrate informational input from several signal transduction pathways.

The escape from competence

All global regulatory systems require mechanisms for escape from the induced state. Sporulation and SOS induction provide two well-known examples of this requirement. The escape from competence uses the MecA/ClpC/ClpP system to degrade ComK. Upon dilution into fresh medium, ComK is degraded over a period of 2 hours. This degradation requires MecA and ClpC, and because competent cells are growth inhibited, loss-of-function mutations in these two genes prevent the resumption of growth after competence (Hahn et al., 1995). Probably the initiation of ComK degradation is due in part to the degradation of ComS. In the absence of ComS, MecA binds to ComK, targeting it for degradation. Also, dilution into fresh medium lowers the concentration of pheromones and reverses the starvation control that acts at several points in the competence regulatory network (e.g., via CodY, Spo0A). Because only 10% to 20% of the cells in a culture express competence, only these cells are growth inhibited. Consequently, when a competent culture is diluted into fresh medium, the majority subpopulation resumes growth within a few minutes, whereas growth in the competence-expressing cells is delayed for about 2 hours, the time it takes to remove ComK. However, ComK is not the only player in this escape scenario. *comGA*, the first gene in the *comG* operon, is uniquely required for the competence-associated growth arrest (Haijema et al., 2001). ComGA resembles traffic ATPases and also has a role in the assembly of the DNA-binding/uptake apparatus (Chen and Dubnau, unpublished data, 2001). A loss-of-function *comGA* mutant resumes growth immediately after dilution into fresh medium, and the resulting cells are slightly filamented and appear to eventually lyse (Haijema et al., 2001). Thus, the ComGA-associated growth arrest comprises a checkpoint to prevent the premature resumption of growth. When ComGA is expressed from an IPTG-regulated promoter during exponential growth, it has no effect, thus demonstrating that its growth inhibiting function is manifested only in the context of competence (Hahn and Dubnau, unpublished data, 2001). A point mutation in the Walker A motif of *comGA* eliminates its ability to function in transformation, but has no effect on its ability to suppress growth in the context of competence. This mutation separates the two functions of ComGA (Haijema et al., 2001).

Why do only 10% to 20% of the cells express competence?

The cells in a well-stirred competent culture, initiated from a single colony, should be genetically and environmentally identical. Thus, the choice of which cells will express *comK* must reflect a stochastic process. It is certainly possible that a deterministic regulatory mechanism is imposed on this underlying stochastic process to ensure the fraction of cells that accumulate ComK does not increase beyond some limit. However, the choice of *which* cells become competent must be stochastic.

The system is delicately poised in the sense that any mutation that increases ComK synthesis in all the cells also results in all or nearly all the cells expressing competence. Thus, mutational inactivation of CodY, Rok, MecA, or ClpC, as well as overexpression of ComS, all result in the entire population expressing competence genes. This suggests that stochastic fluctuations in any of these factors may determine which individual cells will become competent. For instance, although ComS is made in all the cells (Hahn et al., 1994), it is possible that the cells at the upper range of *comS* expression are the ones that are "chosen." The MecA switch and the autoregulatory expression of *comK* may therefore act as an analog-to-digital converter, transducing a graded response of ComS synthesis to pheromone concentration, into a nearly all-or-nothing ComK output. In fact, the use of *gfp* fusions to P*comK* or to late competence gene promoters, reveals that the distribution of fluorescence in the population is indeed nearly all or nothing; 10% to 20% of the cells fluoresce brightly, whereas the rest are dark (Haijema et al., 2001). Because ComS overproduction from a multicopy plasmid results in competence expression in all the cells, something must normally limit ComS accumulation. This limited production is probably not a reflection of a paucity of pheromone production because strains that overproduce pheromones do not exhibit hypercompetence. Perhaps the *comS* synthetic capacity of the cell is limited by its *cis*-regulatory translational and transcriptional sequences and by the stability of ComS. Whatever the mechanism, the balance of ComS synthesis and degradation may be such that the distribution of cellular ComS concentrations in the population allows only 10% to 20% of the cells to exceed a critical threshold, sufficient to throw the MecA switch.

Competence development in the *B. subtilis* system possesses two characteristics that have been shown by theoretical and experimental work to be crucial for systems to exhibit "bistability," the ability to stably exist in alternative states (see, for instance, Becskei et al., 2001, and Ferrell, 2002). The first characteristic is positive autoregulation, which is satisfied because ComK, as noted previously, is required for its own synthesis. The second is cooperativity. In fact, ComK binds to DNA as a tetramer (or more likely as a dimer of dimers) and is therefore expected to act with a high degree of cooperativity (Hamoen et al., 1998). Further exploration of these concepts will require a systematic and quantitative exploration of the responses of ComK synthesis to varying concentrations of each of the regulatory inputs in both mutant and wild-type cells.

The K state

ComK regulates DNA repair genes (Love et al., 1985) and motility genes (Liu and Zuber, 1998), in addition to those needed for transformation. More recently, global gene analysis (Berka et al., 2002; Hamoen et al., 2002; Ogura et al., 2002) confirmed the previously described roles of ComK and further extended the ComK regulon to genes involved in intermediary metabolism, antibiotic production, cell shape, cell division, and cell wall synthesis. A role for ComK in the regulation of cell division is consistent with the observation that competent cells are arrested in cell division and growth (Haijema et al., 2001). Competence, strictly defined, involves the induction of genes required for transformation. Because ComK seems to regulate many genes with no obvious roles in transformability, it was proposed that ComK is the master regulator for a "K state" (Berka et al., 2002) of which competence is but one feature. The K state can be viewed as a global adaptation to stress, which enables the cell to repair DNA damage, to use novel substrates, and to acquire new fitness-enhancing genes by transformation.

LOGIC OF COMPETENCE REGULATION IN *B. SUBTILIS* AND *S. PNEUMONIAE*

The mechanisms of competence regulation differ in these two organisms, and it is instructive to compare the logic of their respective control circuits. In both organisms, quorum-sensing serves to coordinate the onset of competence with population density. In both organisms, the quorum-sensing mechanism drives the synthesis of a competence transcription factor. In *S. pneumoniae*, all the cells in the population achieve the competent state, and accumulation of the competence pheromone probably has much to do with the abrupt and synchronous wave of competence expression that sweeps through the population. This explosive expression apparently relies on the positive autoregulation of pheromone production; the pheromone secretion and synthesis genes are induced by pheromones. In *B. subtilis*, only 10% to 20% of the cells in a culture express competence genes. As noted previously, this limited expression is probably not a reflection of a paucity of pheromones. In *B. subtilis*, as in *S. pneumoniae*, the expression of competence is explosive and positively autoregulated. However, in *B. subtilis* the autoregulatory step involves intracellular synthesis of the transcription factor ComK, rather than pheromone production. This is consistent with the expression of competence in only a fraction of the population. Finally, both organisms possess mechanisms for the escape from competence. In *B. subtilis*, escape relies on the regulated degradation of ComK. In *S. pneumoniae*, it has more recently been reported that the transcription factor COM is unstable, hinting at the existence of a comparable mechanism (Luo and Morrison, 2003).

COMPETENCE REGULATION IN *H. INFLUENZAE*

In *H. influenzae*, competence is low during exponential growth in rich medium and develops spontaneously at the onset of stationary phase, or when cells are transferred to nutrient-limiting conditions (Herriott et al.,

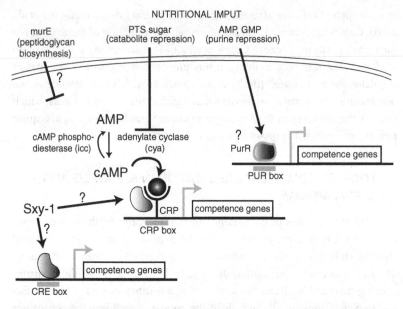

Figure 7.7. Regulation of competence in *Haemophilus influenzae*. Two possible models for the action of Sxy-1 are indicated as described in the text. It has not been proven that the purine repression of competence genes is mediated by PurR acting at PUR boxes. Lines ending in perpendiculars indicate negative inputs.

1970). The expression of competence upon entry into stationary phase is low compared with that achieved following starvation. Competence can also develop following transient exposure to anerobiosis. 3′,5′ cyclic AMP (cAMP) and cAMP receptor protein are absolutely required for the development of competence, and are believed to be primary regulatory factors (Chandler, 1992; Dorocicz et al., 1993). Additional regulator mechanisms were suggested to exist because the addition of cAMP at stationary phase does not further increase the low level of competence in such cultures. The regulation of competence in *H. influenzae* is diagrammed in Fig. 7.7.

Competence and cAMP

The genes encoding adenylate cyclase (*cya*) and cAMP receptor protein (*crp*) are required for competence development (Chandler, 1992; Dorocicz et al., 1993). *cya* or *crp* mutant strains fail to develop competence upon reaching stationary phase or after transfer to a starvation medium. However, a *cya* mutant becomes competent in the presence of exogenously added cAMP

(Dorocicz et al., 1993). Factors that regulate the synthesis of cAMP have an impact on competence. In *Haemophilus* as in *E. coli* (Postma et al., 1993), the phosphoenolpyruvate phosphotransferase (PTS) sugar-import system regulates cAMP synthesis. Components of the PTS system of *Haemophilus* are essential for competence development. Thus, strains with loss-of-function mutations in *crr* (which encodes EIIA) and in *ptsI* (which encodes EI) are markedly lowered in competence in response to both starvation and anerobiosis (Gwinn et al., 1996; Macfadyen et al., 1996).

Sxy-1 (TfoX)

A positive regulator of competence was discovered by two independent approaches. A point mutation was found in *sxy-1*, which increased competence 100- to 1000-fold, whereas a deletion of *sxy-1* resulted in competence deficiency. It was concluded that this gene acts as a positive regulator (Lovett et al., 1994; Redfield, 1991). By sequencing the region upstream of a late competence gene *rec-1*, a divergently transcribed ORF named *tfoX* was identified that is required for transformation (Zulty and Barcak, 1995). These two genes are the same, and in the following discussion, we use the term *sxy-1* for convenience. The expression of plasmid-borne *sxy-1* resulted in constitutively expressed competence (Lovett et al., 1994; Zulty and Barcak, 1995). The transcription of *sxy-1* increased modestly after the addition of cAMP, suggesting that *sxy-1* is itself an early competence gene (Zulty and Barcak, 1995).

It has been proposed that Sxy-1 works directly on the promoters of late competence genes as a positive transcription factor (Karudapuram and Barcak, 1997). In this cascade model, cAMP/Crp would act to stimulate the transcription of *sxy-1*, and Sxy-1 would then activate the late genes. However, a bioinformatics approach has suggested that the proposed Sxy-1-binding sites detected upstream from competence genes are in fact Crp-binding sites (Macfadyen, 2000). It was proposed, therefore, that Sxy-1 is not an independent transcriptional activator of the late competence genes but rather functions together with the cAMP–CRP complex to activate these genes.

Purine repression

It was shown more recently that competence development in *H. influenzae* is regulated by the availability of nucleic acid precursors (MacFadyen et al., 2001). The presence of excess AMP or GMP, but not of dAMP or dGMP reduced competence 200-fold. Potential binding sites for the PurR repressor were detected upstream of one or two competence genes. This is consistent with a model in which the depletion of purine pools signals the need for nucleotides. These observations have been suggested as support for the hypothesis that competence evolved for nucleotide acquisition (DNA as food). It is equally consistent with the idea that depletion of the nucleotide pool is a signal of generally stressful conditions, which threaten DNA damage or advertise the advantage of newly acquired genetic information. It is noteworthy that in *B. subtilis*, depletion of the GTP pool is a signal for both the development of competence and for sporulation (Dworkin and Losick, 2001; Ratnayake-Lecamwasam et al., 2001) and may therefore be a general signal of stress.

Role of peptidoglycan in competence signaling

Point mutations in a peptidoglycan biosynthesis gene induce competence in *H. influenzae* (Ma and Redfield, 2000). Point mutations in *murE*, which encodes the mesodiaminopimelate-adding enzyme of peptidoglycan synthesis, exhibit hypercompetence at all stages of growth. These mutations do not act by increasing the entry of DNA but by causing induction of the competence pathway. It is possible that the state of structural components of the cell surface may function as an important signal for competence development. As noted previously, in *S. pneumoniae* the CiaH/CiaR two-component system regulates competence, probably by affecting the quorum-sensing system. More recently, it has been shown that these proteins also control the expression of genes required for the expression of cell wall polymers (Mascher et al., 2003). It is therefore possible that in both *S. pneumoniae* and *H. influenzae* a cell-surface signal regulates the initiation of competence development.

ACKNOWLEDGMENT

We thank members of our lab for useful discussions. The work quoted from our lab was supported by National Institutes of Health grants GM57720 and GM43756.

REFERENCES

Albano, M., Breitling, R., and Dubnau, D.A. (1989) Nucleotide sequence and genetic organization of the *Bacillus subtilis* comG operon. *J Bacteriol* 171: 5386–5404.

Albano, M., and Dubnau, D.A. (1989) Cloning and characterization of a cluster of linked *Bacillus subtilis* late competence mutations. *J Bacteriol* 171: 5376–5385.

Albano, M., and Dubnau, D. (1999) Unpublished data.

Alloing, G., Martin, B., Granadel, C., and Claverys, J.P. (1998) Development of competence in *Streptococcus pneumoniae*: Pheromone autoinduction and control of quorum sensing by the oligopeptide permease. *Mol Microbiol* 29: 75–83.

Ansaldi, M., Marolt, D., Stebe, T., Mandic-Mulec, I., and Dubnau, D. (2002) Specific activation of the *Bacillus* quorum-sensing systems by isoprenylated pheromone variants. *Mol Microbiol* 44: 1561–1573.

Bacon Schneider, K., Palmer, T.M., and Grossman, A.D. (2002) Characterization of comQ and comX, two genes required for production of ComX pheromone in *Bacillus subtilis*. *J Bacteriol* 184: 410–419.

Becskei, A., Seraphin, B., and Serrano, L. (2001) Positive feedback in eukaryotic gene networks: Cell differentiation by graded to binary response conversion. *EMBO J* 20: 2528–2535.

Berka, R.M., Hahn, J., Albano, M., Draskovic, I., Persuh, M., Cui, X., et al. (2002) Microarray analysis of the *Bacillus subtilis* K-state: Genome-wide expression changes dependent on ComK. *Mol Microbiol* 43: 1331–1345.

Burkholder, W.F., and Grossman, A.D. (2000) Regulation of the initiation of endospore formation in *Bacillus subtilis*. In Shimkets, L.J. (ed). *Prokaryotic development*. Washington, DC: ASM Press, pp. 151–166.

Campbell, E.A., Choi, S.Y., and Masure, H.R. (1998) A competence regulon in *Streptococcus pneumoniae* revealed by genomic analysis. *Mol Microbiol* 27: 929–939.

Chandler, M.S. (1992) The gene encoding cAMP receptor protein is required for competence development in *Haemophilus influenzae* Rd. *Proc Natl Acad Sci U S A* 89: 1626–1630.

Chen, I., and Dubnau, D. (2001) Unpublished data.

Chen, I., and Dubnau, D. (2003) DNA transport during transformation. *Front Biosci* 8: 544–56.

Chen, I., and Gotschlich, E.C. (2001) ComE, a competence protein from *Neisseria gonorrhoeae* with DNA-binding activity. *J Bacteriol* 183: 3160–3168.

Claverys, J.P., Prudhomme, M., Mortier-Barriere, I., and Martin, B. (2000) Adaptation to the environment: *Streptococcus pneumoniae*, a paradigm for recombination-mediated genetic plasticity? *Mol Microbiol* 35: 251–259.

Clifton, S.W., McCarthy, D., and Roe, B.A. (1994) Sequence of the *rec-2* locus of *Hemophilus influenzae*: Homologies to *comE*-ORF3 of *Bacillus subtilis* and *msbA* of *Escherichia coli*. *Gene* 146: 95–100.

Cohan, F.M., Roberts, M.S., and King, E.C. (1991) The potential for genetic exchange by transformation within a natural population of *Bacillus subtilis*. *Evolution* 45: 1383–1421.

D'Souza, C., Nakano, M.M., and Zuber, P. (1994) Identification of *comS*, a gene of the *srfA* operon that regulates the establishment of genetic competence in *Bacillus subtilis*. *Proc Natl Acad Sci U S A* 91: 9397– 9401.

Dahl, M.K., Msadek, T., Kunst, F., and Rapoport, G. (1992) The phosphorylation state of the DegU response regulator acts as a molecular switch allowing either degradative enzyme synthesis or expression of genetic competence in *Bacillus subtilis*. *J Biol Chem* 267: 14509–14514.

Danner, D.B., Deich, R.A., Sisco, K.L., and Smith, H.O. (1980) An eleven-base-pair sequence determines the specificity of DNA uptake in *Haemophilus* transformation. *Gene* 11: 311–318.

Davies, J. (1994) Inactivation of antibiotics and the dissemination of resistance genes. *Science* 264: 375–382.

Dorocicz, I.R., Williams, P.M., and Redfield, R.J. (1993) The *Haemophilus influenzae* adenylate cyclase gene: Cloning, sequence, and essential role in competence. *J Bacteriol* 175: 7142–7149.

Drake, S.L., and Koomey, M. (1995) The product of the *pilQ* gene is essential for the biogenesis of type IV pili in *Neisseria gonorrhoeae*. *Mol Microbiol* 18: 975–986.

Drake, S.L., Sandstedt, S.A., and Koomey, M. (1997) PilP, a pilus biogenesis lipoprotein in *Neisseria gonorrhoeae*, affects expression of PilQ as a high-molecular-mass multimer. *Mol Microbiol* 23: 657–668.

Dubnau, D. (1999) DNA uptake in bacteria. *Annu Rev Microbiol* 53: 217–244.

Dubnau, D., and Lovett, C.M. Jr., C.M.L. (2001) Transformation and recombination. In Hoch, J.A., Losick, R., and Sonenshein, A.L. (eds). *Bacillus subtilis and its relatives: From genes to cells*. Washington, DC: American Society for Microbiology, pp. 453–472.

Dubnau, D., and Roggiani, M. (1990) Growth medium-independent genetic competence mutants of *Bacillus subtilis*. *J Bacteriol* 172: 4048–4055.

Dworkin, J., and Losick, R. (2001) Linking nutritional status to gene activation and development. *Genes Dev* 15: 1051–1054.

Elkins, C., Thomas, C.E., Seifert, H.S., and Sparling, P.F. (1991) Species-specific uptake of DNA by gonococci is mediated by a 10-base-pair sequence. *J Bacteriol* 173: 3911–3913.

Facius, D., and Meyer, T.F. (1993) A novel determinant (*comA*) essential for natural transformation competence in *Neisseria gonorrhoeae*. *Mol Microbiol* 10: 699–712.

Ferrell, J.E., Jr. (2002) Self-perpetuating states in signal transduction: Positive feedback, double-negative feedback and bistability. *Curr Opin Cell Biol* 14: 140–148.

Finkel, S.E., and Kolter, R. (2001) DNA as a nutrient: Novel role for bacterial competence gene homologs. *J Bacteriol* 183: 6288–6293.

Gibbs, C.P., Reimann, B.Y., Schultz, E., Kaufmann, A., Haas, R., and Meyer, T.F. (1989) Reassortment of pilin genes in *Neisseria gonorrhoeae* occurs by two distinct mechanisms. *Nature* 338: 651–652.

Guenzi, E., Gasc, A.M., Sicard, M.A., and Hakenbeck, R. (1994) A two-component signal-transducing system is involved in competence and penicillin susceptibility in laboratory mutants of *Streptococcus pneumoniae*. *Mol Microbiol* 12: 505–515.

Gwinn, M.L., Yi, D., Smith, H.O., and Tomb, J.F. (1996) Role of the two-component signal transduction and the phosphoenolpyruvate: Carbohydrate phosphotransferase systems in competence development of *Haemophilus influenzae* Rd. *J Bacteriol* 178: 6366–6368.

Hahn, J., Bylund, J., Haines, M., Higgins, M., and Dubnau, D. (1995) Inactivation of *mecA* prevents recovery from the competent state and interferes with cell division and the partitioning of nucleoids in *Bacillus subtilis*. *Mol Microbiol* 18: 755–767.

Hahn, J., and Dubnau, D. (1991) Growth stage signal transduction and the requirements for *srfA* induction in development of competence. *J Bacteriol* 173: 7275–7282.

Hahn, J., and Dubnau, D. (2001) Unpublished data.

Hahn, J., Inamine, G., Kozlov, Y., and Dubnau, D. (1993) Characterization of *comE*, a late competence operon of *Bacillus subtilis* required for the binding and uptake of transforming DNA. *Mol Microbiol* 10: 99–111.

Hahn, J., Kong, L., and Dubnau, D. (1994) The regulation of competence transcription factor synthesis constitutes a critical control point in the regulation of competence in *Bacillus subtilis*. *J Bacteriol* 176: 5753–5761.

Hahn, J., Luttinger, A., and Dubnau, D. (1996) Regulatory inputs for the synthesis of ComK, the competence transcription factor of *Bacillus subtilis*. *Mol Microbiol* 21: 763–775.

Haijema, B.J., van Sinderen, D., Winterling, K., Kooistra, J., Venema, G., and Hamoen, L.W. (1996) Regulated expression of the *dinR* and *recA* genes during competence development and SOS induction in *Bacillus subtilis*. *Mol Microbiol* 22: 75–85.

Haijema, B.J., Hahn, J., Haynes, J., and Dubnau, D. (2001) A ComGA-dependent checkpoint limits growth during the escape from competence. *Mol Microbiol* 40: 52–64.

Hamoen, L.W., Eshuis, H., Jongbloed, J., Venema, G., and van Sinderen, D. (1995) A small gene, designated comS, located within the coding region of the fourth amino acid-activation domain of srfA, is required for competence development in *Bacillus subtilis*. *Mol Microbiol* 15: 55–63.

Hamoen, L.W., Haijema, B., Bijlsma, J.J., Venema, G., and Lovett, C.M. (2001) The *Bacillus subtilis* competence transcription factor, ComK, overrides LexA-imposed transcriptional inhibition without physically displacing LexA. *J Biol Chem* 276: 42901–7.

Hamoen, L.W., Kausche, D., Marahiel, M.A., Sinderen, D.V., Venema, G., and Serror, P. The *Bacillus subtilis* transition state regulator AbrB binds to the –35 promoter region of comK. *FEMS Microbiol Lett* 218: 299–304.

Hamoen, L.W., Smits, W.K., Jong Ad, A., Holsappel, S., and Kuipers, O.P. (2002) Improving the predictive value of the competence transcription factor (ComK) binding site in *Bacillus subtilis* using a genomic approach. *Nucleic Acids Res* 30: 5517–5528.

Hamoen, L.W., Van Werkhoven, A.F., Bijlsma, J.J.E., Dubnau, D., and Venema, G. (1998) The competence transcription factor of *Bacillus subtilis* recognizes short A/T-rich sequences arranged in a unique, flexible pattern along the DNA helix. *Genes Dev* 12: 1539–1550.

Hamoen, L.W., Van Werkhoven, A.F., Venema, G., and Dubnau, D. (2000) The pleiotropic response regulator DegU functions as a priming protein in competence development in *Bacillus subtilis*. *Proc Natl Acad Sci U S A* 97: 9246–9251.

Håvarstein, L.S., Coomaraswamy, G., and Morrison, D.A. (1995) An unmodified heptadecapeptide pheromone induces competence for genetic transformation in *Streptococcus pneumoniae*. *Proc Natl Acad Sci U S A* 92: 11140–11144.

Håvarstein, L.S., Gaustad, P., Nes, I.F., and Morrison, D.A. (1996) Identification of the streptococcal competence-pheromone receptor. *Mol Microbiol* 21: 863–869.

Håvarstein, L.S., Hakenbeck, R., and Gaustad, P. (1997) Natural competence in the genus *Streptococcus*: Evidence that streptococci can change pherotype by interspecies recombinational exchanges. *J Bacteriol* 179: 6589–6594.

Håvarstein, L.S., and Morrison, D.A. (1999) Quorum sensing and peptide pheromones in streptococcal competence for genetic transformation. In Winans, S.C. (ed). *Cell–cell signaling in bacteria*. Washington, DC: American Society for Microbiology, pp.9–26.

Healy, J., Weir, J., Smith, I., and Losick, R. (1991) Post-transcriptional control of a sporulation regulatory gene encoding transcription factor sigma H in *Bacillus subtilis*. *Mol Microbiol* 5: 477–487.

Herriott, R.M., Meyer, E.M., and Vogt, M. (1970) Defined nongrowth media for stage II development of competence in *Haemophilus influenzae*. *J. Bacteriol* 101: 517–524.

Hoa, T.T., and Dubnau, D. (2001) Unpublished data.

Hoa, T.T., Tortosa, P., Albano, M., and Dubnau, D. (2002) Rok (YkuW) regulates genetic competence in *Bacillus subtilis* by directly repressing *comK*. *Mol Microbiol* 43: 15–26.

Hoelzer, M.A., and Michod, R.E. (1991) DNA repair and the evolution of transformation in *Bacillus subtilis*. III. Sex with damaged DNA. *Genetics* 128: 215–23.

Hofreuter, D., Odenbreit, S., and Haas, R. (2001) Natural transformation competence in *Helicobacter pylori* is mediated by the basic components of a type IV secretion system. *Mol Microbiol* 41: 379–391.

Hofreuter, D., Odenbreit, S., Henke, G., and Haas, R. (1998) Natural competence for DNA transformation in *Helicobacter pylori*: Identification and genetic characterization of the *comB* locus. *Mol Microbiol* 28: 1027–1038.

Hofreuter, D., Odenbreit, S., Puls, J., Schwan, D., and Haas, R. (2000) Genetic competence in *Helicobacter pylori*: Mechanisms and biological implications. *Res Microbiol* 151: 487–491.

Horwich, A.L., Weber-Ban, E.U., and Finley, D. (1999) Chaperone rings in protein folding and degradation. *Proc Natl Acad Sci U S A* 96: 11033–11040.

Hui, F.M., and Morrison, D.A. (1991) Genetic transformation in *Streptococcus pneumoniae*: Nucleotide sequence analysis shows *comA*, a gene required for competence induction, to be a member of a bacterial ATP-dependent transport protein family. *J Bacteriol* 173: 372–381.

Ikawa, S., Shibata, T., Ando, T., and Saito, H. (1980) Genetic studies on site-specific endodeoxyribonucleases in *Bacillus subtilis*: Multiple modification and restriction systems in transformants of *Bacillus subtilis*. *Mol Gen Genet* 177: 359–368.

Inamine, G.S., and Dubnau, D. (1995) ComEA, a *Bacillus subtilis* integral membrane protein required for genetic transformation, is needed for both DNA binding and transport. *J Bacteriol* 177: 3045–3051.

Karudapuram, S., and Barcak, G.J. (1997) The *Haemophilus influenzae dprABC* genes constitute a competence-inducible operon that requires the product of the *tfoX (sxy)* gene for transcriptional activation. *J Bacteriol* 179: 4815–4820.

Kong, L., and Dubnau, D. (1994) Regulation of competence-specific gene expression by Mec-mediated protein–protein interaction in *Bacillus subtilis*. *Proc Natl Acad Sci U S A* 91: 5793–5797.

Kong, L., Siranosian, K.J., Grossman, A.D., and Dubnau, D. (1993) Sequence and properties of *mecA*, a negative regulator of genetic competence in *Bacillus subtilis*. *Mol Microbiol* 9: 365–373.

Kroll, J.S., Wilks, K.E., Farrant, J.L., and Langford, P.R. (1998) Natural genetic exchange between *Haemophilus* and *Neisseria*: Intergeneric transfer of chromosomal genes between major human pathogens. *Proc Natl Acad Sci U S A* 95: 12381–12385.

Lacks, S., Greenberg, B., and Neuberger, M. (1975) Identification of a deoxyribonuclease implicated in genetic transformation of *Diplococcus pneumoniae*. *J Bacteriol* 123: 222–232.

Lacks, S., and Neuberger, M. (1975) Membrane location of a deoxyribonuclease implicated in the genetic transformation of *Diplococcus pneumoniae*. *J Bacteriol* 124: 1321–1329.

Lacks, S.A. (1999) DNA uptake by transformable bacteria. In Broome-Smith, J.K., Baumberg, S., Stirling, C.J., and Ward, F.B. (eds). *Transport of molecules across microbial membranes*. Cambridge: Cambridge University Press, pp. 138–168.

Lacks, S.A., and Greenberg, B. (2001) Constitutive competence for genetic transformation in *Streptococcus pneumoniae* caused by mutation of a transmembrane histidine kinase. *Mol Microbiol* 42: 1035–1045.

Lazazzera, B.A. (2000) Quorum sensing and starvation: Signals for entry into stationary phase. *Curr Opin Microbiol* 3: 177–182.

Lazazzera, B.A., and Grossman, A.D. (1998) The ins and outs of peptide signaling. *Trends Microbiol* 6: 288–294.

Lazazzera, B.A., Kurtser, I.G., McQuade, R.S., and Grossman, A.D. (1999a) An autoregulatory circuit affecting peptide signaling in *Bacillus subtilis*. *J Bacteriol* 181: 5193–5200.

Lazazzera, B.A., Palmer, T., Quisel, J., and Grossman, A.D. (1999b) Cell density control of gene expression and development in *Bacillus subtilis*. In Dunny, G.M., and Winans, S.C. (eds). *Cell–cell signaling in bacteria*. Washington, DC: American Society of Microbiology Press, pp.27–46.

Lazazzera, B.A., Solomon, J.M., and Grossman, A.D. (1997) An exported peptide functions intracellularly to contribute to cell density signaling in *B. subtilis*. *Cell* 89: 917–925.

Lee, M.S., and Morrison, D.A. (1999) Identification of a new regulator in *Streptococcus pneumoniae* linking quorum sensing to competence for genetic transformation. *J Bacteriol* 181: 5004–5016.

Liu, J., and Zuber, P. (1998) A molecular switch controlling competence and motility: Competence regulatory factors ComS, MecA, and ComK control sigmaD-dependent gene expression in *Bacillus subtilis*. *J Bacteriol* 180: 4243–4251.

Liu, L., Nakano, M., Lee, O.H., and Zuber, P. (1996) Plasmid-amplified *comS* enhances genetic competence and suppresses *sinR* in *Bacillus subtilis*. *J Bacteriol* 178: 5144–5152.

Londono-Vallejo, J.A., and Dubnau, D. (1993) *comF*, a *Bacillus subtilis* late competence locus, encodes a protein similar to ATP-dependent RNA/DNA helicases. *Mol Microbiol* 9: 119–131.

Londono-Vallejo, J.A., and Dubnau, D. (1994a) Mutation of the putative nucleotide binding site of the *Bacillus subtilis* membrane protein ComFA abolishes the uptake of DNA during transformation. *J Bacteriol* 176: 4642–4645.

Londono-Vallejo, J.A., and Dubnau, D. (1994b) Membrane association and role in DNA uptake of the *Bacillus subtilis* PriA analogue ComF1. *Mol Microbiol* 13: 197–205.

Lorenz, M.G., and Wackernagel, W. (1994) Bacterial gene transfer by natural genetic transformation in the environment. *Microbiol Rev* 58: 563–602.

Love, P.E., Lyle, M.J., and Yasbin, R.E. (1985) DNA-damage-inducible (*din*) loci are transcriptionally activated in competent *Bacillus subtilis*. *Proc Natl Acad Sci U S A* 82: 6201–6205.

Lovett, C.M., Jr., O'Gara, T.M., and Woodruff, J.N. (1994) Analysis of the SOS inducing signal in *Bacillus subtilis* using *Escherichia coli* LexA as a probe. *J Bacteriol* 176: 4914–4923.

Lunsford, R.D., and Roble, A.G. (1997) *comYA*, a gene similar to comGA of *Bacillus subtilis*, is essential for competence-factor-dependent DNA transformation in *Streptococcus gordonii*. *J Bacteriol* 179: 3122–3126.

Luo, P., and Morrison, D.A. (2003) Transient association of an alternative sigma factor, ComX, with RNA polymerase during the period of competence for genetic transformation in *Streptococcus pneumoniae*. *J Bacteriol* 185: 349– 358.

Luttinger, A., Hahn, J., and Dubnau, D. (1996) Polynucleotide phosphorylase is necessary for competence development in *Bacillus subtilis*. *Mol Microbiol* 19: 343–356.

Ma, C., and Redfield, R.J. (2000) Point mutations in a peptidoglycan biosynthesis gene cause competence induction in *Haemophilus influenzae*. *J Bacteriol* 182: 3323–3330.

Macfadyen, L.P., Dorocicz, I.R., Reizer, J., Saier, M.H., Jr., and Redfield, R.J. (1996) Regulation of competence development and sugar utilization in *Haemophilus influenzae* Rd by a phosphoenolpyruvate:fructose phosphotransferase system. *Mol Microbiol* 21: 941–952.

Macfadyen, L.P. (2000) Regulation of competence development in *Haemophilus influenzae*. *J Theor Biol* 207: 349–359.

MacFadyen, L.P., Chen, D., Vo, H.C., Liao, D., Sinotte, R., and Redfield, R.J. (2001) Competence development by *Haemophilus influenzae* is regulated by the availability of nucleic acid precursors. *Mol Microbiol* 40: 700–707.

Magnuson, R., Solomon, J., and Grossman, A.D. (1994) Biochemical and genetic characterization of a competence pheromone. *Cell* 77: 207–216.

Martin, B., Garcia, P., Castanie, M.P., and Claverys, J.P. (1995) The *recA* gene of *Streptococcus pneumoniae* is part of a competence-induced operon and controls lysogenic induction. *Mol Microbiol* 15: 367–379.

Mascher, T., Zahner, D., Merai, M., Balmelle, N., De Saizieu, A.B., and Hakenbeck, R. (2003) The *Streptococcus pneumoniae cia* regulon: CiaR target sites and transcription profile analysis. *J Bacteriol* 185: 60–70.

McQuade, R.S., Comella, N., and Grossman, A.D. (2001) Control of a family of phosphatase regulatory genes (*phr*) by the alternate sigma factor sigma-H of *Bacillus subtilis*. *J Bacteriol* 183: 4905–4909.

Mejean, V., and Claverys, J.P. (1988) Polarity of DNA entry in transformation of *Streptococcus pneumoniae*. *Mol Gen Genet* 213: 444–448.

Mejean, V., and Claverys, J.P. (1993) DNA processing during entry in transformation of *Streptococcus pneumoniae*. *J Biol Chem* 268: 5594–5599.

Mongold, J.A. (1992) DNA repair and the evolution of transformation in *Haemophilus influenzae*. *Genetics* 132: 893–898.

Msadek, T., Kunst, F., and Rapoport, G. (1994) MecB of *Bacillus subtilis* is a pleiotropic regulator of the ClpC ATPase family, controlling competence gene expression and survival at high temperature. *Proc Natl Acad Sci U S A* 91: 5788–5792.

Msadek, T., Kunst, F., and Rapoport, G. (1995) A signal transduction network in *Bacillus subtilis* includes the DegS/DegU and ComP/ComA two-component systems. In Hoch, J.A., and Silhavy, T.J. (eds). Two-component signal transduction. Washington, DC: ASM Press.

Nakano, M.M., Hajarizadeh, F., Zhu, Y., and Zuber, P. (2001) Loss-of-function mutations in *yjbD* result in ClpX- and ClpP-independent competence development of *Bacillus subtilis*. *Mol Microbiol* 42: 383–394.

Nakano, M.M., Nakano, S., and Zuber, P. (2002a) Spx (YjbD), a negative effector of competence in *Bacillus subtilis*, enhances ClpC-MecA-ComK interaction. *Mol Microbiol* 44: 1341–1349.

Nakano, M.M., Xia, L., and Zuber, P. (1991) Transcription initiation region of the *srfA* operon which is controlled by the *comP-comA* signal transduction system in *Bacillus subtilis*. *J Bacteriol* 173: 5487–5493.

Nakano, S., Zheng, G., Nakano, M.M., and Zuber, P. (2002b) Multiple pathways of Spx (YjbD) proteolysis in *Bacillus subtilis*. *J Bacteriol* 184: 3664–3670.

Neuwald, A.F., Aravind, L., Spouge, J.L., and Koonin, E.V. (1999) AAA+: A class of chaperone-like ATPases associated with the assembly, operation, and disassembly of protein complexes. *Genome Res* 9: 27–43.

Odenbreit, S., Puls, J., Sedlmaier, B., Gerland, E., Fischer, W., and Haas, R. (2000) Translocation of *Helicobacter pylori* CagA into gastric epithelial cells by type IV secretion. *Science* 287: 1497–1500.

Ogura, M., Liu, L., Lacelle, M., Nakano, M.M., and Zuber, P. (1999) Mutational analysis of ComS: Evidence for the interaction of ComS and MecA in the regulation of competence development in *Bacillus subtilis*. *Mol Microbiol* 32: 799–812.

Ogura, M., Yamaguchi, H., Kobayashi, K., Ogasawara, N., Fujita, Y., and Tanaka, T. (2002) Whole-genome analysis of genes regulated by the *Bacillus subtilis* competence transcription factor ComK. *J Bacteriol* 184: 2344–2351.

Palmen, R., Vosman, B., Buijsman, P., Breek, C.K., and Hellingwerf, K.J. (1993) Physiological characterization of natural transformation in *Acinetobacter calcoaceticus*. *J Gen Microbiol* 139: 295–305.

Pan, Q., Garsin, D.A., and Losick, R. (2001) Self-reinforcing activation of a cell-specific transcription factor by proteolysis of an anti-sigma factor in *B. subtilis*. *Mol Cell* 8: 873–883.

Pearce, B.J., Naughton, A.M., Campbell, E.A., and Masure, H.R. (1995) The *rec* locus, a competence-induced operon in *Streptococcus pneumoniae*. *J Bacteriol* 177: 86–93.

Perego, M. (1999) Self-signaling by Phr peptides modulates *Bacillus subtilis* development. In Dunny, G.M., and Winans, S.C. (eds). *Cell–cell signaling in bacteria*. Washington, DC: American Society of Microbiology Press, pp. 243–258.

Perego, M., Higgins, C.F., Pearce, S.R., Gallagher, M.P., and Hoch, J.A. (1991) The oligopeptide transport system of *Bacillus subtilis* plays a role in the initiation of sporulation. *Mol Microbiol* 5: 173–185.

Perego, M., Spiegelman, G.B., and Hoch, J.A. (1988) Structure of the gene for the transition state regulator *abrB*: Regulator synthesis is controlled by the *spoOA* sporulation gene in *Bacillus subtilis*. *Mol Microbiol* 2: 689–699.

Persuh, M., Dubnau, D. (2000) Unpublished data.

Persuh, M., Dubnau, D. (2001) Unpublished data.

Persuh, M., Mandic-Mulec, I., and Dubnau, D. (2002) A MecA paralog, YpbH, binds ClpC, affecting both competence and sporulation. *J Bacteriol* 184: 2310–2313.

Persuh, M., Turgay, K., Mandic-Mulec, I., and Dubnau, D. (1999) The N- and C-terminal domains of MecA recognize different partners in the competence molecular switch. *Mol Microbiol* 33: 886–894.

Pestova, E.V., Håvarstein, L.S., and Morrison, D.A. (1996) Regulation of competence for genetic transformation in *Streptococcus pneumoniae* by an

COMPETENCE FOR GENETIC TRANSFORMATION

auto-induced peptide pheromone and a two-component regulatory system. *Mol Microbiol* 21: 853–862.

Pestova, E.V., and Morrison, D.A. (1998) Isolation and characterization of three *Streptococcus pneumoniae* transformation-specific loci by use of a *lacZ* reporter insertion vector. *J Bacteriol* 180: 2701–2710.

Peterson, S., Cline, R.T., Tettelin, H., Sharov, V., and Morrison, D.A. (2000) Gene expression analysis of the *Streptococcus pneumoniae* competence regulons by use of DNA microarrays. *J Bacteriol* 182: 6192–6202.

Piazza, F., Tortosa, P., and Dubnau, D. (1999) Mutational analysis and membrane topology of ComP, a quorum-sensing histidine kinase of *Bacillus subtilis* controlling competence development. *J Bacteriol* 181: 4540–4548.

Porstendorfer, D., Drotschmann, U., and Averhoff, B. (1997) A novel competence gene, *comP*, is essential for natural transformation of *Acinetobacter* sp. strain BD413. *Appl Environ Microbiol* 63: 4150–4157.

Postma, P.W., Lengeler, J.W., and Jacobson, G.R. (1993) Phosphoenolpyruvate: carbohydrate phosphotransferase systems of bacteria. *Microbiol Rev* 57: 543–594.

Pozzi, G., Masala, L., Iannelli, F., Manganelli, R., Håvarstein, L.S., Piccoli, L., et al. (1996) Competence for genetic transformation in encapsulated strains of *Streptococcus pneumoniae*: Two allelic variants of the peptide pheromone. *J Bacteriol* 178: 6087–6090.

Provvedi, R., Chen, I., and Dubnau, D. (2001) NucA is required for DNA cleavage during transformation of *Bacillus subtilis*. *Mol Microbiol* 40: 634–644.

Provvedi, R., and Dubnau, D. (1999) ComEA is a DNA receptor for transformation of competent *Bacillus subtilis*. *Mol Microbiol* 31: 271–280.

Puyet, A., Greenberg, B., and Lacks, S.A. (1990) Genetic and structural characterization of EndA. A membrane-bound nuclease required for transformation of *Streptococcus pneumoniae*. *J Mol Biol* 213: 727–738.

Ratnayake-Lecamwasam, M., Serror, P., Wong, K.W., and Sonenshein, A.L. (2001) *Bacillus subtilis* CodY represses early-stationary-phase genes by sensing GTP levels. *Genes Dev* 15: 1093–1103.

Redfield, R.J. (1991) *sxy-1*, A *Haemophilus influenzae* mutation causing greatly enhanced spontaneous competence. *J Bacteriol* 173: 5612–5618.

Redfield, R.J. (1993) Evolution of natural transformation: Testing the DNA repair hypothesis in *Bacillus subtilis* and *Haemophilus influenzae*. *Genetics* 133: 755–761.

Rimini, R., Jansson, B., Feger, G., Roberts, T.C., de Francesco, M., Gozzi, A., et al. (2000) Global analysis of transcription kinetics during competence development in *Streptococcus pneumoniae* using high density DNA arrays. *Mol Microbiol* 36: 1279–1292.

Roberts, M.S., and Cohan, F.M. (1993) The effect of DNA sequence divergence on sexual isolation in *Bacillus*. *Genetics* 134: 401–408.

Roberts, M.S., and Cohan, F.M. (1995) Recombination and migration rates in natural populations of *Bacillus subtilis* and *Bacillus mojavensis*. *Evolution* 49: 1081–1094.

Roggiani, M., and Dubnau, D. (1993) ComA, a phosphorylated response regulator protein of *Bacillus subtilis*, binds to the promoter region of *srfA*. *J Bacteriol* 175: 3182–3187.

Roggiani, M., Hahn, J., and Dubnau, D. (1990) Suppression of early competence mutations in *Bacillus subtilis* by *mec* mutations. *J Bacteriol* 172: 4056–4063.

Rosenthal, A.L., and Lacks, S.D. (1980) Complex structure of the membrane nuclease of *Streptococcus pneumoniae* revealed by two dimensional electrophoresis. *J Mol Biol* 141: 133–146.

Rudner, D.Z., LeDeaux, J.R., Ireton, K., and Grossman, A.D. (1991) The *spoOK* locus of *Bacillus subtilis* is homologous to the oligopeptide permease locus and is required for sporulation and competence. *J Bacteriol* 173: 1388–1398.

Schirmer, E.C., Glover, J.R., Singer, M.A., and Lindquist, S. (1996) HSP100/Clp proteins: A common mechanism explains diverse functions. *Trends Biochem Sci* 21: 289–296.

Schlothauer, T., Nogk, A., Dougan, D., Bukau, B., and Turgay, K. (2003) MecA, an adapter protein necessary for ClpC chaperone activity. *Proc Natl Acad Sci U S A* 100: 2306–11.

Seifert, H.S., Ajioka, R.S., Marchal, C., Sparling, P.F., and So, M. (1988) DNA transformation leads to pilin antigenic variation in *Neisseria gonorrhoeae*. *Nature* 336: 392–395.

Seifert, H.S., Ajioka, R.S., Paruchuri, D., Heffron, F., and So, M. (1990) Shuttle mutagenesis of *Neisseria gonorrhoeae*: Pilin null mutations lower DNA transformation competence. *J Bacteriol* 172: 40–46.

Serror, P., and Sonenshein, A.L. (1996) CodY is required for nutritional repression of *Bacillus subtilis* genetic competence. *J Bacteriol* 178: 5910–5915.

Smeets, L.C., and Kusters, J.G. (2002) Natural transformation in *Helicobacter pylori*: DNA transport in an unexpected way. *Trends Microbiol* 10: 159– 162.

Smith, H.O., Gwinn, M.L., and Salzberg, S.L. (1999) DNA uptake signal sequences in naturally transformable bacteria. *Res Microbiol* 150: 603–616.

Solomon, J., Magnuson, R., Srivastava, A., and Grossman, A.D. (1995) Convergent sensing pathways mediate response to two extracellular competence factors in *Bacillus subtilis*. *Genes Dev* 9: 547–558.

Solomon, J.M., Lazazzera, B.A., and Grossman, A.D. (1996) Purification and characterization of an extracellular peptide factor that affects two developmental pathways in *Bacillus subtilis*. *Genes Dev* 10: 2014–2024.

Strauch, M., Webb, V., Spiegelman, G., and Hoch, J.A. (1990) The SpoOA protein of *Bacillus subtilis* is a repressor of the *abrB* gene. *Proc Natl Acad Sci U S A* 87: 1801–1805.

Suerbaum, S., Smith, J.M., Bapumia, K., Morelli, G., Smith, N.H., Kunstmann, E., Dyrek, I., et al. (1998) Free recombination within *Helicobacter pylori*. *Proc Natl Acad Sci U S A* 95: 12619–12624.

Tomasz, A., and Hotchkiss, R.D. (1964) Regulation of the transformability of pneumococcal cell cultures by macromolecular cell products. *Proc Natl Acad Sci U S A* 51: 480–486.

Tomasz, A., and Mosser, J.L. (1966) On the nature of the pneumococcal activator substance. *Proc Natl Acad Sci U S A* 55: 58–66.

Tomb, J.-F., El-Hajj, H., and Smith, H.O. (1991) Nucleotide sequence of a cluster of genes involved in the transformation of *Haemophilus influenzae* RD. *Gene* 104: 1–10.

Tønjum, T., Freitag, N.E., Namork, E., and Koomey, M. (1995) Identification and characterization of *pilG*, a highly conserved pilus-assembly gene in pathogenic *Neisseria*. *Mol Microbiol* 16: 451–464.

Tortosa, P., Logsdon, L., Kraigher, B., Itoh, Y., Mandic-Mulec, I., and Dubnau, D. (2000) Specificity and genetic polymorphism of the *Bacillus* competence quorum-sensing system. *J Bacteriol* 183: 451–460.

Tran, L.-S.P., Nagai, T., and Itoh, Y. (2000) Divergent structure of the ComQXPA quorum sensing components: Molecular basis of strain-specific communication mechanism in *Bacillus subtilis*. *Mol Microbiol* 37: 1159–1171.

Trautner, T.A., Pawlek, B., Bron, S., and Anagnostopoulos, C. (1974) Restriction and modification in *B. subtilis*. Biological aspects. *Mol Gen Genet* 131: 181–191.

Turgay, K., Hamoen, L.W., Venema, G., and Dubnau, D. (1997) Biochemical characterization of a molecular switch involving the heat shock protein ClpC, which controls the activity of ComK, the competence transcription factor of *Bacillus subtilis*. *Genes Dev* 11: 119–128.

Turgay, K., Hahn, J., Burghoorn, J., and Dubnau, D. (1998) Competence in *Bacillus subtilis* is controlled by regulated proteolysis of a transcription factor. *EMBO J* 17: 6730–6738.

Turgay, K., Persuh, M., Hahn, J., and Dubnau, D. (2001) Roles of the two ClpC ATP binding sites in the regulation of competence and the stress response. *Mol Microbiol* 42: 717–727.

van Sinderen, D., Luttinger, A., Kong, L., Dubnau, D., Venema, G., and Hamoen, L. (1995) *comK* encodes the competence transcription factor, the key regulatory protein for competence development in *Bacillus subtilis*. *Mol Microbiol* 15: 455–462.

IRENA DRASKOVIC AND DAVID DUBNAU

van Sinderen, D., ten Berge, A., Hayema, B.J., Hamoen, L., and Venema, G. (1994) Molecular cloning and sequence of *comK*, a gene required for genetic competence in *Bacillus subtilis*. *Mol Microbiol* 11: 695–703.

van Sinderen, D., and Venema, G. (1994) *comK* acts as an autoregulatory control switch in the signal transduction route to competence in *Bacillus subtilis*. *J Bacteriol* 176: 5762–5770.

Vosman, B., Kooistra, J., Olijve, J., and Venema, G. (1987) Cloning in *Escherichia coli* of the gene specifying the DNA-entry nuclease of *Bacillus subtilis*. *Gene* 52: 175–183.

Vosman, B., Kuiken, G., and Venema, G. (1988) Transformation in *Bacillus subtilis*: Involvement of the 17-kilodalton DNA-entry nuclease and the competence-specific 18-kilodalton protein. *J Bacteriol* 170: 3703–3710.

Ween, O., Gaustad, P., and Håvarstein, L.S. (1999) Identification of DNA binding sites for ComE, a key regulator of natural competence in *Streptococcus pneumoniae*. *Mol Microbiol* 33: 817–827.

Weinrauch, Y., Msadek, T., Kunst, F., and Dubnau, D. (1991) Sequence and properties of *comQ*, a new competence regulatory gene of *Bacillus subtilis*. *J Bacteriol* 173: 5685–5693.

Weinrauch, Y., Penchev, R., Dubnau, E., Smith, I., and Dubnau, D. (1990) A *Bacillus subtilis* regulatory gene product for genetic competence and sporulation resembles sensor protein members of the bacterial two-component signal-transduction systems. *Genes Dev* 4: 860–872.

Wolfgang, M., Lauer, P., Park, H.S., Brossay, L., Hebert, J., and Koomey, M. (1998) PilT mutations lead to simultaneous defects in competence for natural transformation and twitching motility in piliated *Neisseria gonorrhoeae*. *Mol Microbiol* 29: 321–330.

Zulty, J.J., and Barcak, G.J. (1995) Identification of a DNA transformation gene required for *com101A+* expression and supertransformer phenotype in *Haemophilus influenzae*. *Proc Natl Acad Sci U S A* 92: 3616–3620.

Part 3 Biological Consequences of the
Mobile Genome

CHAPTER 8
Phase variation and antigenic variation

Richard Villemur and Eric Déziel

Bacteria owe their ability to thrive in diverse and ever-changing conditions to their extraordinary faculty of adaptation. They have developed many strategies to adjust to new environments. These mechanisms include random modifications within their genome, such as point mutation, duplication, deletion, insertion, and acquisition of new DNA (e.g., lateral gene transfer). These multiple events generate a heterogeneous microbial population containing numerous novel phenotypes. Whenever the environment changes, a subpopulation more apt to survive in these new conditions emerges, thus allowing bacteria to thrive, for example, by acquiring resistance to antibiotics. However, as the intensity, duration, and nature of stress are extremely variable, the optimal response to new environmental conditions may be unpredictable. The means by which bacteria either respond to stress, such as exposure to toxic/inhibitory compounds or starvation, or avoid detection by the host's immune system are crucial for their survival. The spontaneous mutation rate is usually insufficient for allowing an efficient adaptation to these changes. However, certain bacterial populations contain hypermutator strains exhibiting highly increased rates of spontaneous mutations, therefore promoting adaptation to changing environments (Taddei et al., 1997). Nevertheless, this benefit may disappear once adaptation is achieved because the evolved genotype may have accumulated irreversible mutations that are detrimental in other conditions (Giraud et al., 2001a, 2001b).

Alternatively, bacteria have developed adaptation strategies based on DNA rearrangement events restricted to specific genomic regions. These defined loci allow bacteria to generate an array of phenotypic variants, whereas minimizing detrimental effects of random mutations on fitness. One such mechanism named *phase (or phenotypic) variation* is defined as the gain or loss (often represented as ON/OFF switching) of a phenotype resulting from changes in expression of one or multiple gene(s). Phase variation arises from rearrangements, gene replacements, or mutations changing the structure of a transcriptional unit. Such changes can alter the expression or the sequence of a gene. Phase variation occurs generally at high frequency ($>10^{-5}$ per generation) and is usually reversible. When phase variation takes place, a phenotypically heterogeneous bacterial population is generated. As phase variation produces genetic diversity, individuals bearing the most adequate fitness for the new environment are favored by clonal expansion.

An additional mechanism related to phase variation and involving pathogenic bacteria is *antigenic variation*. Pathogens express virulence factors that cause or lead to disease (e.g., the production of toxins or the colonization of the host via adherence to epithelial cells). Bacteria can adhere to host cells using various ways. In particular cases, pili (or fimbriae) mediate attachment to host molecules (glycoproteins or glycolipids). In other instances, bacteria bear surface adhesins or develop into a biofilm. The host reacts to infection by producing antibodies and T cells that recognize microbial antigens such as pilins or adhesins. To survive, microorganisms must evade the host's immune responses. A number of bacteria produce a capsule protecting them against phagocytosis, others mimic host cell proteins to hide from the immune system, and many constantly alter their surface structures and excreted proteins. Production of a modified surface structure that is then no longer recognized by the immune components, which are directed to the original protein, is called *antigenic variation*. Mechanisms of phase variation are sometimes used by microorganisms to create antigenic variation. These inexorable modifications are a tremendous obstacle to the development of effective vaccines because the structure of the primary target is variable.

A limited number of mechanisms have been implicated in most phenotypic and antigenic variation events. DNA rearrangements generating phase variation by switching ON/OFF gene expression include DNA inversion of promoter, slipped strand mispairing, and DNA methylation. Most instances of antigenic variation include mechanisms of DNA recombination or gene conversion between one gene and a series of homologous, but not identical, loci leading to the generation of proteins bearing new antigenicity. With the event of molecular genetics and microbial genome sequencing, we now realize that generation of diversity by phase variation and antigenic variation is widespread in both pathogenic and environmental bacteria, suggesting that this strategy is extensively used by microorganisms to respond and adapt to their environment.

In this chapter, the best understood mechanisms of phenotypic diversification by phase variation and antigenic variation is presented and their functions are discussed.

DNA INVERSION

Many bacteria use DNA inversion to mediate phase variation. Various examples from genes controlling surface proteins or appendages (pilus and flagellum) are presented in this section.

Type 1 fimbriae of *Escherichia coli*

Similar to most pathogenic bacteria, *Escherichia coli* (*E. coli*) expresses adhesion components, such as fimbriae (also known as pili), to attach to host epithelium. Fimbriae are hairlike appendages, 2 to 10 nm in diameter, that extend from the bacterial surface. They are composed of repeating protein subunits termed *fimbrins* that range in size from 14 to 30 kDa. Production of fimbriae is modulated by a global regulatory network that coordinates their expression concomitantly with other genes in response to environmental stimuli. These stimuli inform bacteria that they are appropriately positioned to produce adherent fimbriae for colonization. By expressing adhesins with unique receptor specificities, fimbriae impart tissue tropism, allowing the colonization of different sites within the host. Importantly, the expression of many fimbrial operons is subject to phase variation, including the type 1 fimbriae and the P fimbriae.

Type 1 fimbriae are adhesion pili produced by many species within the enterobacterial family. They play a key role in commensal host–bacterial interactions. Strains of *E. coli* expressing this type of fimbriae, although associated with various clinical conditions, are mostly encountered with urinary tract infections (uropathogenic *E. coli*; Hal Jones et al., 1996; Low et al., 1996). Type 1 fimbriae allow colonization of many host surfaces by mediating attachment to mannose-containing receptors. These appendages play a critical role in the initial establishment and then in persistent infections (Gunther et al., 2002; Hultgren et al., 1996).

Nine genes forming the *fim* operon control the synthesis of type 1 fimbriae in *E. coli*. FimA makes up most of the fimbrial shaft, whereas FimH mediates adhesion to host cell receptors. The *fim* operon is subject to phase variation, based on a mechanism of DNA sequence inversion, at a rate of approximately 10^{-3} to 10^{-4} per cell per generation (Donnenberg and Welch, 1996). It involves the reversible inversion of a 314-bp DNA element located

ON phase

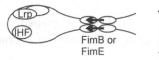

314-bp inversion region

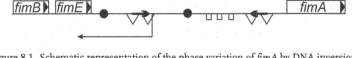

● : IHF binding sites
□ : Lrp binding sites
▽ : FimB, FimE binding sites

FimB or FimE

OFF phase

RICHARD VILLEMUR AND ERIC DÉZIEL

Figure 8.1. Schematic representation of the phase variation of *fimA* by DNA inversion regulating the expression of the FimA fimbrin in *Escherichia coli*. The reversible 314-bp DNA element contains the *fimA* promoter and the transcriptional start site *(thin arrows)*. IR, inverted repeats *(thick arrows)*; closed triangles, direction of the coding frame. (Adapted from Blomfield, 2001.)

upstream of the gene, and contains both the *fimA* promoter and the transcriptional start site (Fig. 8.1). During the ON phase, the *fimA* promoter is oriented to transcribe the gene, whereas during the OFF phase the promoter is oriented in the reverse direction and the transcription of *fimA* cannot take place. The reversible sequence contains a 9 bp-inverted repeat (TTGGGGCCA, named IRL and IRR) located at both ends, and its inversion is promoted by the product of the *fimB* and *fimE* genes located upstream of *fimA* (Henderson et al., 1999; Low et al., 1996). FimB and FimE are analogous to members of the invertase family of site-specific recombinases (Dybvig, 1993; van de Putte and Goosen, 1992). These two enzymes have similar sequences (48% identical; Klemm, 1986), and each possesses the conserved motif RHRY that forms the catalytic tetrad of recombinase family members (Dorman and Higgins, 1987; Eisenstein et al., 1987; Nunes-Duby et al., 1998). FimB promotes the inversion in both directions, whereas FimE is only active in catalyzing the ON to OFF phase change. When *fimB* and *fimE* are co-expressed, FimE activity dominates and the switch turns to the OFF phase (bacteria are afimbriate; Blomfield, 2001; Gally et al., 1993, 1996; Kulasekara and Blomfield, 1999; McClain et al., 1991).

FimB and FimE bind to half-sites on either side of the 9-bp inverted repeats (Gally et al., 1996). FimE preferentially promotes the ON-to-OFF conversion, due to different affinity for the binding site. Although the sequences flanking both 9 bp-inverted repeats are similar, they are not identical, suggesting different binding potential (Blomfield, 2001; Gally et al., 1996). In the OFF mode, FimE may have a low affinity for both sequences (Kulasekara and Blomfield, 1999). Moreover, *fimE* expression is controlled by the orientation of the 314-bp region. In the OFF orientation, the *fimA* promoter is located near *fimE*, which results in the production of a FimE antisense transcript that may regulate *fimE* expression (Pallesen et al., 1989). Finally, the FimE sequence may confer different affinity to the 9-bp inverted repeats by substituting an arginine for a lysine at position 59, conveying a FimB-like switching character to FimE (Smith and Dorman, 1999). However, control of *fimA*, *fimB*, and *fimE* gene expression is even more complex as *trans*- and *cis*-acting factors are also involved. For instance, the −10 promoter region of *fimA* overlaps the inverted repeats, such that factors binding to this region may block the accessibility for recombination factors binding to the inverted repeats. In addition, expression from the *fimA* promoter and the inversion promoted by FimB are mutually exclusive events (Dove and Dorman, 1996; O'Gara and Dorman, 2000).

Both the DNA gyrase and the DNA topology of the promoter also influence the *fim* switch (Dove and Dorman, 1994). Many global regulatory factors are known to control the expression of the type 1 fimbriae operon in *E. coli*: the integration host factor (IHF), the leucine-responsive regulatory protein (Lrp), the histone-like nucleoid structuring protein (H-NS), and the stationary-phase sigma factor RpoS. For example, the 314-bp sequence contains three Lrp-binding sites and one IHF-binding site, and an additional IHF-binding site is located upstream of the *fimE* gene. IHF, a DNA-binding protein, attaches to specific sequences, thus locally inducing DNA bending. The IHF–DNA complex facilitates the interaction between proteins bound on each site of the bending location, and therefore, promotes site-specific recombination. It has been suggested that Lrp and IHF facilitate the recombination event by bending the DNA, thus aligning the inverted repeats (Blomfield, 2001; Henderson et al., 1999; Low et al., 1996).

Flagella of *Salmonella typhimurium*

Synthesis of the flagellin of *Salmonella enterica* serovar Typhimurium also undergoes phase variation by a DNA inversion mechanism. Flagellin is the major component of the flagellar filament and is expressed from either one of two genes, named *fljB* (or H1) and *fliC* (or H2), located on different chromosomal positions. Both proteins, although highly similar, diverge in the central core, resulting in distinct flagella antigenicity. The expression of

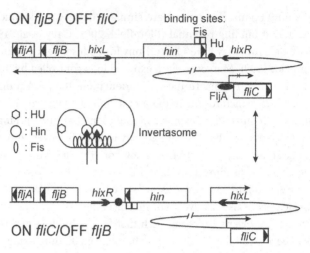

ON *fljB* / OFF *fliC*

Figure 8.2. Schematic representation of the phase variation of *fljB/fliC* by DNA inversion regulating the expression of the FljB and FliC flagellins in *Salmonella typhimurium*. The reversible 996-bp DNA segment contains the *hin* gene and the *fljBA* promoter, and is flanked by two inverted homologous sequences, *hixL* and *hixR*. The schematic representation of the invertasome is illustrated as proposed by Merickel et al. (1998). Closed triangles, direction of the coding frame.

both genes is susceptible to phase variation, which occurs every 10^3 to 10^5 generations to produce cells with a new antigenic structure. This is a typical example where a phase variation mechanism, meaning the ON/OFF switch of gene expression, can produce antigenic variation (MacNab, 1996).

The *fljB* gene is flanked by the *hin* gene, which encodes a site-specific recombinase (or invertase), and *fljA*, whose product acts as a repressor of the *fliC* gene (Fig. 8.2). *fljB* and *fljA* are co-expressed from the same promoter allowing the simultaneous production of FljB and repression of the other flagellin FliC. In *S. enterica*, which has a similar *fljB* and *fliC* arrangement, Bonifield and Hughes (2003) showed that FljA prevents production of FliC protein at the posttranscriptional level. They proposed a model in which FljA regulates both *fliC* transcription and translation via interactions with the 5′-UTR of the transcript.

Phase variation is induced by the inversion of a 996-bp segment containing the *hin* gene and the *fljBA* promoter. When the inversion occurs, the expression of *fljBA* is turned OFF, allowing the expression of *fliC* (ON) because of the absence of its repressor (Henderson et al., 1999; MacNab, 1996). The 996-bp segment is flanked by two inverted homologous sequences, called *hixL* and *hixR*, to which Hin invertase binds and then catalyzes the homologous recombination. A second protein called Fis is also needed to stimulate

Hin activity. Homodimers of Fis bind to these two sites separated by 48 bp within the 996-bp segment, being positioned on opposite sides of the DNA helix (Johnson et al., 1987). In addition, the histonelike protein HU is required for high rates of recombination by assisting in DNA looping between *hixL* and Fis-binding sites. The interaction of Hin, Fis, and HU, with their respective binding sites, generates a complex called an *invertasome* and requires supercoiled DNA. This complex holds the invertible DNA segment in a topological arrangement so Hin can catalyze the cleavage and rearrangement of the exchanged DNA strands (Heichman and Johnson, 1990; Kanaar et al., 1990; Merickel et al., 1998).

Type IVB pilus of *E. coli*

Enteropathogenic strains of *E. coli* bear the IncI1 plasmid R64, which encodes for a thin-type IVB pilus involved in conjugation. Fourteen genes designated *pilI* through *pilV* compose the R64 *pil* locus, among which the *pilS* encodes for the type IVB pilin. PilV, a minor component of the pilus (Yoshida et al., 1998), determines the recipient specificity by recognition of LPS molecules on recipient bacterial cells (Ishiwa and Komano, 2000).

A region named *shufflon* at the C-terminal end of the *pilV* gene is predisposed to antigenic variation by DNA inversion (Ishiwa and Komano, 2000; Komano et al., 1994). This shufflon is composed of four segments, designated A, B, C, and D, which are separated and flanked by imperfect 19-bp repeat sequences. These repeats are grouped in four distinct types; each type differs from one to three nucleotides. There are seven possible *pilV* C-terminal sequences resulting from different DNA inversion combinations of four segments, independently or in groups (Fig. 8.3A). The inversion frequency of these segments differs greatly and is dependent on the four distinct types of repeat sequences (Gyohda et al., 1997; 2004; Gyohda and Komano, 2000). The Rci recombinase, genetically located downstream of *pilV*, cuts at the end of the 19-bp repeats generating a 5′ protruding 7-bp staggered cut. Two recombinases bind to 12-bp sequences flanking this 7-bp sequence. The 5′ end of the binding site is a conserved sequence of the 19-bp repeat, whereas the 3′ end is not conserved among inverted repeats. It is the affinity of the Rci recombinase for the nonconserved sequence that determines the inversion frequency (Gyohda et al., 2002; 2004).

Campylobacter and the S layer

The S layer (for surface layer), found in more than 300 different species of Bacteria and Archaea, is a uniformly arranged structure coating the cell

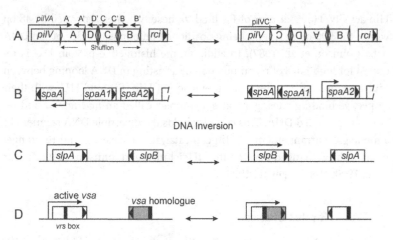

Figure 8.3. Schematic representation various DNA inversion mechanisms causing phase variation or antigenic variation. Closed triangles, direction of the coding frame. (A) *pilV* shufflon in *Escherichia coli*. Illustrated is an example of the result of one shuffling event. Open triangles represent the seven 19-bp repeat sequences. Thin arrows above *pilV* represent the location and direction of the proper coding frame in the shufflon when fused to the 5' end region. (B) DNA inversion of the *sapA* promoter in front of a different *sapA* cassette in *Campylobacter fetus*. (C) DNA inversion of the *slpB* gene cassette in front of the active promoter in *Lactobacillus acidophilus*. (D) DNA inversion of *vsa* gene cassettes in *Mycoplasma pulmonis*. The inversion results in the replacement of the equivalent region of the expressed *vsa* gene by a silent *vsa* locus.

envelope. A single protein, the S protein, autoassembles by noncovalent attachment to form the S layer (Bahl et al., 1997). Although no definitive function has been assigned to the S layer, it has been associated with cell shape, protective coats, molecular sieves, ion traps, and cell surface interactions, as well as virulence (Sleytr et al., 1993).

Campylobacter fetus interferes with the reproduction of ungulates, such as pig and ruminants, and is also a human opportunistic pathogen. The S layer of wild-type strains of *C. fetus* is an essential virulence factor. The S proteins composing the S layer are subject to antigenic variation conferring protection against the host immune responses. *C. fetus* strains are classified as type A or B, depending on the S protein, either type A (SapA) or type B (SapB), interacting with LPS. SapA and SapB differ mainly in the 184-amino acid N-terminal sequence where the LPS-binding site resides (Dworkin and Blaser, 1997a; Dworkin et al., 1995a, 1995b). No shift from one type to the other has been observed in any *C. fetus* strains (Bahl et al., 1997; Boot and Pouwels, 1996; Dworkin and Blaser, 1997a). However, variations within the same group A or B have been demonstrated.

In wild-type A strains, eight to nine full-length and functional *sapA* gene cassettes, located in a 93-kb chromosomic region, can encode S proteins varying in size from 97 to 149 kDa. However, only one of these genes can be expressed at a time. Two regions are found in each gene cassette: One beginning 74 bp upstream of the start codon extends up to the first 184 amino acids, and the other is located downstream of the coding sequence. DNA inversion between the active *sapA* gene and one of the other *sapA* gene cassettes brings the promoter in close vicinity to a different cassette and allows the expression of a new S-protein variant (Fig. 8.3B). The inverted region lies within the 5' conserved region, just upstream of the start codon. It contains a sequence very similar (GCTGGTGA; seven of eight bp) to the *E. coli* RecBCD recognition (Chi) site. This region is then followed by three pentameric (ATTTT) repeats, and 7-bp inverted repeats with a single nucleotide spacer (Dworkin and Blaser, 1996, 1997a, 1997b). The *sapA* DNA inversion mechanism differs from most other DNA inversion mechanisms using specific recombinases in that it requires the generalized (RecA-dependent) homologous recombination pathway (Dworkin et al., 1997).

Assorted DNA inversion mechanisms

Similarly to what is observed in *C. fetus*, *Lactobacillus acidophilus* uses DNA inversion to modulate the expression of two gene cassettes, slpA and slpB, encoding its S proteins. Both genes, oriented in opposite direction and separated by 3.0 kb, encode S proteins that diverge in their N-terminal and middle part but are nearly identical at the C-terminal. Only one cassette at a time has an active promoter (Fig. 8.3C). However, in contrast to the *sapA* system where the promoter moves in front of a new gene cassette, it is the gene cassette that moves downstream of the active promoter. The inversion may involve a Din-type invertase as a consensus recognition site is present within the region where recombination occurs (Boot and Pouwels, 1996).

Mycoplasma bovis contains several copies of the *vsp* genes (for variable surface proteins). The N-terminal region of all Vsp proteins is highly conserved, whereas the remaining of the Vsp sequence displays divergence with in-frame tandem repeats. Antigenic variation is suggested not only by the observation of recombination between these genes, but also by spontaneous deletions and insertions in the repeat units, all that varying the antigenicity of the Vsp proteins (Lysnyansky et al., 1999). The expression of the *vsp* gene family shows phase variation by DNA inversion similar to the *sap* system. Only one *vsp* gene is expressed at the time in *M. bovis*, whereas the others are silent. The inversion involves moving the active promoter in front of a silent *vsp* allele. Therefore, the formerly expressed gene is switched OFF and the recipient gene is switched ON. The recombination event occurs at a 35-bp conserved region present upstream of all *vsp* genes. Frequency of Vsp variation occurs at a rate of 10^{-2} to 10^{-3} per cell per generation (Lysnyansky et al., 2001). A similar arrangement is present in the *vsp*-related genes (or *avg* and *Vpma*) of *M. agalactiae* (Flitman-Tene et al., 2000; Glew et al. 2002). Similarly, the *vsa* gene family of *M. pulmonis* encoding highly variable V-1 surface lipoproteins undergoes DNA inversion. A single full-length copy of the *vsa* gene is expressed, whereas the other *vsa* loci are silent and encode only for the C-terminal part, which is composed of repeated units. By DNA inversion, the C-terminal region of a *vsa* silent locus replaces the equivalent region of the expressed *vsa* gene, thus generating a protein with a different antigenicity (Fig. 8.3D). The proposed mechanism of inversion involves a staggered cleavage reaction at the *vrs* box, a conserved 34-bp sequence present upstream of the C-terminal region of the expressed Vsa and also found upstream of all silent *vsa* alleles. The reannealing of the staggered DNA (6 bp) upon inversion is not homologous and creates unmatched base pairs. Upon DNA repair or DNA replication, point mutations can occur that change the amino acid residues and add to the antigenic variation, or introduce a stop codon resulting in a truncated nonfunctional protein (phase variation, OFF phase; Bhugra et al., 1995; Shen et al., 2000). Adjacent to the *M. bovis vsp*, *M. pulmonis vsa*, and *M. agalactiae avg* and *Vpma* loci, an ORF bearing homology with members of the lambda integrase family of tyrosine site-specific recombinases could be responsible for this DNA inversion mechanism (Glew et al., 2002; Ron et al., 2002).

The same type of DNA inversion observed with the *M. pulmonis vsa* genes occurs with the genes encoding for the type IV pilin, *tfpQ* and *tfpI*, of *Moraxella lacunata* and *Moraxella bovis*. The gene for the Piv recombinase responsible for the inversion is proximal to the silent gene (Rozsa and Marrs, 1991; Tobiason et al., 1999). DNA inversion has been described or proposed to explain phase variation or antigenic variation observed with other systems, such as the mannose-resistant/*Proteus*-like (MR/P) fimbriae of *Proteus mirabilis* (*mrpA/mrpI*; Li et al., 2002; Zhao et al., 1997) and the flagellin genes (*flaA* and *flaB*) of *Proteus mirabilis* (Murphy and Belas, 1999).

GENE CONVERSION

Neisseria gonorrhoeae

Neisseria gonorrhoeae and *Neisseria meningitidis* cause gonorrhoea and meningitis, respectively, in humans. Because of the increasing incidence of antibiotic-resistant strains, the development of a vaccine to prevent the spread of these diseases would represent a significant advancement. Unfortunately, *Neisseria* has developed strategies to escape the host immune responses by constantly changing its surface antigens or by mimicking host-tissue epitopes.

Several components of the outer membrane of *N. gonorrhoeae* are required for the initial steps of the infection, and most of them demonstrate phase variation and/or antigenic variation. The attachment to epithelial cells depends on type IV pili. Although the pilin subunit PilE represents the major component of the pilus, the PilC protein, which is located at the tip of the pilus, is a key player in pilus assembly, epithelial cell adherence, cell specificity, and competence for DNA transformation (Rudel et al., 1992, 1995a, 1995b; Seifert, 1996). Additional outer membrane proteins, called Opa (for opacity), are involved in various adherence and invasion functions, such as interbacterial adhesion and interaction with human epithelial and phagocytic cells. In *N. meningitidis*, another outer membrane protein, Opc, is implicated in epithelial and endothelial cell binding (Meyer et al., 1994; Nassif, 1999; Seifert, 1996). These proteins vary either by phase variation (ON/OFF expression) or antigenic variation (gene rearrangement) mechanisms, as is detailed in the following section.

Pilin antigenic variation in *Neisseria*

The type IV pilin is encoded by *pilE*. However, the chromosome of *N. gonorrhoeae* contains many other *pil* loci; the majority of them are transcriptionally inactive and called "silent" loci or *pilS* (Haas and Meyer, 1986). The *pilS* loci lack a promoter and most of the coding sequence for the PilE N-terminal hydrophobic region. Alignment between sequences of *pilE* and the *pilS* alleles shows semivariable and hypervariable regions known as *minicassettes* (mc1–mc6; Howell-Adams et al., 1996; Meyer et al., 1994; Seifert, 1996). The most hypervariable region is immunodominant and flanked by two conserved regions, cys1 and cys2, both containing a cysteine residue. A disulfide bound between these two residues generates the hypervariable loop of the pilin. Pilin antigenic variation in *Neisseria* occurs at a high frequency (approximately 10^{-2}–10^{-3} *pilE* variants per total *pilE;* Serkin and Seifert, 1998), and involves exchanges of small DNA sequences (30–450 nt) of the variable region between *pilE* and *pilS* by a mechanism of homologous recombination involving crossing over at small conserved regions (8–40 nt; Howell-Adams and Seifert, 1999). The result does not alter the function of the pilus but changes its antigenicity. Because *N. gonorrhoeae* contains more than 15 *pilS* alleles, rearrangement with *pilE* can theoretically produce more than 10^7 different variants from one bacterial cell. The type IV pili are prime targets of the host immune system, thus by varying its antigenicity, *N. gonorrhoeae* can escape recognition (Meyer et al., 1994). This DNA exchange phenomenon is not completely understood; nevertheless, RecA (Koomey et al., 1987), as well as RecO and RecQ, recombinases are required, suggesting the involvement of the RecF-like pathway of homologous recombination (Mehr and Seifert, 1998). Furthermore, most of the DNA exchange events are nonreciprocal, meaning that the *pilS* gene copies are not modified after *pilE* variation. This type of recombination is usually called *gene conversion*. However, as mentioned by Seifert (1996, page 434), "the gene conversion describes the result of a recombination reaction, not the mechanism used to obtain the result." Several molecular models could explain this apparent gene conversion mechanism:

a. Interchromosomal homologous recombination – during a normal infection, a *N. gonorrhoeae* population is composed of several dead cells due in part to host immune responses. If the DNA released from these lysed bacteria is taken up by a viable *N. gonorrhoeae*, homologous recombination could occur between a silent locus and *pilE*.

b. Intrachromosomal homologous recombination between a *pilS* copy and the *pilE* during the process of autolysis, and the transformation of this variant *pilE* into a sibling cell.

c. Homologous recombination between daughter chromosomes during DNA replication when multiple chromosomes exist in rapidly growing cells.

d. A putative mechanism that allows unidirectional repair of the *pilS* locus by removal of the *pilE*-derived sequence (Hill and Grant, 2002; Meyer et al., 1994; Seifert, 1996).

Howell-Adams and Seifert (2000) proposed an elegant gene conversion model that would involve three recombination events. The first recombination is a cross-over with a short conserved sequence between *pilE* and one *pilS* allele to generate an extrachromosomal circular *pilE/pilS* hybrid. The two other recombinations involve a *pilE* recipient (from chromosomal replication for instance) and the *pilE/pilS* hybrid. A first cross-over would occur between another short conserved sequence in the hybrid and the recipient *pilE*, followed by a second cross-over between an extended sequence upstream of *pilE* and *pilE/pilS* hybrid, resulting in the insertion of the *pilS* variant sequence in the *pilE* recipient (Fig. 8.4). This model predicts that the recombination events with the short sequences should be RecA independent, but that the last recombination at the extended sequence must be RecA dependent.

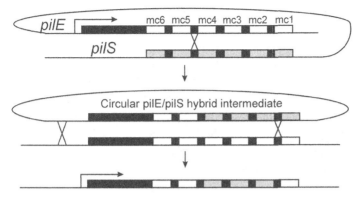

Figure 8.4. Schematic representation of gene conversion of *pilE/pilS* in *Neisseria gonorrhoeae* (see text). Conserved regions are in black, and the variable regions are white and gray. (Adapted from Howell-Adams and Seifert, 2000.)

Additional pilin-specific factors are required to achieve pilin antigenic variation in *N. gonorrhoeae* such as the recombination-dependent growth gene, *rdg*C (Mehr et al., 2000), the cys2 sequence (Howell-Adams et al., 1996), and the appropriate spacing between cys1 and cys2 (Howell-Adams and Seifert, 1999). Finally, the Sma/Cla repeat located in the *pilE* 3′-untranscribed sequence that binds proteins is also required at the *pilE* locus for efficient recombination with the silent loci (Wainwright et al., 1994, 1997).

Phase variation of pilin in *Neisseria*

The prepilin peptidases (PilD) process the PilE synthesized as a precursor (prepilin) by removing the leader sequence (first seven amino acids) at the P+ site in order to generate the mature pilin (Freitag et al., 1995). Next to the leader sequence is a hydrophobic and conserved region essential for pilin polymerization and pilus assembly. A phase variation mechanism allowing piliated bacteria (Pil+; ON) to switch to a nonpiliated phenotype (OFF) involves an alternative processing of the prepilin. This procedure implicates cleavage of the N-terminal of the precursor protein 39 amino acids downstream (at the Ps site) from the initial cleavage site, generating truncated pilins that are then secreted (soluble pilin or S-pilin) instead of forming pili (Meyer et al., 1990). These S-pilin bacteria are less piliated than the Pil+ cells but harbor a nonpiliated phenotype (Seifert, 1996). The precise mechanism of this variation is undetermined, but the control of the reversible switch from the P+ to PS sites seems associated with a reassortment of variable sequences in the *pilE* (Haas and Meyer, 1987; Haas et al., 1987). In contrast, the recombination event of antigenic variation can sometimes generate larger than normal pilin called *L-pilin*. This pilin is not assembled and is ultimately secreted but, in contrast to S-pilin cells, L-pilin cells are not piliated and display a true Pil⁻ phenotype (OFF phase). The L-pilin phenotype is unstable and often reverts to the piliated phenotype (ON phase) probably by a mechanism of deletion of the extra sequence in *pilE* (Meyer et al., 1994). In summary, two distinct phase variation mechanisms in *N. gonorrhoeae* generate the reversible switch between the Pil⁺ phenotype and the Pil⁻ phenotype by producing soluble pilin. The soluble pilin may act as a decoy to the immune system (Rytkonen et al., 2001).

Antigenic variation in *Borrelia hermsii*

The spirochete *Borrelia hermsii (B. Hermsii)* causes a tick-borne relapsing fever. A change in *B. hermsii* serotype accompanies each new cycle of fever. For instance, up to 24 antigenically distinct serotypes have been identified in the progeny of a single cell (Stoenner et al., 1982). The serotype changes are caused by antigenic variation in the lipoprotein of the outer surface. These variable major proteins (Vmps) are composed of Vsp (variable small proteins) and Vlp (the variable long proteins), encoded by multigene families displaying 39% to 78% nucleotide identity (Restrepo et al., 1992). However, only one *vsp/vlp* gene is expressed, determining the spirochete serotype, whereas the others are silent (Barbour, 1990; Barbour and Restrepo, 2000; Haake, 2000).

B. hermsii contains linear plasmids (or minichromosomes) carrying the different *vsp/vlp* copies (Barbour, 1993). Activation of a different *vsp/vlp* gene is caused by the replacement of the active gene by a silent one at the expression site and the loss of the formerly expressed *vsp/vlp* gene. One potential DNA rearrangement mechanism involves a nonreciprocal, unidirectional recombination event (gene conversion) between two different plasmids (interplasmidic recombination; Barbour et al., 1991a; Kitten and Barbour, 1990, 1992; Kitten et al., 1993; Meier et al., 1985; Plasterk et al., 1985). In contrast to *Neisseria* where parts of the sequences are exchanged, a complete gene set is switched in front of the promoter (Fig. 8.5A). The precise mechanism is not yet elucidated but appears to involve 2-kb DNA segments containing 1-kb elements with inverted repeats at their termini (Barbour et al., 1991b). The second mechanism involves an intraplasmidic recombination between the active copy and a silent copy. Recombination occurs between direct repeats located at the 5' end of the expressed and silent genes, creating a nonreplicative circular DNA and causing the loss of the formerly active

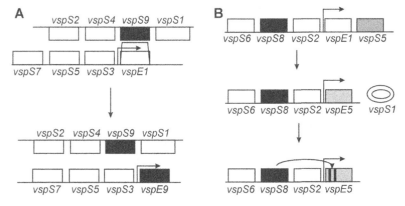

Figure 8.5. Schematic representation of gene conversion involved in antigenic variation of Vsp and Vlp in *Borrelia hermsii*. *vspS* gene cassettes are silent, and the *vspE* gene is expressed. The *vspS* and *vspE* gene numbering is hypothetical. Panel A illustrates gene conversion between two linear plasmids, and panel B, gene deletion followed by multiple point mutations by gene conversion. In panel **B**, the *vspS8* cassette serves as the DNA template for the postswitch mutations of the expressed gene.

copy (Fig. 8.5B; Restrepo et al., 1994). Interestingly, following the intraplasmidic recombination, multiple point mutations occur at the 5' end of the newly expressed copy. These postswitching mutations might take place by gene conversion between the expressed gene and a DNA template (one of the silent copy) located upstream of the expressed gene (Restrepo and Barbour, 1994). This mechanism resembles the somatic mutations used by B and T cells to further tune immunoglobulin and T-cell receptor affinity following a DNA rearrangement. Finally, transcriptional control of two plasmidic potential sites of *vsp/vlp* expression has been observed in *B. hermsii* HS1 (Carter et al., 1994). Upon expression of one site, the other is silent. In addition, the expression pattern follows the alternate mammal–tick life cycle of *B. hermsii*. This switch appears to be temperature sensitive as *B. hermsii* grown at 23°C produces tick-associated Vsp (site 1), whereas *B. hermsii* grown at 37°C, produces mammal-associated Vsp (site 2; Barbour et al., 2000; Schwan and Hinnebusch, 1998).

Antigenic variation occurs in other vector-borne pathogens and also involves major surface proteins, such as Vsp/Vlp in *B. turicatae*, VlsE in *B. burgdorferi* (Lyme disease), MSP2 in *Anaplasma marginale* (anaplasmosis), VSG in *Trypanosoma brucei* (african trypanosomiasis), PfEMP1 *Plasmodium falciparum* (malaria), and VESA1 *Babesia bovis* (babesiosis). Multigenic

PHASE VARIATION AND ANTIGENIC VARIATION

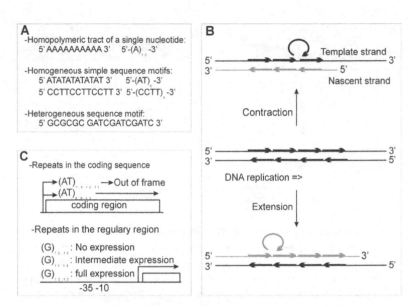

A
-Homopolymeric tract of a single nucleotide:
5' AAAAAAAAAA 3'   5'-(A)$_{1.0}$ -3'

-Homogeneous simple sequence motifs:
5' ATATATATAT 3'   5'-(AT)$_5$ -3'
5' CCTTCCTTCCTT 3' 5'-(CCTT)$_3$ -3'

-Heterogeneous sequence motif:
5' GCGCGC GATCGATCGATC 3'

C
-Repeats in the coding sequence

┌─►(AT)$_{7,8,10,11}$ ─►Out of frame
└─►(AT)$_{6,9,12}$ ──────────►
   ┌──────────────┐
   │ coding region │
   └──────────────┘

-Repeats in the regulary region

(G)$_{10,16}$: No expression
(G)$_{11,14}$: Intermediate expression
(G)$_{12,13}$: full expression ┌──────┐
                                 └──────┘
     -35 -10

B

5' ─────────────➤──➤──⟳ Template strand ─────3'
3' ─────────────◄──◄──◄──── 5'
                         Nascent strand

Contraction ↑

5' ─────────➤──➤──➤──➤─── 3'
3' ─────────◄──◄──◄──◄─── 5'

DNA replication =>

Extension ↓

5' ────────⟳──➤──➤──➤──── 3'
3' ────────◄──◄──◄──◄──── 5'

Figure 8.6. Schematic representation of slipped-strand mispairing (SSM). Panel A illustrates examples of short DNA repeats. Panel **B** illustrates the proposed mechanism of SSM during DNA replication, and panel C, the consequences of SSM on gene expression. Contraction occurs when a bulge is generated in the template strand. Extension occurs with a bulge forming in the nascent strand. Thick arrows represent repeats.

families encode all these proteins. DNA rearrangements as described in *B. hermsii* have the potential to produce antigenic variation (Barbour and Restrepo, 2000; Deitsch et al., 1997; Haake, 2000; Penningon et al., 1999).

PHASE VARIATION BY SLIPPED-STRAND MISPAIRING

A number of specific, highly mutable genes regulated by phase variation are associated with short DNA sequence repeats. These repeats can be composed of simple homopolymeric tracts of a single nucleotide, or multimeric repeats of homogeneous, heterogeneous, or degenerate repeat sequence motifs (Fig. 8.6A). These sequences are called short sequence repeats (SSRs) or variable number of tandem repeats. Genomic regions containing these repeats are subject to transient mispairing during DNA replication or DNA repair, leading to variation in the number of repeats, a process called *slipped-strand mispairing* (SSM). Genes associated with a variable simple repeat sequence have been termed *contingency genes* (Moxon et al., 1994). The growing availability of total microbial genome sequences reveals

RICHARD VILLEMUR AND ERIC DÉZIEL

that such contingency loci are widespread in many microorganisms, suggesting that SSM is a central mechanism involved in host adaptation and pathogenesis.

The general process of SSM is illustrated in Fig. 8.6B. Throughout DNA replication, a slippage occurs during leading- or lagging-strand synthesis; when the DNA polymerase reaches a repeat sequence and repeat units are unpaired, it results in a chromosomal bulge. Reduction in the repeat sequence length occurs when the bulge develops on the template strand, whereas extension arises when the polymerase slips backward, producing a bulge on the nascent strand (Henderson et al., 1999). DNA conformation such as triple-stranded DNA (H-DNA) appears to be involved (Belland, 1991). Genes involved in mismatched repair (*polI* and *mutS*) were shown to influence the number of tetra- or dinucleotide repeats in contingency reporter genes in *Haemophilus influenzae* (Bayliss et al., 2002).

Variations in these repeats can influence the expression of a contingency gene at different levels, depending on the position of the repeats in and around the gene unit:

1. *The repeats occur within the protein-coding sequence.* Upon transcription, the RNA transcript encodes a polypeptide in or out of phase, leading to the expression of longer or truncated forms of the protein (Fig. 8.6C). Because SSM can lead to contraction or expansion of the repeat sequence length, the expression switch of the encoded protein is reversible.

2. *The repeat sequence is located within the regulatory region.* The contingency gene expression can be affected; SSM can modify the spacing between −35 and −10 promoter sequences, leading to a significant increase or decrease in transcription (Fig. 8.6C). It could modulate the level of expression of individual contingency loci. The variation in the repeat's length may interfere with the interaction between the different transcriptional factors or promote the formation of specific structures in DNA or RNA.

Phase variation by SSM in *N. gonorrhoeae* and *N. meningitidis*

Phase variation by SSM occurs in *N. gonorrhoeae* and *N. meningitides* for several genes resulting in antigenic variation. As many as 12 *Opa* alleles are constitutively expressed in *N. gonorrhoeae* strains, and all possess CTCTT repeats near the start codon that vary in number. A multiple of three (6, 9, 12...) repeats produces an in-frame gene and an Opa$^+$ phenotype (ON phase), whereas 4, 5, 7, 8 . . . repeats generate an out-of-frame gene and an Opa⁻ phenotype (OFF phase). Combined with the possibility of recombination between *opa* alleles as described earlier, a very high level of antigenic variability generates an ever-changing population adding to the challenge faced by the immune system fighting this pathogen. *N. meningitidis* also possesses multiple *opa* genes that are subject to phase variation by SSM (Murphy et al., 1989; Stern et al., 1986). ON/OFF switch by SSM also occurs in *pilC1* and *pilC2*, encoding for the PilC adhesin, causing the PilC⁻/PilC⁺ phenotypes. In both genes, a homopolymeric tract (poly G) is present near the start codon (Jonsson et al., 1991).

The *opc* locus of *N. meningitidis* is susceptible to phase variation by SSM at the transcriptional level. The number of residues in a C-stretch downstream of the promoter's −10 region fluctuates in an SSM-dependent manner, thus varying the distance between the promoter and the transcriptional start, and modulating the level of expression of this locus. The presence of 12 or 13 cytidine residues leads to an efficient expression (ON phase) of *opc*, whereas 11 or 14 residues cause an intermediate expression (on phase), and finally, no expression is observed when ≤ 10 or ≥ 15 residues are present (OFF phase; Sarkari et al., 1994). The same type of phase variation mechanism is observed in the *porA* gene encoding the class 1 outer membrane protein; a G-tract located between the −10 and −35 domains contains either 11, 10, or 9 contiguous guanidine residues, resulting in high level, medium level, or no expression of the gene, respectively (van der Ende et al., 1995).

N. gonorrhoeae and *N. meningitides* produce highly immunogenic lipooligosaccharides (LOS), which are involved in bacterium–bacterium attachment (via the Opa proteins). LOS form a collection of heterogeneous and highly variable glycolipids that are organized in a complex branched formation. This complex displays a significant variation both within and between gonococcal strains in the number and size of expressed LOS components, as well as in the relative concentrations of the constituting glycolipids (Danaher et al., 1995). Homopolymeric tracts have been found in the coding sequences of genes encoding for enzymes involved in LOS synthesis, such as *pgtA*, *lgtA*, *lgtC*, *lgtD*, and *lgtG* in *N. gonorrhoeae* (Banerjee et al., 1998, 2002; Gotschlich, 1994; Yang and Gotschlich, 1996), and *pglA* and *siaD* in *N. meningitidis* (Hammerschmidt et al., 1996; Jennings et al., 1998).

Searching bacterial genomes for indications of phase variation and antigenic variation

Phase variation and antigenic variation are observed in numerous bacteria, and the mechanisms used are similar. The complete genomic sequences of an increasing number of bacteria reveal that a tremendous amount of genes with known and unknown functions contain sequence motifs and characteristics, such as homo- and heteronucleotide tracts, that make them likely candidates for differential expression by phase or antigenic variation. This section summarizes analyses of some microbial genomes regarding phase variation and antigenic variation.

Neisseria

Comparison of three *Neisseria* genomes identified sequence repeats in several loci candidates for phase variation by an SSM mechanism. They include 68 phase-variable genes in *N. meningitidis* strain Z2491, 83 candidates in *N. gonorrhoeae* strain FA1090, and 82 candidates in *N. meningitidis* strain MC58 (Snyder et al., 2001). Combined, these loci include 18 *opa* genes, 14 known genes, and 68 strong candidates adding to more than 100 identified genes whose expression is, or could be regulated, by SSM phase variation. Most phase-variable genes either encode surface exposed proteins or are involved in the biosynthesis or modification of surface structures, which supports a role in virulence and/or immune evasion (Parkhill et al., 2000a). Other potential genes are involved in bacteriocin-related activity, restriction/modification, transcription of glycosyltransferases, toxin-related products, or in other metabolic activities (Saunders et al., 2000).

Mycoplasma

Analysis of four *Mycoplasma* genomes (*M. pulmonis, Ureaplasma urealyticum, M. genitalium,* and *M. pneumoniae*) revealed that several kinds of repeats are present in these bacteria, suggesting that these genomes undergo various types of DNA rearrangements (Rocha and Blanchard, 2002). Many of the potentially affected genes code for surface-exposed proteins, which are usually subject to antigenic and phase variation. Mechanisms proposed for these rearrangements are (1) variation in protein size due to insertion or deletion of repeated elements in the structural gene, and (2) presence of multigene families and three SSRs, such as nucleotide tracts in the promoter region or the coding region. Combinations of two or three mechanisms are encountered, resulting in tremendous possible phenotype variations in *Mycoplasma*. We already discussed the Vsa/Vsp surface proteins whose expression is regulated by a DNA inversion mechanism. In addition, homo- and heteronucleotide tracts are detected in several genes in these four genomes, suggesting a SSM rearrangement mechanism. These tracts are positioned either in the intergenic sequence, presumably in the promoter region, or within the coding sequence. Among the genes putatively regulated by phase variation, most encode lipoproteins or restriction/modification systems, or are ORFs with unknown functions (Rocha and Blanchard, 2002). These types of tracts are present upstream of genes encoding the pMGA family of hemagglutinin (adhesin) proteins in *M. gallisepticum* (GAA repeats; Liu et al., 2002), the variable lipoproteins (Vlp) in *M. hyorhinis* (Yogev et al., 1991), and the variable surface protein (Vmm) in *M. mycoides* (Persson et al., 2002). They are also found in the coding region of the single-copy gene encoding the variable adherence-associated (Vaa) antigen in *M. hominis* (poly A; Zhang and Wise, 1997) and in the *pvpA* gene of *M. gallisepticum*, which encodes a putative variable cytadhesin protein (Boguslavsky et al., 2000).

Helicobacter pylori

Analysis of the *Helicobacter pylori* strain 26695 genome revealed 17 genes containing homopolymeric tracts of C or G residues, or CT or AG dinucleotide repeats within coding sequences, and poly (A) or poly (G) tracts in potential promoter region of 18 different loci. These genes encode for either surface structures, such as outer membrane proteins (OMPs), or enzymes involved in the production of lipopolysaccharides, such as glycosyltransferases and fucosyltransferase (HP379). A homopolymeric tract is present in both the coding sequence and the promoter region of five OMP genes, suggesting a complex fine-tuning of expression by phase variation (Tomb et al., 1997; Wang et al., 2000). Finally, antigenic variation may occur between members of the OMP family because they share sequences, through recombinational events, such as homologous recombination or gene conversion mechanisms, which could lead to a mosaic organization (Tomb et al., 1997).

Haemophilus

Study of the *Haemophilus influenzae* strain Rd genome uncovered tetranucleotide tracts in 12 different genes, all within coding regions (Bayliss et al., 2001). A dinucleotide tract is also present into the intergenic region between *hifA* and *hifB* genes, where the promoter for both genes resides, and interferes with the transcription of *hifA* and *hifB* (van Ham et al., 1993). Heteropolynucleotide tracts (2 to 6 nt) are also present in other *H. influenzae* strains (van Belkum et al., 1998). Functions ascribed to these genes include lipooligosaccharide (LOS) biosynthesis, production of adhesins (pili), iron acquisition, and restriction/modification system. An increased number of repeats in a tract augments the rate of phase variation (De Bolle et al., 2000). This changing rate could allow *H. influenzae* to adapt more readily to new environments. Tracts are also present within *lob1* and *lob2A*, both involved in LOS biosynthesis in *H. sommus* (McQuiston et al., 2000; Wu et al., 2000).

Campylobacter

The genome of *Campylobacter jejuni* NCTC11168 (Parkhill et al., 2000b) contains 30 genes associated with an homopolymeric tract, most being intragenic with repeats varying from 8 to 13 G or C residues. A majority of these tracts are located in loci involved in biosynthesis and modification of surface proteins, such as lipooligopolysaccharide biosynthesis locus, flagellin modification locus, and capsular polysaccharide synthesis genes (Faguy, 2000; Karlyshev et al., 2002; Linton et al., 2001; Parkhill et al., 2000b), but many of these genes have unknown functions.

PHASE VARIATION BY AN EPIGENETIC MECHANISM: DNA METHYLATION

P fimbriae (PAP) of *E. coli*

Some bacteria have developed an epigenetic (genotype not modified) phase variation mechanism to switch on and off the expression of cell surface pili-adhesin complexes. P fimbriae encoded by the *pap* (pyelonephritis-associated pilus) operon are important virulence factors of uropathogenic *E. coli*. The fimbrial shaft is composed mostly of the PapA fimbrin, whereas the PapE, PapF, and PapG fimbrins form the fibrillar tip. PapG acts as adhesin and binds to specific receptors on host target cells (Hultgren et al., 1996). Among the 11 genes of the *pap* operon, *papI* and *papB* are regulatory genes that are transcribed in opposite directions. PapB is a DNA-binding protein that binds upstream of the *papI* promoter and stimulates *papI* expression. PapI stimulates the expression of both *papB* and the *pap* operon. At elevated concentration, PapB represses the expression of the *pap* operon by binding to its own promoter (Blomfield, 2001; Donnenberg and Welch, 1996; Hal Jones et al., 1996; Hultgren et al., 1996).

P fimbriae operon expression undergoes phase variation due to differences in the DNA methylation pattern in the regulatory sequences located between the *papI* and *papB* genes (Blomfield, 2001; Henderson et al., 1999; Hernday et al. 2004). This intergenic sequence includes two regions each containing three Lrp-binding domains and one GATC site that overlaps on one of the Lrp-binding domains. The GATC site in each region is differentially methylated than the other by the deoxyadenosine methylase (Dam), depending on the phase variation status (Fig. 8.7). When transcription of the *pap* operon is OFF, the GATC-I (proximal to the *pap* operon) is methylated but the GATC-II (distal to the *pap* operon) site is not. This methylation pattern allows Lrp molecules to bind to the three Lrp-binding sites (designated sites

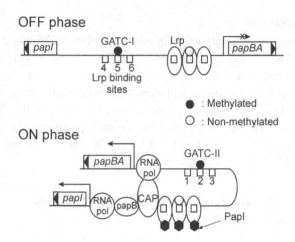

Figure 8.7. Schematic representation of the phase variation of the *pap* operon by DNA methylation. (Adapted from Low et al., 1996.)

1, 2, and 3) containing the GATC-II site, thus inhibiting the *papBA* transcription (Nou et al., 1995). Lrp is a regulatory factor with pleiotropic stimulating or inhibiting effects on the expression of many genes. In contrast, during the ON phase, the GATC methylation pattern is reversed and Lrp molecules bind now to at least two of the three Lrp-binding sites (designated 4, 5, and 6), allowing transcription of the *papBA* genes to proceed, and thus *papI*. PapI can in turn bind to Lrp proteins, which promotes the preferential binding of Lrp to sites 4, 5, and 6. Hernday et al. (2003) showed that PapI increases the affinity of Lrp for the sequence ACGATC, which is found in both GATC sites. PapI plays therefore a key role in the OFF/ON switching mechanism. The PapI/Lrp complex can also stimulate the expression of *papI*, and consequently, the PapB regulatory protein. Basal expression of *papI* or *papB* may be responsible for initiating higher levels of PapI production (Blomfield, 2001). Methylation-dependent phase variation is also involved in *sfa* and *daa* fimbrial *E. coli* operons and in *Salmonella typhimurium* plasmid-encoded fimbriae (Pef; Nicholson and Low, 2000).

   Methylation pattern changes during DNA replication when hemimethylated DNA is generated. In this context, a hemimethylated GATC-I allows the PapI/Lrp complex to bind before full methylation is accomplished. Consequently, when *E. coli* grows, two subpopulations are present, depending on the methylation pattern and the affinity of the PapI/Lrp complex. In the OFF phase, for instance, increasing level of Dam favors the methylation of GATC-I site, and therefore, the OFF mode. However, inhibition of Dam favors also the OFF mode because the methylation of GATC-II site is required for the expression of P fimbriae operon (ON mode; Hernday et al., 2002, 2003; Low et al., 1996).

Finally, Xia et al. (2000) showed a direct interaction between the type 1 fimbriae (*fim*) and pyelonephritis-associated pili (*pap*) adhesin gene clusters in uropathogenic *E. coli*. PapB inhibits phase transition via FimB by blocking the *fim* switch through binding to several regions in the type 1 fimbrial operon. In addition, PapB increases the rate of transition from the ON to the OFF phase by increasing the expression of *fimE*.

Antigen 43 (Ag43) of *E. coli*

Another example of methylation-dependent phase variation is the expression of the so-called "Antigen 43," a major outer membrane protein of *E. coli*, which is encoded by *agn43* (also called *flu*; Owen et al., 1996). Ag43 differential production observed via colony morphology and autoaggregation (Hasman et al., 1999, 2000; Henderson et al., 1997) is involved in biofilm formation (Danese et al., 2000; Kjaergaard et al., 2000).

The expression of *agn43* varies by phase variation at a rate $1-2 \times 10^{-3}$ (Owen et al., 1996) and implicates a methylation-dependent mechanism mediated by Dam, and OxyR, the latter being a LysR-type global transcriptional regulator (Henderson et al., 1999). Three GATC Dam methylation sites are located within the *agn43* regulatory sequence upstream. Methylation of at least two of these sites prevents OxyR binding to this DNA segment and allows the gene transcription to take place (ON phase; Henderson et al., 1999; Waldron et al., 2002). Upon DNA replication, hemimethylated DNA occurs, allowing OxyR to bind and cover the three GATC sites (Correnti et al., 2002; Wallecha et al., 2002). This binding interferes with RNA polymerase activity, preventing the transcription of the gene (OFF phase). OxyR can exist in both a reduced and an oxidized form, the latter containing two oxidized cysteines forming a disulfide bridge. It was shown that only the reduced form represses *agn43* (Haagmans and van der Woude, 2000; Henderson and Owen, 1999; Schembri et al. 2003). On the contrary, Wallecha et al. (2003) showed that both redox forms can repress agn43.

It was observed that the presence of type I fimbriae abolishes Ag43-mediated cell autoaggregation and affects *agn43* phase variation through the action of OxyR (Hasman et al., 1999; Schembri and Klemm, 2001). OxyR plays a role in cellular oxidative stress response and monitors the cellular thiol–disulfide status (Aslund and Beckwith, 1999; Storz and Imlay, 1999; Zheng et al., 1998) that is present in type I fimbriae. Therefore, the phase variation of type 1 fimbriae can affect the phase variation occurrence of Ag43.

Consequently, the expression of three different surface structures regulated by phase variation in *E. coli*: P fimbriae (Pap), type 1 fimbriae (Fim), and Ag43 have an influence on each others.

PHASE VARIATION IS INFLUENCED BY ENVIRONMENTAL CONDITIONS

Phase variation is conventionally viewed as an unregulated, random event, in which all phases are present simultaneously and one (or few) phenotype is favored by environmental selection. However, the rates of switching between the ON and the OFF phases can vary greatly and be influenced by the environmental conditions of the microbial cells.

Mutational processes in bacteria appear to be under environmental control, or at least susceptible to stress responses. Stressful conditions favor phenotypic variability in order to increase the survival rate (Rocha et al., 2002). As discussed previously, global regulators are often required to control the switching events, such as in the phase variation of *E. coli* type 1 and Pap pili. Expression of these factors is dictated by the physiological state of bacteria leading to the modulation of both frequency and timing of phase variation.

One key event in bacterial life is the transition between feast and famine. Starvation and stress regulatory responses overlap considerably (Hengge-Aronis, 2002). For example, specific amino acid availability influences the expression of the Lrp gene. As mentioned previously, Lrp positively regulates the *fim* switch controlling *E. coli* type 1 pili expression (Gally et al., 1994; Roesch and Blomfield, 1998). Transcription of *fimB*, the site-specific recombinase catalyzing the inversion of the *fim* segment, is strongly repressed as *E. coli* enter stationary phase. RpoS, the stress/stationary phase sigma factor performs this repression (Dove et al., 1997). In fact, *fim* switch phase variation is regulated by a complex network of regulatory factors: Lrp, RpoS, PapB, the DNA-binding and DNA-bending protein IHF, and the nucleoid-associated protein H-NS (Blomfield, 2001; Henderson et al., 1999). Moreover, the growth medium composition, osmolarity, pH, and temperature all modulate independently, sometimes synergistically, the action of FimB and FimE, the mediators of the *fim* switch. Culturing *E. coli* on minimal medium at $\geq 37^{\circ}$C favors the OFF-to-ON switch, reflecting the preference for virulence factor expression at the body temperature of mammal hosts (Gally et al., 1993). In contrast, in rich medium at temperatures $< 37^{\circ}$C, the FimB-promoted inversion occurs at 10^{-3} to 10^{-4} per cell per generation, whereas

FimE-promoted inversion to the OFF phase takes place at a much higher frequency (0.3 per cell per generation). Likewise, a low pH combined with a high
osmolarity promotes *fimE* expression and represses *fimB*, which favors a shift
to the phase OFF position and results in the loss of type 1 pili (Schwan et al.,
2002). The expression of *E. coli* P fimbriae operon also can reversibly switch
between the ON and OFF phases at 37°C but stays OFF at 23°C. The H-NS
protein inhibits the OFF-to-ON transition and can prevent methylation of the
pap GATC sites at 23°C but not at 37°C (White-Ziegler et al., 1998). The *rimJ*
gene coding for the ribosomal protein S5N-terminal acetyltransferase also
plays a role in *pap* operon thermoregulation. Insertional mutations in *rimJ*
disrupt this regulation, allowing *papBA* transcription at low temperatures. It
also increases the switching rate to the phase ON state, indicating that RimJ
inhibits the phase OFF-to-ON transition (White-Ziegler et al., 1990; White-
Ziegler and Low, 1992). Moreover, the repressive effect of rich medium on
papBA transcription is also relieved by an in-frame deletion of *rimJ*, con-
firming that RimJ modulates *papBA* transcription in response to multiple
environmental stimuli (White-Ziegler et al., 2002).

Finally, as discussed previously, *S. typhimurium* flagella undergo DNA
inversion-mediated phase variation. The site-specific recombinase Hin
catalyzes this event with the help of the Fis DNA-binding protein (Merickel
et al., 1998). The production of Fis is intensely upregulated during the
transition from the stationary to the exponential phase of growth (Ball
et al., 1992), suggesting that cell populations may undergo increased phase
variation rate when favorable growth conditions and rapid environmental
changes are encountered.

PHASE VARIATION IN ECOLOGICAL INTERACTIONS

Most studies about phase variation have been performed on human
pathogens, usually in connection to adaptation to the host environment. However, phenotypic switching is typical of all bacteria occupying heterogeneous
ecological niches, and is linked to adaptation to different environmental situations and to sudden changes in the ecosystem. Phase variation predominantly
affects surface components of bacterial cells, such as membrane antigens,
flagellum, and fimbriae, causing morphological alterations in colonies that
allow easy detection. We present here some studies where phenotypic variation is involved in ecological adaptation processes. Many of these studies are
preliminary and have not been primarily concerned about the mechanisms
of phenotypic variation.

Transition between association with a host or free-living growth

Phase variation phenomena have been observed among bacteria alternating their lifestyle between *colonization* (e.g., symbiotic, parasitic, and pathogenic interactions) and *dissemination* (free-living growth).

Photorhabdus and *Xenorhabdus* spp. are symbiotic gram-negative bacteria carried respectively in the gut of entomopathogenic nematodes of the families Heterorhabditidae and Steinernematidae. The nematode and bacteria act in concert to kill a variety of insect preys, probably via a combination of insecticidal toxin action and septicemia. Bacteria are essential for the growth and reproduction of these nematodes (Forst et al., 1997). These symbionts can undergo highly variable and reversible phenotypic switching: the forms normally isolated from the symbiotic infective-stage nematodes are the primary variants (referred to as phase I), whereas phase II variants are usually generated *in vitro* by prolonged growth in stationary phase or in a low osmolarity medium (Boemare and Akhurst, 1988; Krasomil-Osterfeld, 1995). Both variants of *Xenorhabdus* and *Photorhabdus* can be distinguished by several features, such as secreted enzymes and crystal protein production, pilus and capsule synthesis, bioluminescence, and motility, all of which are involved in the symbiotic association with nematode hosts. Phase II variants do not carry these properties or display a reduced form; therefore, they are unable to support nematode growth and reproduction (Forst et al., 1997). Accumulating evidence indicate that phenotypic variation is regulated by several genes (O'Neill et al. 2002). For instance, maintenance of the phase II variant phenotype in *Photorhabdus* requires the *hexA* gene (Joyce and Clarke, 2003). The precise role of phase II variants is unknown, but they may be better adapted to survive as free-living organisms in the soil, in part because they have an improved growth capacity under starvation conditions compared with phase I variants (Smigielski et al., 1994). This may allow bacteria to survive in the absence of a host and eventually to disperse to new hosts.

A similar situation is observed in *Listeria monocytogenes*, a ubiquitous and highly resistant gram-positive bacterium predominantly found in decaying moist vegetable and fecal material (Fenlon, 1999). Soil may serve as a reservoir for dormant *L. monocytogenes*. This facultative pathogen can cause severe infections and is responsible for several food-borne epidemics. This bacterium produces phenotypic variants that form rough colonies when plated on agar. These rough (R) variants arise spontaneously in cultures and are isolated from mice colonized by wild-type, smooth (S) variants (Zachar and Savage, 1979). R variants are stable, but they can revert to the S form *in vitro* and sometimes during colonization of infected animals. *L. monocytogenes* is obviously well adapted to grow in animal hosts, but it also persists in the environment. Interestingly, the R variants are less virulent to mice and their capacity to survive in the environment is enhanced compared with the S variants (Lenz and Portnoy, 2002). Although it is not clear how frequently environmental isolates actually contain R-type variants, the cycling between smooth and rough phases seems to accompany transition between parasitic and saprophytic growth. The superior ability of R variants to colonize abiotic surfaces supports this model (Monk et al. 2004).

Involvement of phenotypic variation in biofilm formation

The use of phase variation for the transition between attached and free-living/planktonic modes of growth is probably a widespread behavior among microorganisms. Most bacterial species can produce a specialized and highly adapted form of surface growth called biofilm. These biofilm bacteria produce an extracellular polysaccharide (EPS) matrix into which they are growing as microcolonies (Costerton et al., 1995; Stoodley et al., 2002). Interestingly, bacteria use the biofilm state both as a survival mechanism outside the host and as a phenotype of virulence.

Vibrio cholerae, the causative agent of cholera, resides within aquatic habitats during interepidemic periods as a natural inhabitant of coastal and estuarine environments, both as free-living bacteria and in association with other members of the aquatic flora. In fact, *V. cholerae* inhabits more than one ecological niche in a variety of aquatic habitats. Thus, long-term survival in aquatic environments is essential for its spread, and in these conditions, bacteria are predominantly attached to surfaces. Biofilm growth is beneficial because surfaces adsorb and thus concentrate nutrients diluted in the fluid phase. In addition, the degradation of biotic surfaces, such as chitin, by attached bacteria releases assimilable sources of carbon and nitrogen. Bacteria in biofilms are also extremely resistant to biocides and to harsh environmental conditions, such as desiccation (Costerton et al., 1995; Gilbert et al., 2002). Therefore, the surface mode of growth is apparently preferred by *V. cholerae* in natural aquatic habitats.

The production of EPS is critical for the development of a mature biofilm. In response to nutrient starvation, *V. cholerae* shifts from its normal translucent colony morphology to a rugose colony morphology. This rugose form develops a thicker and more differentiated biofilm than the smooth form, due to the production of a glucose- and galactose-rich EPS. The *vsp* gene cluster responsible for EPS synthesis is partially characterized, but the putative mechanism of phase variation has not yet been determined (Yildiz and Schoolnik, 1999; Yildiz et al., 2001). Nevertheless, microarray expression profiling studies showed that smooth-to-rugose conversion has a profound impact on the bacteria with 124 genes differentially regulated between the two morphotypes (Yildiz et al. 2004). Because rugose variants form thicker biofilms than smooth variants, they may have a survival advantage in the environment outside the human host.

A similar rugose phenotype is observed in *Salmonella typhimurium*, the rdar (red, dry and rough) morphotype. Although the regulation of this phase transition is not completely elucidated, point mutations in the promoter of *agfD* are implicated. This response regulator controls the production of the extracellular matrix components that define this aggregative and multicellular phenotype (Anriany et al., 2001; Romling et al., 1998, 2000).

In contrast, nosocomial infections caused by coagulase-negative *Staphylococcus epidermidis* correlate with the ability to elaborate a biofilm on catheters and other biomedical devices, and resistance to host defenses (Arciola et al., 2001). Development of a mature *S. epidermidis* biofilm might take place in a multistage manner in which rapid initial attachment to an inert synthetic surface is followed by cellular accumulation (Gotz, 2002; Heilmann et al., 1996). Initial adherence to biomaterials is principally mediated by polysaccharide adhesin (PS/A), which has a capsular function and thus protects bacteria from host responses. Moreover, accumulation of bacteria results from the elaboration of polysaccharide intercellular adhesin (PIA). The PIA and PS/A are chemically closely related and are both encoded by the *ica* (intercellular adhesin) operon (McKenney et al., 1998).

Production of PIA, the key virulence factor of *S. epidermidis*, is subject to phase-variable regulation and is modulated by various environmental conditions (Ziebuhr et al., 1997). The primary mechanism responsible for biofilm phenotypic variation is downregulation of *ica* operon expression by mutation (Handke et al. 2004). Moreover, at least part (one-fourth to one-third) of the phase variation process involves alternating insertion and excision of an insertion sequence (IS) element (IS256) in *icaA* or *icaC* of the *ica* operon (Ziebuhr et al., 1999). This mechanism is relevant to biofilm formation in *S. epidermidis* obtained from infections (Cho et al., 2002) because the element IS257 has been found inserted at the *ica* locus of clinical *S. epidermidis* isolates (Rohde et al., 2001). Interestingly, IS256 insertions outside the *ica* locus also participate in phenotypic switching (Conlon et al. 2004). A comparable IS-mediated mechanism of phenotypic switching controls the production of EPS adhesin in *Pseudoalteromonas atlantica*. Insertion and precise excision of the 1.2-kb IS492 at an essential locus for EPS synthesis regulate the life cycle of this biofilm-forming marine bacterium (Bartlett et al., 1988; Bartlett and Silverman, 1989).

Bacteria growing as biofilms exhibit significant phenotypic and biochemical differences from their planktonic counterparts (Stoodley et al., 2002). It is increasingly evident that this different state of growth reflects global changes in gene expression compared with freely floating cells (Prigent-Combaret et al., 1999; Whiteley et al., 2001). For example, the comparison of total proteins between planktonic *Pseudomonas aeruginosa* cells and mature biofilm cells demonstrates a sixfold or greater change in expression level of more than 800 proteins, which represent more than 50% of the proteome (Sauer et al., 2002). How such gene expression reprogramming occurs is a controversial issue. Phenotypic variants displaying multiple differences from typical planktonic bacteria, including overadherence and defective motility, play a role in initiation of biofilm development in *P. aeruginosa* (Déziel et al., 2001). Phenotypic diversity, determined by phase variation, may ensure bacteria initiating the formation of a biofilm are already present when environmental conditions are favorable to biofilm development. Thus, surface growth may represent a selective environment where biofilm initiator variants are favored. The biofilm genetic program could therefore be the result of a phase variation mechanism affecting a global regulator of gene expression.

CONCLUSION

The diversity of situations that triggers reversible phenotypic switching points toward a widespread occurrence of phase variation and antigenic variation mechanisms in all environmental conditions. DNA methylation, DNA inversion, DNA recombination, gene conversion, and slipped-strand mispairing are the basic known mechanisms of DNA rearrangements leading to phase variation and antigenic variation. Microbial pathogens have developed these mechanisms to adapt to new environmental conditions and evade immune responses of the hosts they colonize. A dramatic consequence lies in the difficulty of developing effective vaccines against these pathogens. For example, as mentioned in this chapter, *N. gonorrhoeae* is using phase variation to generate up to 10^7 different variants of its main surface antigen, the pilin PilE. Furthermore, *B. hermsii* uses a mechanism of somatic mutations similar to what is seen in B and T lymphocytes that can lead to an infinite possibility of variants. Nevertheless, the increasing availability of complete microbial genome sequences and the advent of proteomic studies will facilitate the identification of new phase variation and antigenic variation motifs.

This will be highly valuable to select more suitable surface molecule targets for the development of innovative vaccines.

ACKNOWLEDGMENTS

The authors greatly appreciate the assistance of Nathalie Arbour in reviewing the manuscript.

REFERENCES

Anriany, Y. A., Weiner, R. M., Johnson, J. A., De Rezende, C. E., and Joseph, S. W. (2001). *Salmonella enterica* serovar Typhimurium DT104 displays a rugose phenotype. *Appl. Environ. Microbiol.* **67**, 4048–4056.

Arciola, C. R., Baldassarri, L., and Montanaro, L. (2001). Presence of *icaA* and *icaD* genes and slime production in a collection of staphylococcal strains from catheter-associated infections. *J. Clin. Microbiol.* **39**, 2151–2156.

Aslund, F., and Beckwith, J. (1999). Bridge over troubled waters: Sensing stress by disulfide bond formation. *Cell* **96**, 751–753.

Bahl, H., Scholz, H., Bayan, N., Chami, M., Leblon, G., Gulik-Krzywicki, T., Shechter, E., et al. (1997). Molecular biology of S-layers. *FEMS Microbiol. Rev.* **20**, 47–98.

Ball, C. A., Osuna, R., Ferguson, K. C., and Johnson, R. C. (1992). Dramatic changes in Fis levels upon nutrient upshift in *Escherichia coli*. *J. Bacteriol.* **174**, 8043–8056.

Banerjee, A., Wang, R., Supernavage, S. L., Ghosh, S. K., Parker, J., Ganesh, N. F., Wang, P. G., et al. (2002). Implications of phase variation of a gene (*pgtA*) encoding a pilin galactosyl transferase in gonococcal pathogenesis. *J. Exp. Med.* **196**, 147–162.

Banerjee, A., Wang, R., Uljon, S. N., Rice, P. A., Gotschlich, E. C., and Stein, D.C. (1998). Identification of the gene (*lgtG*) encoding the lipooligosaccharide beta chain synthesizing glucosyl transferase from *Neisseria gonorrhoeae*. *Proc. Natl. Acad. Sci. U S A* **95**, 10872–10877.

Barbour, A. G. (1990). Antigenic variation of a relapsing fever *Borrelia* species. *Annu. Rev. Microbiol.* **44**, 155–171.

Barbour, A. G. (1993). Linear DNA of *Borrelia* species and antigenic variation. *Trends Microbiol.* **1**, 236–239.

Barbour, A. G., Burman, N., Carter, C. J., Kitten, T., and Bergstrom, S. (1991a). Variable antigen genes of the relapsing fever agent *Borrelia hermsii* are activated by promoter addition. *Mol. Microbiol.* **5**, 489–493.

RICHARD VILLEMUR AND ERIC DÉZIEL

Barbour, A. G., Carter, C. J., Burman, N., Freitag, C. S., Garon, C. F., and Bergstrom, S. (1991b). Tandem insertion sequence-like elements define the expression site for variable antigen genes of *Borrelia hermsii*. *Infect. Immun.* **59**, 390–397.

Barbour, A. G., Carter, C. J., and Sohaskey, C. D. (2000). Surface protein variation by expression site switching in the relapsing fever agent *Borrelia hermsii*. *Infect. Immun.* **68**, 7114–7121.

Barbour, A. G., and Restrepo, B. I. (2000). Antigenic variation in vector-borne pathogens. *Emerg. Infect. Dis.* **6**, 449–457.

Bartlett, D. H., and Silverman, M. (1989). Nucleotide sequence of IS492, a novel insertion sequence causing variation in extracellular polysaccharide production in the marine bacterium *Pseudomonas atlantica*. *J. Bacteriol.* **171**, 1763–1766.

Bartlett, D. H., Wright, M. E., and Silverman, M. (1988). Variable expression of extracellular polysaccharide in the marine bacterium *Pseudomonas atlantica* is controlled by genome rearrangement. *Proc. Natl. Acad. Sci. U S A* **85**, 3923–3927.

Bayliss, C. D., Field, D., and Moxon, E. R. (2001). The simple sequence contingency loci of *Haemophilus influenzae* and *Neisseria meningitidis*. *J. Clin. Invest.* **107**, 657–662.

Bayliss, C. D., van de Ven, T., and Moxon, E. R. (2002). Mutations in *polI* but not *mutSLH* destabilize *Haemophilus influenzae* tetranucleotide repeats. *EMBO J.* **21**, 1465–1476.

Belland, R. J. (1991). H-DNA formation by the coding repeat elements of neisserial *opa* genes. *Mol. Microbiol.* **5**, 2351–2360.

Bhugra, B., Voelker, L. L., Zou, N., Yu, H., and Dybvig, K. (1995). Mechanism of antigenic variation in *Mycoplasma pulmonis*: Interwoven, site-specific DNA inversions. *Mol. Microbiol.* **18**, 703–714.

Blomfield, I. C. (2001). The regulation of *pap* and type 1 fimbriation in *Escherichia coli*. *Adv. Microb. Physiol.* **45**, 1–49.

Boemare, N. E., and Akhurst, R. J. (1988). Biochemical and physiological characterization of colony form variants in *Xenorhabdus* spp. (Enterobacteriaceae). *J. Gen. Microbiol.* **134**, 751–761.

Boguslavsky, S., Menaker, D., Lysnyansky, I., Liu, T., Levisohn, S., Rosengarten, R., Garcia, M., et al. (2000). Molecular characterization of the *Mycoplasma gallisepticum pvpA* gene which encodes a putative variable cytadhesin protein. *Infect. Immun.* **68**, 3956–3964.

Bonifield, H. R., and Hughes, K. T. (2003). Flagellar phase variation in *Salmonella enterica* is mediated by a posttranscriptional control mechanism. *J. Bacteriol.* **185**, 3567–3574.

Boot, H. J., and Pouwels, P. H. (1996). Expression, secretion and antigenic variation of bacterial S-layer proteins. *Mol. Microbiol.* **21**, 1117–1123.

Carter, C. J., Bergstrom, S., Norris, S. J., and Barbour, A. G. (1994). A family of surface-exposed proteins of 20 kilodaltons in the genus *Borrelia*. *Infect. Immun.* **62**, 2792–2799.

Cho, S. H., Naber, K., Hacker, J., and Ziebuhr, W. (2002). Detection of the *icaADBC* gene cluster and biofilm formation in *Staphylococcus epidermidis* isolates from catheter-related urinary tract infections. *Int. J. Antimicrob. Agents* **19**, 570–575.

Conlon, K. M., Humphreys, H., and O'Gara, J. P. (2004). Inactivations of *rsbU* and *sarA* by IS256 represent novel mechanisms of biofilm phenotypic variation in *Staphylococcus epidermidis*. *J. Bacteriol* **186**, 6208–6219.

Correnti, J., Munster, V., Chan, T., and van der Woude, M. (2002). Dam-dependent phase variation of Ag43 in *Escherichia coli* is altered in a *seqA* mutant. *Mol. Microbiol.* **44**, 521–532.

Costerton, J. W., Lewandowski, Z., Caldwell, D. E., Korber, D. R., and Lappin-Scott, H. M. (1995). Microbial biofilms. *Annu. Rev. Microbiol.* **49**, 711–745.

Danaher, R. J., Levin, J. C., Arking, D., Burch, C. L., Sandlin, R., and Stein, D. C. (1995). Genetic basis of *Neisseria gonorrhoeae* lipooligosaccharide antigenic variation. *J. Bacteriol.* **177**, 7275–7279.

Danese, P. N., Pratt, L. A., Dove, S. L., and Kolter, R. (2000). The outer membrane protein, antigen 43, mediates cell-to-cell interactions within *Escherichia coli* biofilms. *Mol. Microbiol.* **37**, 424–432.

De Bolle, X., Bayliss, C. D., Field, D., van de Ven, T., Saunders, N. J., Hood, D. W., and Moxon, E. R. (2000). The length of a tetranucleotide repeat tract in *Haemophilus influenzae* determines the phase variation rate of a gene with homology to type III DNA methyltransferases. *Mol. Microbiol.* **35**, 211–222.

Deitsch, K. W., Moxon, E. R., and Wellems, T. E. (1997). Shared themes of antigenic variation and virulence in bacterial, protozoal, and fungal infections. *Microbiol. Mol. Biol. Rev.* **61**, 281–293.

Déziel, E., Comeau, Y., and Villemur, R. (2001). Initiation of biofilm formation by *Pseudomonas aeruginosa* 57RP correlates with emergence of hyperpiliated and highly adherent phenotypic variants deficient in swimming, swarming, and twitching motilities. *J. Bacteriol.* **183**, 1195–1204.

Donnenberg, M. S., and Welch, R. A. (1996). Virulence determinants of uropathogenic *Escherichia coli*. In H. L. Mobley and J. W. Warren (Eds.), *Urinary tract infections: Molecular pathogenesis and clinical management*, pp. 135–174. Washington, DC: ASM Press.

Dorman, C. J., and Higgins, C. F. (1987). Fimbrial phase variation in *Escherichia coli*: Dependence on integration host factor and homologies with other site-specific recombinases. *J. Bacteriol.* **169**, 3840–3843.

Dove, S. L., and Dorman, C. J. (1994). The site-specific recombination system regulating expression of the type 1 fimbrial subunit gene of *Escherichia coli* is sensitive to changes in DNA supercoiling. *Mol. Microbiol.* **14**, 975–988.

Dove, S. L., and Dorman, C. J. (1996). Multicopy *fimB* gene expression in *Escherichia coli*: Binding to inverted repeats *in vivo*, effect on *fimA* gene transcription and DNA inversion. *Mol. Microbiol.* **21**, 1161–1173.

Dove, S. L., Smith, S. G., and Dorman, C. J. (1997). Control of *Escherichia coli* type 1 fimbrial gene expression in stationary phase: A negative role for Rpos. *Mol. Gen. Genet.* **254**, 13–20.

Dworkin, J., and Blaser, M. J. (1996). Generation of *Campylobacter fetus* S-layer protein diversity utilizes a single promoter on an invertible DNA segment. *Mol. Microbiol.* **19**, 1241–1253.

Dworkin, J., and Blaser, M. J. (1997a). Molecular mechanisms of *Campylobacter fetus* surface layer protein expression. *Mol. Microbiol.* **26**, 433–440.

Dworkin, J., and Blaser, M. J. (1997b). Nested DNA inversion as a paradigm of programmed gene rearrangement. *Proc. Natl. Acad. Sci. U S A* **94**, 985–990.

Dworkin, J., Shedd, O. L., and Blaser, M. J. (1997). Nested DNA inversion of *Campylobacter fetus* S-layer genes is *recA* dependent. *J. Bacteriol.* **179**, 7523–7529.

Dworkin, J., Tummuru, M. K., and Blaser, M. J. (1995a). A lipopolysaccharide-binding domain of the *Campylobacter fetus* S-layer protein resides within the conserved N terminus of a family of silent and divergent homologs. *J. Bacteriol.* **177**, 1734–1741.

Dworkin, J., Tummuru, M. K., and Blaser, M. J. (1995b). Segmental conservation of *sapA* sequences in type B *Campylobacter fetus* cells. *J. Biol. Chem.* **270**, 15093–15101.

Dybvig, K. (1993). DNA rearrangements and phenotypic switching in prokaryotes. *Mol. Microbiol.* **10**, 465–471.

Eisenstein, B. I., Sweet, D. S., Vaughn, V., and Friedman, D. I. (1987). Integration host factor is required for the DNA inversion that controls phase variation in *Escherichia coli. Proc. Natl. Acad. Sci. U S A* **84**, 6506–6510.

Faguy, D. M. (2000). The controlled chaos of shifty pathogens. *Curr. Biol.* **10**, R498–501.

Fenlon, D. R. (1999). *Listeria monocytogenes* in the natural environment. In E. T. Ryser and E. H. Marth (Eds.), *Listeria, listeriosis, and food safety*, vol. 1, pp. 21–38. New York: Marcel Dekker.

Flitman-Tene, R., Levisohn, S., Lysnyansky, I., Rapoport, E., and Yogev, D. (2000). A chromosomal region of *Mycoplasma agalactiae* containing *vsp*-related genes undergoes *in vivo* rearrangement in naturally infected animals. *FEMS Microbiol. Lett.* **191**, 205–212.

Forst, S., Dowds, B., Boemare, N., and Stackebrandt, E. (1997). *Xenorhabdus* and *Photorhabdus* spp.: Bugs that kill bugs. *Annu. Rev. Microbiol.* **51**, 47–72.

Freitag, N. E., Seifert, H. S., and Koomey, M. (1995). Characterization of the *pilF-pilD* pilus-assembly locus of *Neisseria gonorrhoeae*. *Mol. Microbiol.* **16**, 575–586.

Gally, D. L., Bogan, J. A., Eisenstein, B. I., and Blomfield, I. C. (1993). Environmental regulation of the *fim* switch controlling type 1 fimbrial phase variation in *Escherichia coli* K-12: Effects of temperature and media. *J. Bacteriol.* **175**, 6186–6193.

Gally, D. L., Leathart, J., and Blomfield, I. C. (1996). Interaction of FimB and FimE with the *fim* switch that controls the phase variation of type 1 fimbriae in *Escherichia coli* K-12. *Mol. Microbiol.* **21**, 725–738.

Gally, D. L., Rucker, T. J., and Blomfield, I. C. (1994). The leucine-responsive regulatory protein binds to the *fim* switch to control phase variation of type 1 fimbrial expression in *Escherichia coli* K-12. *J. Bacteriol.* **176**, 5665–5672.

Gilbert, P., Maira-Litran, T., McBain, A. J., Rickard, A. H., and Whyte, F. W. (2002). The physiology and collective recalcitrance of microbial biofilm communities. *Adv. Microb. Physiol.* **46**, 202–256.

Giraud, A., Matic, I., Tenaillon, O., Clara, A., Radman, M., Fons, M., and Taddei, F. (2001a). Costs and benefits of high mutation rates: Adaptive evolution of bacteria in the mouse gut. *Science* **291**, 2606–2608.

Giraud, A., Radman, M., Matic, I., and Taddei, F. (2001b). The rise and fall of mutator bacteria. *Curr. Opin. Microbiol.* **4**, 582–585.

Glew, M. D., Marenda, M., Rosengarten, R., and Citti, C. (2002). Surface diversity in *Mycoplasma agalactiae* is driven by site-specific DNA inversions within the *vpma* multigene locus. *J. Bacteriol.* **184**, 5987–5998.

Gotschlich, E. C. (1994). Genetic locus for the biosynthesis of the variable portion of *Neisseria gonorrhoeae* lipooligosaccharide. *J. Exp. Med.* **180**, 2181–2190.

Gotz, F. (2002). *Staphylococcus* and biofilms. *Mol. Microbiol.* **43**, 1367–1378.

Gunther, N. W., Snyder, J. A., Lockatell, V., Blomfield, I., Johnson, D. E., and Mobley, H. L. (2002). Assessment of virulence of uropathogenic *Escherichia coli* type 1 fimbrial mutants in which the invertible element is phase-locked on or off. *Infect. Immun.* **70**, 3344–3354.

Gyohda, A., Funayama, N., and Komano, T. (1997). Analysis of DNA inversions in the shufflon of plasmid R64. *J. Bacteriol.* **179**, 1867–1871.

Gyohda, A., Furuya, N., Kogure, N., and Komano, T. (2002). Sequence-specific and non-specific binding of the Rci protein to the asymmetric recombination sites of the R64 shufflon. *J. Mol. Biol.* **318**, 975–983.

Gyohda, A., Furuya, N., Ishiwa, A., Zhu, S., and Komano T. (2004). Structure and function of the shufflon in plasmid R64. *Adv. Biophysics* **38**, 183–213.

Gyohda, A., and Komano, T. (2000). Purification and characterization of the R64 shufflon-specific recombinase. *J. Bacteriol.* **182**, 2787–2792.

Haagmans, W., and van der Woude, M. (2000). Phase variation of Ag43 in *Escherichia coli*: Dam-dependent methylation abrogates OxyR binding and OxyR-mediated repression of transcription. *Mol. Microbiol.* **35**, 877–887.

Haake, D. A. (2000). Spirochaetal lipoproteins and pathogenesis. *Microbiology* **146**, 1491–1504.

Haas, R., and Meyer, T. F. (1986). The repertoire of silent pilus genes in *Neisseria gonorrhoeae*: Evidence for gene conversion. *Cell* **44**, 107–115.

Haas, R., and Meyer, T. F. (1987). Molecular principles of antigenic variation in *Neisseria gonorrhoeae*. *Antonie Van Leeuwenhoek* **53**, 431–434.

Haas, R., Schwarz, H., and Meyer, T. F. (1987). Release of soluble pilin antigen coupled with gene conversion in *Neisseria gonorrhoeae*. *Proc. Natl. Acad. Sci. U S A* **84**, 9079–9083.

Hal Jones, C., Dodson, K., and Hultgren, S. J. (1996). Structure, function, and assembly of adhesive P pili. In H. L. Mobley and J. W. Warren (Eds.), *Urinary tract infections: Molecular pathogenesis and clinical management*, pp. 175–219. Washington, DC: ASM Press.

Hammerschmidt, S., Hilse, R., van Putten, J. P., Gerardy-Schahn, R., Unkmeir, A., and Frosch, M. (1996). Modulation of cell surface sialic acid expression in *Neisseria meningitidis* via a transposable genetic element. *EMBO J.* **15**, 192–198.

Handke, L. D., Conlon, K. M., Slater, S. R., Elbaruni, S., Fitzpatrick, F., Humphreys, H., Giles, W. P., Rupp, M. E., Fey, P. D., and O'Gara, J. P. (2004). Genetic and phenotypic analysis of biofilm phenotypic variation in multiple *Staphylococcus epidermidis* isolates. *J. Med. Microbiol.* **53**, 367–374.

Hasman, H., Chakraborty, T., and Klemm, P. (1999). Antigen-43-mediated autoaggregation of *Escherichia coli* is blocked by fimbriation. *J. Bacteriol.* **181**, 4834–4841.

Hasman, H., Schembri, M. A., and Klemm, P. (2000). Antigen 43 and type 1 fimbriae determine colony morphology of *Escherichia coli* K-12. *J. Bacteriol.* **182**, 1089–1095.

Heichman, K. A., and Johnson, R. C. (1990). The Hin invertasome: Protein-mediated joining of distant recombination sites at the enhancer. *Science* **249**, 511–517.

Heilmann, C., Schweitzer, O., Gerke, C., Vanittanakom, N., Mack, D., and Gotz, F. (1996). Molecular basis of intercellular adhesion in the biofilm-forming *Staphylococcus epidermidis*. *Mol. Microbiol.* **20**, 1083–1091.

Henderson, I. R., Meehan, M., and Owen, P. (1997). Antigen 43, a phase-variable bipartite outer membrane protein, determines colony morphology and autoaggregation in *Escherichia coli* K- 12. *FEMS Microbiol. Lett.* **149**, 115–120.

Henderson, I. R., and Owen, P. (1999). The major phase-variable outer membrane protein of *Escherichia coli* structurally resembles the immunoglobulin A1 protease class of exported protein and is regulated by a novel mechanism involving Dam and OxyR. *J. Bacteriol.* **181**, 2132–2141.

Henderson, I. R., Owen, P., and Nataro, J. P. (1999). Molecular switches – The ON and OFF of bacterial phase variation. *Mol. Microbiol.* **33**, 919–932.

Hengge-Aronis, R. (2002). Signal transduction and regulatory mechanisms involved in control of the sigma(S) (RpoS) subunit of RNA polymerase. *Microbiol. Mol. Biol. Rev.* **66**, 373–395.

Hernday, A. D., Braaten, B. A., and Low, D. A. (2003). The mechanism by which DNA adenine methylase and PapI activate the *pap* epigenetic switch. *Mol. Cell* **12**, 947–957.

Hernday, A., Braaten, B., and Low, D. (2004). The intricate workings of a bacterial epigenetic switch. *Adv. Exp. Med. Biol.* **547**, 83–89.

Hernday, A., Krabbe, M., Braaten, B., and Low, D. A. (2002). Self-perpetuating epigenetic pili switches in bacteria. *Proc. Nat. Acad. Sc. USA.* **99**, 16470–16476 Suppl. 4.

Hill, S. A., and Grant, C. C. (2002). Recombinational error and deletion formation in *Neisseria gonorrhoeae*: A role for RecJ in the production of *pilE* (L) deletions. *Mol. Genet. Genomics* **266**, 962–972.

Howell-Adams, B., and Seifert, H. S. (1999). Insertion mutations in *pilE* differentially alter gonococcal pilin antigenic variation. *J. Bacteriol.* **181**, 6133–6141.

Howell-Adams, B., and Seifert, H. S. (2000). Molecular models accounting for the gene conversion reactions mediating gonococcal pilin antigenic variation. *Mol. Microbiol.* **37**, 1146–1158.

Howell-Adams, B., Wainwright, L. A., and Seifert, H. S. (1996). The size and position of heterologous insertions in a silent locus differentially affect pilin recombination in *Neisseria gonorrhoeae*. *Mol. Microbiol.* **22**, 509–522.

Hultgren, S. J., Hal Jones, C., and Normark, S. (1996). Bacterial adhesins and their assembly. In R. Curtiss III, J. L. Ingraham, E. C. C. Lin, K. B. Low, B. Magasanik, W. S. Reznikoff, M. Riley, et al. (Eds.), *Escherichia coli and Salmonella. Cellular and molecular biology,* 2nd ed., pp. 2730–2756. Washington, DC: ASM Press.

Ishiwa, A., and Komano, T. (2000). The lipopolysaccharide of recipient cells is a specific receptor for PilV proteins, selected by shufflon DNA rearrangement, in liquid matings with donors bearing the R64 plasmid. *Mol. Gen. Genet.* **263**, 159–164.

Jennings, M. P., Virji, M., Evans, D., Foster, V., Srikhanta, Y. N., Steeghs, L., van der Ley, P., et al. (1998). Identification of a novel gene involved in pilin glycosylation in *Neisseria meningitidis*. *Mol. Microbiol.* **29**, 975–984.

Johnson, R. C., Glasgow, A. C., and Simon, M. I. (1987). Spatial relationship of the Fis binding sites for Hin recombinational enhancer activity. *Nature* **329**, 462–465.

Jonsson, A. B., Nyberg, G., and Normark, S. (1991). Phase variation of gonococcal pili by frameshift mutation in *pilC*, a novel gene for pilus assembly. *EMBO J.* **10**, 477–488.

Joyce, S. A., and Clarke, D. J. (2003). A *hexA* homologue from *Photorhabdus* regulates pathogenicity, symbiosis and phenotypic variation. *Mol. Microbiol.* **47**, 1445–1457.

Kanaar, R., Klippel, A., Shekhtman, E., Dungan, J. M., Kahmann, R., and Cozzarelli, N. R. (1990). Processive recombination by the phage Mu Gin system: Implications for the mechanisms of DNA strand exchange, DNA site alignment, and enhancer action. *Cell* **62**, 353–366.

Karlyshev, A. V., Linton, D., Gregson, N. A., and Wren, B. W. (2002). A novel paralogous gene family involved in phase-variable flagella-mediated motility in *Campylobacter jejuni*. *Microbiology* **148**, 473–480.

Kitten, T., and Barbour, A. G. (1990). Juxtaposition of expressed variable antigen genes with a conserved telomere in the bacterium *Borrelia hermsii*. *Proc. Natl. Acad. Sci. U S A* **87**, 6077–6081.

Kitten, T., and Barbour, A. G. (1992). The relapsing fever agent *Borrelia hermsii* has multiple copies of its chromosome and linear plasmids. *Genetics* **132**, 311–324.

Kitten, T., Barrera, A. V., and Barbour, A. G. (1993). Intragenic recombination and a chimeric outer membrane protein in the relapsing fever agent *Borrelia hermsii*. *J. Bacteriol.* **175**, 2516–2522.

Kjaergaard, K., Schembri, M. A., Ramos, C., Molin, S., and Klemm, P. (2000). Antigen 43 facilitates formation of multispecies biofilms. *Environ. Microbiol.* **2**, 695–702.

Klemm, P. (1986). Two regulatory *fim* genes, *fimB* and *fimE*, control the phase variation of type 1 fimbriae in *Escherichia coli*. *EMBO J.* **5**, 1389–1393.

Komano, T., Kim, S. R., Yoshida, T., and Nisioka, T. (1994). DNA rearrangement of the shufflon determines recipient specificity in liquid mating of IncI1 plasmid R64. *J. Mol. Biol.* **243**, 6–9.

Koomey, M., Gotschlich, E. C., Robbins, K., Bergstrom, S., and Swanson, J. (1987). Effects of *recA* mutations on pilus antigenic variation and phase transitions in *Neisseria gonorrhoeae*. *Genetics* **117**, 391–398.

Krasomil-Osterfeld, K. (1995). Influence of osmolarity on phase shift in *Photorhabdus luminescens*. *Appl. Environ. Microbiol.* **61**, 3748–3749.

Kulasekara, H. D., and Blomfield, I. C. (1999). The molecular basis for the specificity of *fimE* in the phase variation of type 1 fimbriae of *Escherichia coli* K-12. *Mol. Microbiol.* **31**, 1171–1181.

Lenz, L. L., and Portnoy, D. A. (2002). Identification of a second *Listeria secA* gene associated with protein secretion and the rough phenotype. *Mol. Microbiol.* **45**, 1043–1056.

Li, X., Lockatell, C. V., Johnson, D. E., and Mobley, H. L. (2002). Identification of MrpI as the sole recombinase that regulates the phase variation of MR/P fimbria, a bladder colonization factor of uropathogenic *Proteus mirabilis*. *Mol. Microbiol.* **45**, 865–874.

Linton, D., Karlyshev, A. V., and Wren, B. W. (2001). Deciphering *Campylobacter jejuni* cell surface interactions from the genome sequence. *Curr. Opin. Microbiol.* **4**, 35–40.

Liu, L., Panangala, V. S., and Dybvig, K. (2002). Trinucleotide GAA repeats dictate pMGA gene expression in *Mycoplasma gallisepticum* by affecting spacing between flanking regions. *J. Bacteriol.* **184**, 1335–1339.

Low, D., Braaten, B., and van der Woude, M. (1996). Fimbriae. In R. Curtiss III, J. L. Ingraham, E. C. C. Lin, K. B. Low, B. Magasanik, W. S. Reznikoff, M. Riley, M. Schaechter, and H. E. Umbarger (Eds.), *Escherichia coli and Salmonella. Cellular and molecular biology*, 2nd ed., pp. 146–157. Washington, DC: ASM Press.

Lysnyansky, I., Ron, Y., and Yogev, D. (2001). Juxtaposition of an active promoter to *vsp* genes via site-specific DNA inversions generates antigenic variation in *Mycoplasma bovis*. *J. Bacteriol.* **183**, 5698–5708.

Lysnyansky, I., Sachse, K., Rosenbusch, R., Levisohn, S., and Yogev, D. (1999). The vsp locus of Mycoplasma bovis: Gene organization and structural features. *J. Bacteriol.* **181**, 5734–5741.

MacNab, R. M. (1996). Flagella and motility. In R. Curtiss III, J. L. Ingraham, E. C. C. Lin, K. B. Low, B. Magasanik, W. S. Reznikoff, M. Riley, et al. (Eds.), *Escherichia coli and Salmonella. Cellular and molecular biology*, 2nd ed., pp. 123–145. Washington, DC: ASM Press.

McClain, M. S., Blomfield, I. C., and Eisenstein, B. I. (1991). Roles of *fimB* and *fimE* in site-specific DNA inversion associated with phase variation of type 1 fimbriae in *Escherichia coli*. *J. Bacteriol.* **173**, 5308–5314.

McKenney, D., Hubner, J., Muller, E., Wang, Y., Goldmann, D. A., and Pier, G. B. (1998). The *ica* locus of *Staphylococcus epidermidis* encodes production of the capsular polysaccharide/adhesin. *Infect. Immun.* **66**, 4711–4720.

McQuiston, J. H., McQuiston, J. R., Cox, A. D., Wu, Y., Boyle, S. M., and Inzana, T. J. (2000). Characterization of a DNA region containing 5'-(CAAT)(n)-3' DNA sequences involved in lipooligosaccharide biosynthesis in *Haemophilus somnus*. *Microb. Pathog.* **28**, 301–312.

Mehr, I. J., Long, C. D., Serkin, C. D., and Seifert, H. S. (2000). A homologue of the recombination-dependent growth gene, *rdgC*, is involved in gonococcal pilin antigenic variation. *Genetics* **154**, 523–532.

Mehr, I. J., and Seifert, H. S. (1998). Differential roles of homologous recombination pathways in *Neisseria gonorrhoeae* pilin antigenic variation, DNA transformation and DNA repair. *Mol. Microbiol.* **30**, 697–710.

Meier, J. T., Simon, M. I., and Barbour, A. G. (1985). Antigenic variation is associated with DNA rearrangements in a relapsing fever *Borrelia. Cell* **41**, 403–409.

Merickel, S. K., Haykinson, M. J., and Johnson, R. C. (1998). Communication between Hin recombinase and Fis regulatory subunits during coordinate activation of Hin-catalyzed site-specific DNA inversion. *Genes Dev* **12**, 2803–2816.

Meyer, T. F., Gibbs, C. P., and Haas, R. (1990). Variation and control of protein expression in *Neisseria. Annu. Rev. Microbiol.* **44**, 451–477.

Meyer, T. F., Pohlner, J., and van Putten, J. P. (1994). Biology of the pathogenic *Neisseriae. Curr. Top. Microbiol. Immunol.* **192**, 283–317.

Monk, I. R., Cook, G. M., Monk, B. C., and Bremer, P. J. (2004). Morphotypic conversion in *Listeria monocytogenes* biofilm formation: biological significance of rough colony isolates. *Appl. Environ. Microbiol.* **70**, 6686–6694.

Moxon, E. R., Rainey, P. B., Nowak, M. A., and Lenski, R. E. (1994). Adaptive evolution of highly mutable loci in pathogenic bacteria. *Curr. Biol.* **4**, 24–33.

Murphy, C. A., and Belas, R. (1999). Genomic rearrangements in the flagellin genes of *Proteus mirabilis. Mol. Microbiol.* **31**, 679–690.

Murphy, G. L., Connell, T. D., Barritt, D. S., Koomey, M., and Cannon, J. G. (1989). Phase variation of gonococcal protein II: Regulation of gene expression by slipped-strand mispairing of a repetitive DNA sequence. *Cell* **56**, 539–547.

Nassif, X. (1999). Interaction mechanisms of encapsulated meningococci with eucaryotic cells: What does this tell us about the crossing of the blood–brain barrier by *Neisseria meningitidis? Curr. Opin. Microbiol.* **2**, 71–77.

Nicholson, B., and Low, D. (2000). DNA methylation-dependent regulation of Pef expression in *Salmonella typhimurium. Mol. Microbiol.* **35**, 728–742.

Nou, X., Braaten, B., Kaltenbach, L., and Low, D. A. (1995). Differential binding of Lrp to two sets of *pap* DNA binding sites mediated by Pap I regulates Pap phase variation in *Escherichia coli. EMBO J.* **14**, 5785–5797.

Nunes-Duby, S. E., Kwon, H. J., Tirumalai, R. S., Ellenberger, T., and Landy, A. (1998). Similarities and differences among 105 members of the Int family of site-specific recombinases. *Nucleic Acids Res.* **26**, 391–406.

O'Gara, J. P., and Dorman, C. J. (2000). Effects of local transcription and H-NS on inversion of the *fim* switch of *Escherichia coli*. *Mol. Microbiol.* **36**, 457–466.

O'Neill, K. H., Roche, D. M., Clarke, D. J., and Dowds, B. C. (2002). The *ner* gene of *Photorhabdus*: effects on primary-form-specific phenotypes and outer membrane protein composition. *J. Bacteriol.* **184**, 3096–3105.

Owen, P., Meehan, M., de Loughry-Doherty, H., and Henderson, I. (1996). Phase-variable outer membrane proteins in *Escherichia coli*. *FEMS Immunol. Med. Microbiol.* **16**, 63–76.

Pallesen, L., Madsen, O., and Klemm, P. (1989). Regulation of the phase switch controlling expression of type 1 fimbriae in *Escherichia coli*. *Mol. Microbiol.* **3**, 925–931.

Parkhill, J., Achtman, M., James, K. D., Bentley, S. D., Churcher, C., Klee, S. R., Morelli, G., et al. (2000a). Complete DNA sequence of a serogroup A strain of *Neisseria meningitidis* Z2491. *Nature* **404**, 502–506.

Parkhill, J., Wren, B. W., Mungall, K., Ketley, J. M., Churcher, C., Basham, D., Chillingworth, T., et al. (2000b). The genome sequence of the food-borne pathogen *Campylobacter jejuni* reveals hypervariable sequences. *Nature* **403**, 665–668.

Penningon, P. M., Cadavid, D., Bunikis, J., Norris, S. J., and Barbour, A. G. (1999). Extensive interplasmidic duplications change the virulence phenotype of the relapsing fever agent *Borrelia turicatae*. *Mol. Microbiol.* **34**, 1120–1132.

Persson, A., Jacobsson, K., Frykberg, L., Johansson, K. E., and Poumarat, F. (2002). Variable surface protein Vmm of *Mycoplasma mycoides* subsp. mycoides small colony type. *J. Bacteriol.* **184**, 3712–3722.

Plasterk, R. H., Simon, M. I., and Barbour, A. G. (1985). Transposition of structural genes to an expression sequence on a linear plasmid causes antigenic variation in the bacterium *Borrelia hermsii*. *Nature* **318**, 257–263.

Prigent-Combaret, C., Vidal, O., Dorel, C., and Lejeune, P. (1999). Abiotic surface sensing and biofilm-dependent regulation of gene expression in *Escherichia coli. J. Bacteriol.* **181**, 5993–6002.

Restrepo, B. I., and Barbour, A. G. (1994). Antigen diversity in the bacterium *B. hermsii* through "somatic" mutations in rearranged *vmp* genes. *Cell* **78**, 867–876.

Restrepo, B. I., Carter, C. J., and Barbour, A. G. (1994). Activation of a *vmp* pseudogene in *Borrelia hermsii*: An alternate mechanism of antigenic variation during relapsing fever. *Mol. Microbiol.* **13**, 287–299.

Restrepo, B. I., Kitten, T., Carter, C. J., Infante, D., and Barbour, A. G. (1992). Subtelomeric expression regions of *Borrelia hermsii* linear plasmids are highly polymorphic. *Mol. Microbiol.* **6**, 3299–3311.

Rocha, E. P., and Blanchard, A. (2002). Genomic repeats, genome plasticity and the dynamics of *Mycoplasma* evolution. *Nucleic Acids Res.* **30**, 2031–2042.

Rocha, E. P., Matic, I., and Taddei, F. (2002). Over-representation of repeats in stress response genes: A strategy to increase versatility under stressful conditions? *Nucleic Acids Res.* **30**, 1886–1894.

Roesch, P. L., and Blomfield, I. C. (1998). Leucine alters the interaction of the leucine-responsive regulatory protein (Lrp) with the fim switch to stimulate site-specific recombination in *Escherichia coli*. *Mol. Microbiol.* **27**, 751–761.

Rohde, H., Knobloch, J. K., Horstkotte, M. A., and Mack, D. (2001). Correlation of biofilm expression types of *Staphylococcus epidermidis* with polysaccharide intercellular adhesin synthesis: Evidence for involvement of *icaADBC* genotype-independent factors. *Med. Microbiol. Immunol. (Berl)* **190**, 105–112.

Romling, U., Rohde, M., Olsen, A., Normark, S., and Reinkoster, J. (2000). AgfD, the checkpoint of multicellular and aggregative behaviour in *Salmonella typhimurium* regulates at least two independent pathways. *Mol. Microbiol.* **36**, 10–23.

Romling, U., Sierralta, W. D., Eriksson, K., and Normark, S. (1998). Multicellular and aggregative behaviour of *Salmonella typhimurium* strains is controlled by mutations in the *agfD* promoter. *Mol. Microbiol.* **28**, 249–264.

Ron, Y., Flitman-Tene, R., Dybvig, K., and Yogev, D. (2002). Identification and characterization of a site-specific tyrosine recombinase within the variable loci of *Mycoplasma bovis*, *Mycoplasma pulmonis* and *Mycoplasma agalactiae*. *Gene* **292**, 205–211.

Rozsa, F. W., and Marrs, C. F. (1991). Interesting sequence differences between the pilin gene inversion regions of *Moraxella lacunata* ATCC 17956 and *Moraxella bovis* Epp63. *J. Bacteriol.* **173**, 4000–4006.

Rudel, T., Facius, D., Barten, R., Scheuerpflug, I., Nonnenmacher, E., and Meyer, T. F. (1995a). Role of pili and the phase-variable PilC protein in natural competence for transformation of *Neisseria gonorrhoeae*. *Proc. Natl. Acad. Sci. U S A* **92**, 7986–7990.

Rudel, T., Scheurerpflug, I., and Meyer, T. F. (1995b). *Neisseria* PilC protein identified as type-4 pilus tip-located adhesin. *Nature* **373**, 357–359.

Rudel, T., van Putten, J. P., Gibbs, C. P., Haas, R., and Meyer, T. F. (1992). Interaction of two variable proteins (PilE and PilC) required for pilus-mediated adherence of *Neisseria gonorrhoeae* to human epithelial cells. *Mol. Microbiol.* **6**, 3439–3450.

Rytkonen, A., Johansson, L., Asp, V., Albiger, B., and Jonsson, A. B. (2001). Soluble pilin of *Neisseria gonorrhoeae* interacts with human target cells and tissue. *Infect. Immun.* **69**, 6419–6426.

Sarkari, J., Pandit, N., Moxon, E. R., and Achtman, M. (1994). Variable expression of the Opc outer membrane protein in *Neisseria meningitidis* is caused by size variation of a promoter containing poly-cytidine. *Mol. Microbiol.* **13**, 207–217.

Sauer, K., Camper, A. K., Ehrlich, G. D., Costerton, J. W., and Davies, D. G. (2002). *Pseudomonas aeruginosa* displays multiple phenotypes during development as a biofilm. *J. Bacteriol.* **184**, 1140–1154.

Saunders, N. J., Jeffries, A. C., Peden, J. F., Hood, D. W., Tettelin, H., Rappuoli, R., and Moxon, E. R. (2000). Repeat-associated phase variable genes in the complete genome sequence of *Neisseria meningitidis* strain MC58. *Mol. Microbiol.* **37**, 207–215.

Schembri, M. A., and Klemm, P. (2001). Coordinate gene regulation by fimbriae-induced signal transduction. *EMBO J.* **20**, 3074–3081.

Schembri M. A., Hjerrild L., Gjermansen M., and Per Klemm P. (2003). Differential expression of the *Escherichia coli* autoaggregation factor antigen 43. *J. Bacteriol.* **185**, 2236–2242.

Schwan, T. G., and Hinnebusch, B. J. (1998). Bloodstream- versus tick-associated variants of a relapsing fever bacterium. *Science* **280**, 1938–1940.

Schwan, W. R., Lee, J. L., Lenard, F. A., Matthews, B. T., Beck, M. T. (2002). Osmolarity and pH growth conditions regulate *fim* gene transcription and type 1 pilus expression in uropathogenic *Escherichia coli*. *Infect Immun.* **70**, 1391–1402.

Seifert, H. S. (1996). Questions about gonococcal pilus phase and antigenic variation. *Mol. Microbiol.* **21**, 433–440.

Serkin, C. D., and Seifert, H. S. (1998). Frequency of pilin antigenic variation in *Neisseria gonorrhoeae*. *J. Bacteriol.* **180**, 1955–1958.

Shen, X., Gumulak, J., Yu, H., French, C. T., Zou, N., and Dybvig, K. (2000). Gene rearrangements in the *vsa* locus of *Mycoplasma pulmonis*. *J. Bacteriol.* **182**, 2900–2908.

Sleytr, U. B., Messner, P., Pum, D., and Sara, M. (1993). Crystalline bacterial cell surface layers. *Mol. Microbiol.* **10**, 911–916.

Smigielski, A. J., Akhurst, R. J., and Boemare, N. E. (1994). Phase variation in *Xenorhabdus nematophilus* and *Photorhabdus luminescens*: Differences in respiratory activity and membrane energization. *Appl. Environ. Microbiol.* **60**, 120–125.

Smith, S. G., and Dorman, C. J. (1999). Functional analysis of the FimE integrase of *Escherichia coli* K-12: Isolation of mutant derivatives with altered DNA inversion preferences. *Mol. Microbiol.* **34**, 965–979.

Snyder, L. A., Butcher, S. A., and Saunders, N. J. (2001). Comparative whole-genome analyses reveal over 100 putative phase-variable genes in the pathogenic *Neisseria* spp. *Microbiology* **147**, 2321–2332.

Stern, A., Brown, M., Nickel, P., and Meyer, T. F. (1986). Opacity genes in *Neisseria gonorrhoeae*: Control of phase and antigenic variation. *Cell* **47**, 61–71.

Stoenner, H. G., Dodd, T., and Larsen, C. (1982). Antigenic variation of *Borrelia hermsii. J. Exp. Med.* **156**, 1297–1311.

Stoodley, P., Sauer, K., Davies, D. G., and Costerton, J. W. (2002). Biofilms as complex differentiated communities. *Annu. Rev. Microbiol.* **56**, 187–209.

Storz, G., and Imlay, J. A. (1999). Oxidative stress. *Curr. Opin. Microbiol.* **2**, 188–194.

Taddei, F., Radman, M., Maynard-Smith, J., Toupance, B., Gouyon, P. H., and Godelle, B. (1997). Role of mutator alleles in adaptive evolution. *Nature* **387**, 700–702.

Tobiason, D. M., Lenich, A. G., and Glasgow, A. C. (1999). Multiple DNA binding activities of the novel site-specific recombinase, Piv, from *Moraxella lacunata. J. Biol. Chem.* **274**, 9698–9706.

Tomb, J. F., White, O., Kerlavage, A. R., Clayton, R. A., Sutton, G. G., Fleischmann, R. D., Ketchum, K. A., et al. (1997). The complete genome sequence of the gastric pathogen *Helicobacter pylori. Nature* **388**, 539–547.

van Belkum, A., Scherer, S., van Alphen, L., and Verbrugh, H. (1998). Short-sequence DNA repeats in prokaryotic genomes. *Microbiol. Mol. Biol. Rev.* **62**, 275–293.

van de Putte, P., and Goosen, N. (1992). DNA inversions in phages and bacteria. *Trends Genet.* **8**, 457–462.

van der Ende, A., Hopman, C. T., Zaat, S., Essink, B. B., Berkhout, B., and Dankert, J. (1995). Variable expression of class 1 outer membrane protein in *Neisseria meningitidis* is caused by variation in the spacing between the −10 and −35 regions of the promoter. *J. Bacteriol.* **177**, 2475–2480.

van Ham, S. M., van Alphen, L., Mooi, F. R., and van Putten, J. P. (1993). Phase variation of *H. influenzae* fimbriae: Transcriptional control of two divergent genes through a variable combined promoter region. *Cell* **73**, 1187–1196.

Wainwright, L. A., Frangipane, J. V., and Seifert, H. S. (1997). Analysis of protein binding to the Sma/Cla DNA repeat in pathogenic *Neisseriae. Nucleic Acids Res.* **25**, 1362–1368.

Wainwright, L. A., Pritchard, K. H., and Seifert, H. S. (1994). A conserved DNA sequence is required for efficient gonococcal pilin antigenic variation. *Mol. Microbiol.* **13**, 75–87.

Waldron, D. E., Owen, P., and Dorman, C. J. (2002). Competitive interaction of the OxyR DNA-binding protein and the Dam methylase at the antigen 43 gene regulatory region in *Escherichia coli. Mol. Microbiol.* **44**, 509–520.

319

Wallecha, A., Correnti, J., Munster, V., and van der Woude M. (2003). Phase variation of Ag43 is independent of the oxidation state of OxyR. *J. Bacteriol.* **185**, 2203–2209.

Wallecha, A., Munster, V., Correnti, J., Chan, T., and van der Woude, M. (2002). Dam- and OxyR-dependent phase variation of *agn43*: Essential elements and evidence for a new role of DNA methylation. *J. Bacteriol.* **184**, 3338–3347.

Wang, G., Ge, Z., Rasko, D. A., and Taylor, D. E. (2000). Lewis antigens in *Helicobacter pylori*: Biosynthesis and phase variation. *Mol. Microbiol.* **36**, 1187–1196.

White-Ziegler, C. A., Angus Hill, M. L., Braaten, B. A., van der Woude, M. W., and Low, D. A. (1998). Thermoregulation of *Escherichia coli pap* transcription: H-NS is a temperature-dependent DNA methylation blocking factor. *Mol. Microbiol.* **28**, 1121–1137.

White-Ziegler, C. A., Black, A. M., Eliades, S. H., Young, S., and Porter, K. (2002). The N-acetyltransferase RimJ responds to environmental stimuli to repress *pap* fimbrial transcription in *Escherichia coli*. *J. Bacteriol.* **184**, 4334–4342.

White-Ziegler, C. A., Blyn, L. B., Braaten, B. A., and Low, D. A. (1990). Identification of an *Escherichia coli* genetic locus involved in thermoregulation of the *pap* operon. *J. Bacteriol.* **172**, 1775–1782.

White-Ziegler, C. A., and Low, D. A. (1992). Thermoregulation of the *pap* operon: Evidence for the involvement of RimJ, the N-terminal acetylase of ribosomal protein S5. *J. Bacteriol.* **174**, 7003–7012.

Whiteley, M., Bangera, M. G., Bumgarner, R. E., Parsek, M. R., Teitzel, G. M., Lory, S., and Greenberg, E. P. (2001). Gene expression in *Pseudomonas aeruginosa* biofilms. *Nature* **413**, 860–864.

Wu, Y., McQuiston, J. H., Cox, A., Pack, T. D., and Inzana, T. J. (2000). Molecular cloning and mutagenesis of a DNA locus involved in lipooligosaccharide biosynthesis in *Haemophilus somnus*. *Infect. Immun.* **68**, 310–319.

Xia, Y., Gally, D., Forsman-Semb, K., and Uhlin, B. E. (2000). Regulatory crosstalk between adhesin operons in *Escherichia coli*: Inhibition of type 1 fimbriae expression by the PapB protein. *EMBO J.* **19**, 1450–1457.

Yang, Q. L., and Gotschlich, E. C. (1996). Variation of gonococcal lipooligosaccharide structure is due to alterations in poly-G tracts in *lgt* genes encoding glycosyl transferases. *J. Exp. Med.* **183**, 323–327.

Yildiz, F. H., Liu, X. S., Heydorn, A., and Schoolnik, G. K. (2004). Molecular analysis of rugosity in a *Vibrio cholerae* O1 El Tor phase variant. *Mol. Microbiol.* **53**, 497–515.

Yildiz, F. H., Dolganov, N. A., and Schoolnik, G. K. (2001). VpsR, a member of the response regulators of the two-component regulatory systems, is required for expression of *vps* biosynthesis genes and EPS(ETr)-associated phenotypes in *Vibrio cholerae* O1 El Tor. *J. Bacteriol.* **183**, 1716–1726.

Yildiz, F. H., and Schoolnik, G. K. (1999). *Vibrio cholerae* O1 El Tor: Identification of a gene cluster required for the rugose colony type, exopolysaccharide production, chlorine resistance, and biofilm formation. *Proc. Natl. Acad. Sci. U S A* **96**, 4028–4033.

Yogev, D., Rosengarten, R., Watson-McKown, R., and Wise, K. S. (1991). Molecular basis of *Mycoplasma* surface antigenic variation: A novel set of divergent genes undergo spontaneous mutation of periodic coding regions and 5′ regulatory sequences. *EMBO J.* **10**, 4069–4079.

Yoshida, T., Furuya, N., Ishikura, M., Isobe, T., Haino-Fukushima, K., Ogawa, T., and Komano, T. (1998). Purification and characterization of thin pili of IncI1 plasmids ColIb-P9 and R64: Formation of PilV-specific cell aggregates by type IV pili. *J. Bacteriol.* **180**, 2842–2848.

Zachar, Z., and Savage, D. C. (1979). Microbial interference and colonization of the murine gastrointestinal tract by *Listeria monocytogenes*. *Infect. Immun.* **23**, 168–174.

Zhang, Q., and Wise, K. S. (1997). Localized reversible frameshift mutation in an adhesin gene confers a phase-variable adherence phenotype in *mycoplasma*. *Mol. Microbiol.* **25**, 859–869.

Zhao, H., Li, X., Johnson, D. E., Blomfield, I., and Mobley, H. L. (1997). *In vivo* phase variation of MR/P fimbrial gene expression in *Proteus mirabilis* infecting the urinary tract. *Mol. Microbiol.* **23**, 1009–1019.

Zheng, M., Aslund, F., and Storz, G. (1998). Activation of the OxyR transcription factor by reversible disulfide bond formation. *Science* **279**, 1718–1721.

Ziebuhr, W., Heilmann, C., Gotz, F., Meyer, P., Wilms, K., Straube, E., and Hacker, J. (1997). Detection of the intercellular adhesion gene cluster (*ica*) and phase variation in *Staphylococcus epidermidis* blood culture strains and mucosal isolates. *Infect. Immun.* **65**, 890–896.

Ziebuhr, W., Krimmer, V., Rachid, S., Lossner, I., Gotz, F., and Hacker, J. (1999). A novel mechanism of phase variation of virulence in *Staphylococcus epidermidis*: Evidence for control of the polysaccharide intercellular adhesin synthesis by alternating insertion and excision of the insertion sequence element IS256. *Mol. Microbiol.* **32**, 345–356.

CHAPTER 9

Pathogenicity Islands

Bianca Hochhut and Jörg Hacker

(323)

Infections caused by microbial pathogens are a major global health problem not only in developing countries, but also in the industrial world. Similarly, there are numerous diseases of animals and plants due to bacterial infections. Pathogenic bacteria can be found among various bacterial species, and the determination of the factors that are responsible for virulence of a certain bacterium has long been one of the main interests in microbial research. In the early 1990s, new insights into the evolution of bacterial pathogens were gained by the development of the concept of *pathogenicity islands* (PAIs), which are the topic of this chapter. Since then, the advances in genomics have led into a new area of pathogen research, with increasing knowledge of completely sequenced bacterial genomes.

The prokaryotic genome can be generally divided into a core gene pool encompassing those genes that encode essential functions, such as DNA replication, cell division, nucleotide turnover, and key metabolic pathways, and a flexible gene pool containing genes that are only required under certain environmental conditions. The core genes are normally encoded in stable regions of the chromosome and exhibit a relatively homogeneous G+C content. In contrast, genes of the flexible gene pool are often found on mobile genetic elements and are preferentially transmitted between different organisms by natural transformation (the uptake of naked DNA), phage-mediated transduction, or conjugation (the unidirectional transfer of DNA from a donor to a recipient via intimate cell-to-cell contact).

The exploitation of bacterial genome sequences has shown that genes encoding toxins, adhesins, secretion systems, invasins, or other virulence associated factors are often found on transposons, plasmids, or bacteriophages and can be horizontally transferred not only between members of the same species, but also between more distantly related species. Barksdale and

Pappenheimer (1954) discovered that corynephage β of *Corynebacterium diphtheriae* carries the gene-encoding diphtheria toxin, and subsequently, a variety of toxins has been shown to be part of phage genomes, including cholera toxin of *Vibrio cholerae*, enterobacterial shiga toxin, cytotoxin in *Pseudomonas aeruginosa*, and neurotoxin in *Clostridium botulinum* (Cheetham and Katz, 1995; Waldor and Mekalanos, 1996). Other important virulence-associated genes have been located on plasmids in gram-negative and gram-positive bacteria. Examples are the well-characterized virulence plasmids of *Yersinia* and *Shigella* spp. (Iriarte and Cornelis, 1999; Parsot and Sansonetti, 1999), the tetanus neurotoxin-encoding plasmid pCL1 of *Clostridium tetani* (Finn et al., 1984), and the virulence plasmids pXO1 and pXO2 of *Bacillus anthracis* (Mikesell et al., 1983; Thorne, 1985).

When uropathogenic *Escherichia coli* (*E. coli*) isolates were examined, no virulence-associated plasmids or phages were detected. Instead, virulence genes were found to be clustered in large regions of the chromosome that were not present in the chromosome of nonpathogenic isolates, such as laboratory strain K-12, and that were termed PAIs (Hacker et al., 1990). Based on their features, PAIs are believed to have been acquired by horizontal gene transfer. Since the first discovery of PAIs in the genome of uropathogenic *E. coli*, similar structures have been identified in the genomes of other pathogenic enterobacteria, and subsequently, in the genomes of various gram-negative and even gram-positive bacteria that are pathogens of humans, animals, and plants. PAIs are now regarded as a subgroup of genomic islands that are present in the majority of bacterial genomes and may encode accessory functions that have been laterally transferred among bacterial populations.

In this chapter, common features of PAIs are described and the role of these elements in bacterial virulence are shown with examples of relatively well-studied PAIs. Furthermore, integration, deletion, and mobilization mechanisms of PAIs are discussed. Finally, the impact of PAIs and genomic islands, in general, in microbial evolution are described.

THE CONCEPT OF PATHOGENICITY ISLANDS

After the discovery of PAIs in the genome of uropathogenic *E. coli* strains, they have also been identified in the genomes of other pathogenic enterobacteria, such as enteropathogenic and enterohemorrhagic *E. coli*, *Salmonella enterica*, *Yersinia*, and *Shigella* spp. (Hacker and Kaper, 2000). Based on the data of the genome structure of pathogenic enterobacteria, the concept of PAIs was developed. A simplified model of a PAI is shown in Fig. 9.1.

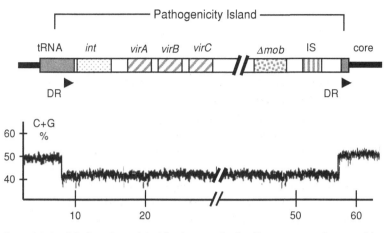

Figure 9.1. Model of a pathogenicity island (PAI). The thin line represents the core of the bacterial chromosome and PAI-specific sequences are represented by boxes (*top*). Island-associated genes are shown as shaded boxes, and the flanking direct repeats (DRs) are indicated by arrows. The G+C content of the PAI normally differs from the G+C value of the core genome (*bottom*). Abbreviations: *int*, integrase gene; *vir*, virulence-associated gene; Δ*mob*, truncated mobility genes; IS, insertion sequence. (Adapted from Hacker and Carniel, 2001.)

Generally, PAIs represent distinct pieces of DNA, which have most of the following features in common (Hacker et al., 1997; Hacker and Kaper, 2000):

1. PAIs occupy relatively large regions in the genome and can cover between 10 and more than 100 kbp, which may reflect the introduction of large pieces of DNA into a new host by horizontal gene transfer. Some strains also carry insertions of smaller pieces of DNA that encode virulence factors and that have been termed *islets* in contrast to the larger islands (Groisman and Ochman, 1997).

2. PAIs are present in the genome of pathogenic bacteria but less often present in the genome of nonpathogenic members of the same or a closely related species.

3. PAIs carry genes encoding one or more virulence factors (e.g., adhesins, invasions, type III and type IV secretion systems, toxins, and iron uptake systems). In addition, some PAIs encode regulators that play a role in the expression of virulence-associated genes.

4. PAI specific DNA sequences often display a G+C content and codon usage that differ from those of the rest of the genome. This is a further

indication that PAIs have been acquired by horizontal gene transfer, but it has to be kept in mind that differences in the G+C content of PAIs and core genome will not be observed if the genome of the donor had a similar or identical G+C content as the recipient.

5. PAIs are distinct genetic elements that are often flanked by direct repeats (DRs). These sequences may have been generated during integration of PAI-specific DNA regions into the recipient's chromosome via site-specific recombination.

6. PAIs are often associated with tRNA genes. The 3' end of tRNA loci are preferred target sites for the integration of foreign DNA (Reiter et al., 1989; Williams, 2002). Because many bacteriophages have their specific attachment sites within tRNA genes, it can be speculated that PAIs, or parts of PAIs, may represent phage-derived elements.

7. PAIs often carry functional or cryptic genes encoding mobility factors such as integrases, transposases, or parts of insertion sequence (IS) elements. Most PAIs do not represent homogeneous elements but rather exhibit mosaic-like structures that have been generated by multistep processes, including DNA rearrangements via IS elements.

8. PAIs often tend to be unstable DNA regions. Deletion of PAIs may occur via recombination between the DRs at their ends, between IS elements, or between other regions of homologous sequences between PAIs. In addition, a PAI of *Yersinia pseudotuberculosis* has been shown to be able to translocate from one tRNA target site to another within the host chromosome (Buchrieser et al., 1998), and other PAIs can be mobilized and transmitted by bacteriophages (Karaolis et al., 1999; Lindsay et al., 1998; O'Shea and Boyd, 2002).

DISTRIBUTION OF PATHOGENICITY ISLANDS

Although the concept of PAIs has been originally defined for human enterobacterial pathogens, the growing knowledge of microbial genome sequences has revealed that the presence of PAIs is not restricted to a particular group of pathogens; rather, they occur in the genomes of various human, animal, and even plant pathogens (Hacker and Kaper, 2000; Table 9.1). By now, more than 50 bacterial genome sequences have been determined, and more than 100 are currently being sequenced. Comparison of the genomic organization of closely related species led to the identification of several DNA regions that exhibit most of the features common to pathogenicity islands. Most of the characterized PAIs so far have been found in members of the *Enterobacteriaceae*, which may partly reflect the fact that this group has been intensively

Table 9.1. *Pathogenicity islands (PAIs) and their functions*

Organism	Description	Functions	Size (kbp)	Junction	Integrase	Insertion site
Escherichia coli 536 (UPEC)	PAI I$_{536}$	α Hemolysin	75.8	DR 16 bp	CP4-like (cryptic)	selC
E. coli 536 (UPEC)	PAI II$_{536}$	α Hemolysin, P fimbriae (Prf)	102	DR 18 bp	P4-like	leuX
E. coli 536 (UPEC)	PAI III$_{536}$	S fimbriae	76.8	DR 46 bp	Sfx-like	thrW
E. coli 536 (UPEC)	PAI V$_{536}$	Capsule	> 75 bp	DR 23 bp	P4-like	pheV
E. coli J96 (UPEC)	PAI I$_{J96}$	α Hemolysin, P fimbriae (Pap)	170	?	?	pheV
E. coli J96 (UPEC)	PAI II$_{J96}$	α Hemolysin, P fimbriae (Prs), cytotoxic necrotizing factor 1	110	DR 135 bp	P4-like	pheU
E. coli CFT073 (UPEC)	PAI I$_{CFT073}$	α Hemolysin, P fimbriae (Pap)	58	DR 9 bp	?	near metV/dadX
E. coli CFT073 (UPEC)	PAI II$_{CFT073}$	P fimbriae (Pap)	71	no DR	?	pheU ?
E. coli AL862	PAI I$_{AL862}$	AfaE-VIII adhesin	61	DR 136 bp (imperfect)	P4-like	pheU

(*cont.*)

(327)

Table 9.1 (*cont.*)

Organism	Description	Functions	Size (kbp)	Junction	Integrase	Insertion site
E. coli K1	kps PAI	Capsule	n.d.	?	?	pheV
E. coli E2348/69 (EPEC)	LEE	Type III secretion, invasion	35	No DR	no	selC
E. coli EPEC2	LEE	Type III secretion, invasion	35	No DR	?	pheU
E. coli 83/39 (REPEC)	LEE	Type III secretion, invasion, putative adhesin, enterotoxin	59.5	No DR	P4-like	pheU
E. coli 84/110–1 (REPEC)	LEE	Type III secretion, invasion	~85	DR 23 bp (imperfect)	P4-like	pheV
E. coli O157:H7 (EHEC)	LEE	Type III secretion, invasion	43	No DR	CP4-like	selC
E. coli EHEC	LPA	Serine protease (EspI), vitamin B_{12} receptor (BtuB), adhesin	33	No DR	CP4-like	selC
E. coli ETEC	Tia-PAI	Invasion	46		?	selC
Pathogenic *E. coli*	HPI (PAI IV_{536})	Yersiniabactin synthesis, transport	31–43	No DR ?	P4-like (cryptic?)	asnT

Organism	PAI	Functions	Size (kb)	Direct repeats	Integrase/phage	tRNA/insertion site
Yersinia enterocolitica	HPI	Yersiniabactin synthesis, transport	43	DR 17 bp	P4-like (cryptic)	asnT
Y. pseudotuberculosis	HPI	Yersiniabactin synthesis, transport	36	DR 17 bp	P4-like	asnT, U, W
Y. pestis	HPI (pgm locus)	Yersiniabactin synthesis, transport, hemin uptake	102	IS100DR 17 bp	P4-like	asnT
Shigella flexneri	SHI-1	Enterotoxin, protease	46.6	DR 22 bp (imperfect)	P4-like	pheV
S. flexneri	SHI-2	Aerobactin synthesis, transport	23–30		CP4-like	selC
S. flexneri	SRL-PAI	Ferric dicitrate transport, antibiotic resistances	66	DR 14 bp	Yes	serX
Salmonella enterica	SPI-1	Type III secretion, invasion into epithelial cells, apoptosis	40	No DR	No	between fhlA/mutS

(cont.)

329

Table 9.1 (cont.)

Organism	Description	Functions	Size (kbp)	Junction	Integrase	Insertion site
S. enterica	SPI-2	Type III secretion, invasion into monocytes	40	No DR	No	valV
S. enterica	SPI-3	Invasion, survival in monocytes	17	No DR	No	selC
S. enterica	SPI-4	Invasion, survival in monocytes	25	No DR	No	Putative tRNA gene
S. enterica	SPI-5	SPI-1 effector protein (SopB)	7			serT
Vibrio cholerae	VPI	Type IV pilus (TCP), regulator	39.5	DR 30 bp	CP4–57 like	ssrA
Helicobacter pylori	cag PAI	Type IV secretion, Cag-antigen	37	DR 31 bp	No	glr
Dichelobacter nodosus	vap region	Vap-antigens	12	DR 19 bp	CP4-like	serV
Neisseria gonorrhoeae	atlA locus	Serum resistance, cytotoxin	60–70	?	?	?
Bordetella pertussis	ptx-ptl locus	Pertussis toxin		No DR	No	tRNA (Asp)
Erwinia amylovora	Hrp PAI	Type III secretion, effectors	~60	?	Yes	pheV

330

Organism	Island	Function	Size (kb)	DR		Insertion site
Pseudomonas syringae	Hrp PAI	Type III secretion, effectors	~35	?	No	tRNA (Leu)
Agrobacterium tumefaciens	T-DNA	Crown gall tumor induction, opine production	20	DR 25 bp	No	Plasmid
Staphylococcus aureus	SaPI1	Toxic shock syndrome toxin-1	15.2	DR 17 bp	Yes	Near tyrB
S. aureus	SaPI3	Enterotoxin serotypes B, K, Q	15.9	DR 17 bp	Yes	?
Enterococcus faecalis	PPI1	Cytolysin, surface protein (Esp)	~150	DR 10 bp	?	Intergenic
Streptococcus pneumoniae	PPI1	Iron uptake system	27.2	No DR	Recombinase	yefA
Listeria ivanovii	vis gene cluster	Internalins, sphingomyelinase	18	No DR	No	

studied on a genomic level, but also indicates that PAIs seem to have played a key role in the evolution of enterobacterial pathogens. Analysis of *S. enterica* and uropathogenic *E. coli* isolates revealed that their respective pathogenicity is based on the presence of several PAIs in their genome (Dobrindt et al., 2002a; Groisman et al., 1999; Hacker et al., 1999). Enterobacteria also show frequent gene transfer by bacteriophages and plasmids, which are regarded as one source of PAIs (Hacker and Kaper, 1999; Lee, 1996). However, PAIs have also been described in bacteria that exhibit natural competence such as *Helicobacter pylori*, *Neisseria gonorrhoeae*, and even *Streptococcus pneumoniae* with the pathogenicity island PPI-1 that encodes an ABC transporter involved in iron uptake and virulence (Brown et al., 2001; Censini et al., 1996; Dillard and Seifert, 2001). In these organisms, acquisition of large pieces of foreign DNA is not believed to occur frequently because they normally tend to introduce smaller pieces of DNA into their genome.

In summary, PAIs can occur in gram-negative and gram-positive bacterial pathogens where their encoded virulence factors modulate the specific pathogenicity of their respective host. So far, the individual PAIs do not seem to be closely related to each other, and therefore, no families or classes have been defined. Instead, the term "pathogenicity island" is used to describe different types of elements with similar features. However, the comparison of the nucleotide sequence of those islands that have been completely analyzed revealed that some PAIs possess blocks (e.g., the integrase genes that are discussed later) of similarity to each other, which may suggest a modular evolution of these elements.

PATHOGENICITY ISLAND-ENCODED VIRULENCE FACTORS

PAIs are characterized by the fact that they commonly carry more than one virulence factor. Therefore, acquisition of PAIs has also been determined as an "evolution in quantum leaps" (Groisman and Ochman, 1996). Besides the fact that the traits of PAIs increase the virulence potential of their respective host, there is no other common theme for the encoded factors. Frequently, PAIs encode adhesins that play a critical role for the colonization of the eukaryotic host. Examples are the PAIs of uropathogenic *E. coli*, which carry the genes for P- or P-related fimbriae and S-fimbriae gene clusters (Hacker et al., 1999), as well as the *V. cholerae*-specific VPI (*Vibrio* pathogenicity island; Karaolis et al., 1998). This island encodes a type IV pilus, which is not only required for colonization of the host, but also is the receptor for the cholera toxin-encoding CTXφ phage (Waldor and Mekalanos, 1996). Furthermore, there is one report that VPI can be transmitted as a phage with TcpA as the major subunit of the pilus and the phage coat protein

(Karaolis et al., 1999). Dual functions of phage-specific proteins as virulence factors have also been reported for the gene products of *ace* and *zot* of CTXφ (Waldor and Mekalanos, 1996).

PAIs also encode secretion systems that are involved in delivering virulence factors to the surface of the host cell or directly into the host cell. From the five types of secretion systems that have been described in gram-negative bacteria, type III and IV systems are most closely associated with PAIs. Type III systems are structurally related to core components of the flagellar machine and have been proposed to act like a needle and syringe to inject effector proteins into host cells (reviewed in Galan and Collmer, 1999). Type III secretion systems have been found on SPI-1 and SPI-2 of *Salmonella enterica* (Groisman et al., 1999), the LEE island of *E. coli* EHEC and EPEC isolates (reviewed in Kaper et al., 1999), and on the Hrp PAI of several plant pathogens, such as *Erwinia amylovora* and *Pseudomonas syringae* (reviewed in Kim and Alfano, 2002). The proteins that comprise the type III secretion apparatus are conserved among the different secretion systems, but sequence and host cell activities of the secreted proteins differ according to the respective pathogen and its lifestyle (Galan and Collmer, 1999). Type IV secretion systems are believed to be built from core components of conjugation machines that are involved in the transfer of DNA from a donor to a recipient cell (Christie, 2001). The *cag*-PAI of *H. pylori*, which is prevalently found in "type I" strains of *H. pylori* that are more often associated with severe disease than "type II" strains, encodes a functional type IV secretion system that delivers the CagA protein into host cells (Segal et al., 1999). In *Bordetella pertussis*, a type IV secretion system is essential for translocation of pertussis toxin into mammalian cells (Burns, 1999).

The presence of PAIs can provide their hosts with the ability to invade epithelial cells or to modulate host cell activities due to invasins, modulins, and effectors encoded on PAIs. These proteins are often substrates of a specific secretion system and are located on the same PAI as their respective delivery system. In some cases, however, effectors may be located elsewhere on the chromosome as, for example, the SPI-5-associated SopB protein, which is secreted by the SPI-1-encoded type III secretion system.

As mentioned, toxins are frequently located on mobile elements such as bacteriophages and plasmids. Therefore, it was not too surprising that PAIs can also carry toxin-encoding genes. Examples are the α-hemolysins of UPEC isolates that belong to the group of pore-forming toxins and that result in lysis of eukaryotic cells after insertion into the eukaryotic cell membrane (Dobrindt and Hacker, 1999). Similarly, the more recently described PAI of *Enterococcus faecalis* encodes a cytolysin (Shankar et al., 2002). Other PAI-encoded toxins are cytotoxic necrotizing factor 1 of *E. coli* and pertussis toxin that acts by

ADP-ribosylating small G proteins (Burns, 1999; Hacker et al., 1997). The toxic shock syndrome toxin 1 of *Staphylococcus aureus* is a superantigen that binds to receptors on T cells, thereby activating these cells in an unspecific manner (Lindsay et al., 1998).

Finally, iron uptake systems are often associated with PAIs because pathogenic microbes have to scavenge this essential nutrient in competition with their host. Therefore, acquisition of iron is required to multiply in the ecological niche of the eukaryotic host and a prerequisite of the infection process. Bacteria have developed at least two strategies for iron uptake. Some microbes express receptors for iron carriers such as hemoglobin, lactoferrin, or transferrin. Another way is the synthesis and secretion of siderophores, small compounds that bind iron with high affinity. Yersiniabactin (Ybt) and aerobactin are siderophore systems that were found to be parts of PAIs. The so-called *high pathogenicity island* (HPI) that was first discovered in *Yersinia* spp. with an increased virulence for mice or humans, consists of a core element that comprises the Ybt-encoding genes and additional loci at one of the ends specific for particular species (reviewed in Carniel, 2001). In *Y. pestis*, the *pgm* locus is located next to the HPI sequences. Interestingly, HPI-related elements are widely spread among enterobacteria and cannot only be found in defined pathotypes of *E. coli*, in *Citrobacter* spp. or *Klebsiella* spp., but also in commensal *E. coli* isolates. This suggests that HPI is rather a "fitness island," increasing the ability to grow in the host rather than a true PAI. Aerobactins are often plasmid encoded and associated with the colicin ColV. In *Shigella flexneri*, the *aer* genes are located on a PAI (SHI-2), which is located next to *selC*. On this PAI, there are also regions with sequence homology to ColV-specific genes, suggesting that SHI-2 may have been evolved by integration of a plasmid encoding the aerobactin system and colicin biosynthesis (Moss et al., 1999; Vokes et al., 1999).

PATHOGENICITY ISLAND-ENCODED REGULATORS AND REGULATION OF ISLAND-ENCODED GENES

The expression of PAI-associated virulence genes is often modulated by island-encoded regulator proteins and chromosomally encoded factors. In addition, PAIs may encode regulator proteins, which influence expression of genes not encoded on the island, but elsewhere in the genome.

This is the case for the *V. cholerae* VPI-encoded ToxT, a transcriptional activator of the AraC family. ToxT is not only involved in the expression of the *tcp* gene cluster, but also controls cholera toxin production by activating expression of *ctxAB* located on CTXφ (reviewed in Klose, 2001). This is another indication for the tight linkage between the filamentous phage encoding the toxin and the VPI encoding the TCP pilus in *V. cholerae*. ToxT itself is regulated by at least two two-component systems: the TcpP/H system encoded by the island, and the chromosomally encoded ToxS/T system that is present in pathogenic and nonpathogenic *Vibrio* spp.

The regulation of the SPI-1-encoded factors in *S. enterica* has been investigated in great detail (reviewed in Lucas and Lee, 2000). Expression of the type III system and the secreted effectors depends on factors encoded within and outside SPI-1. HilA and InvF are two island-associated regulator proteins belonging to the OmpR/ToxR and AraC families of transcriptional activators, respectively. HilA positively controls the components of the secretion apparatus and InvF expression, whereby InvF activates expression of the *sip* genes, resulting in the production of the island-specific secreted proteins. *hilA* itself is not only regulated by HilC and HilD, two AraC-like proteins that are encoded by SPI-1, but also by PhoP and SirA, which are chromosomally encoded. PhoP and SirA are the response regulators of two-component systems that respond to environmental signals such as Mg^{2+} concentration, osmolarity, pH, and oxygen.

The expression of the LEE-encoded type III secretion system and the effector proteins in EHEC and EPEC strains depends on the island-encoded H-NS-like protein Ler (Mellies et al., 1999). In addition, Ler regulates the expression of a fimbrial gene cluster that is located outside the island. In both EHEC and EPEC strains, *ler* expression is modulated by integration host factor and quorum sensing. In EPEC, *ler* is also controlled by the AraC-like regulator protein PerA that is encoded by the large virulence plasmid typically found in EPEC, whereas in EHEC, the alternative sigma factor RpoS is involved in transcription of some LEE-specific genes.

Finally, regulation of PAI-encoded genes by regulators encoded by another PAI can be observed in the case of fimbrial gene cluster expression in UPEC strains and was termed *cross-talk activation* (Morschhäuser et al., 1994). In UPEC strain 536, PAI II_{536} carries the genes for P-related fimbriae, including two regulatory proteins. In addition to the P-fimbrial genes, they can also activate the S-fimbrial gene cluster that is found on PAI III_{536}.

In summary, regulation of PAI-associated genes is generally well integrated into the global regulation mechanisms of the bacterial cell and coordinately expressed with genes of the core chromosome.

INTEGRATION SITES OF PATHOGENICITY ISLANDS

The majority of the so far described PAIs in gram-negative bacteria is associated with tRNA genes or similar genes, such as *ssrA* in case of VPI in *V. cholerae* that encode small regulatory RNAs (Table 9.1). The 3' ends of tRNA genes are also preferred integration sites of bacteriophages, which suggests that PAIs may have evolved from lysogenic phages. It has been speculated as to why tRNA genes seem to represent hot spots for site-specific integration systems. First, tRNA genes are generally highly conserved sequences, which may play a role for the host range of mobile genetic elements. Second, tRNA genes exhibit symmetric sequences facilitating the binding of integrases or additional proteins that mediate the recombination events. Third, some phages carry tRNA genes in their chromosome that may have represented hot spots of recombination with the bacterial chromosome and may have been precursors of the integration systems that we now find in phages or other chromosomally integrating elements. Because integration occurs in most cases into the 3' end of the tRNA gene, the respective promoter regions normally remain functional and expression of the tRNA genes is not altered. However, in some cases, deletion of the integrated element results in interruption of the tRNA gene. Therefore, integration seem to occur preferentially into those tRNA genes that are either nonessential (e.g., *selC*) or are present in more than one copy on the chromosome (e.g., *pheV* and *pheU*). Also, the tRNA genes that function as integration sites are normally not part of an operon.

Some tRNA genes, such as *selC* encoding a selenocysteine-specific tRNA, actually seem to be hot spots of integration of foreign DNA. *selC* is associated with PAI I_{536} in uropathogenic *E. coli*, the LEE or LPA islands in EHEC, SHI-2 of *S. flexneri*, and SPI-3 of *S. enterica* (Table 9.1). *selC* is also the integration site of coliphage ϕR73, which is again indicative that PAIs may be related to phages. Interestingly, the islands of the different *E. coli* pathotypes and the SHI-2 island are not only integrated into the same region of *selC*, they also encode nearly identical integrase genes of the CP4 family. This suggests a modular evolution of the different islands with a precursor carrying the conserved integrase module.

The two genes encoding for a phenylalanine-specific tRNA, *pheV* and *pheU*, that are targeted by several elements are other hot spots of integration. For example, the two PAIs of UPEC strain J96 are associated with *pheV* and *pheU*, respectively. In some *E. coli* isolates, the LEE island is integrated into *pheU* or *pheV*, and PAIV in UPEC strain 536 is associated with *pheV*. The latter have quite similar integration modules with related integrases. Similarly, a 500-kbp "symbiotic island" of *Mesorhizobium loti* that mediates the ability to fix nitrogen is integrated into *pheV* (Sullivan and Ronson, 1998), and the enterobacterial conjugative element CTn*scr*94 that carries genes for sucrose uptake and metabolism also integrates site specifically into *pheV* or *pheU* (Hochhut et al., 1997). However, in contrast to the different elements using *selC* as an attachment site, which were found to be integrated into the same spot and all carry highly conserved integrase genes, the *pheV/U*-specific elements integrate into different regions of their target gene and encode more diverse integrases, indicating that usage of *pheV/U* as an integration site may have evolved independently in the respective elements.

As tRNA genes represent preferred target sites for the integration, a systematic screening of tRNA genes for the presence of foreign DNA can result in the identification of new PAIs. For example, SPI III of *S. enterica* or SHI-2 of *S. flexneri* have been isolated by this method.

In contrast to the PAIs of gram-negative bacteria, PAIs of gram positives seem to integrate more frequently into intergenic regions rather than into tRNA genes.

As already mentioned, PAIs are distinct pieces of DNA integrated into the backbone of the core chromosome with boundaries that are often flanked by direct repeats of varying length of 9 to up to 136 bp with one of the repeats often located within a tRNA gene. In related strains not carrying the island, there is only one copy of the DR sequence. Duplication is believed to be a result of site-specific recombination between the chromosome and the element as it has been described for lysogenic phages or other chromosomally integrating elements. This is supported by the presence of phage-related integrase genes in numerous PAIs, which also suggests that the mechanism of integration and excision of PAIs corresponds to that of phages.

INSTABILITY OF PATHOGENICITY ISLANDS

Certain PAIs show the tendency for deletion of particular PAI-specific sequences or of the complete element. Deletions within an island that occurred over longer time intervals are believed to have been important for optimization of the element's structure, such as immobilization of the element and reduction of the element's size. In contrast to these processes of "macroevolution" that are described in more detail later, some PAIs (e.g., the *cag*-PAI) show variations in their sequence in shorter time ranges. In some derivatives of the *cag*-PAI, IS605-mediated rearrangements and a tendency to reduce their size can be observed. Furthermore, loss of the complete element due to recombination events via flanking regions (rather than by excision through the DRs) has been described. It can be speculated that these events reflect an ongoing adaptation between *H. pylori* and its human host (reviewed in Odenbreit and Haas, 2002).

Deletion of complete PAIs has also been observed in uropathogenic *E. coli* strains, when colonies with a nonhemolytic phenotype were examined

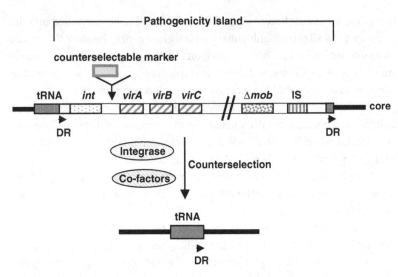

Figure 9.2. The method of island probing. For the quantification of deletion events, a counterselectable marker (e.g., *sacB* from *Bacillus subtilis*) is introduced into the pathogenicity island. Under specific conditions, those colonies that have lost the island can be isolated and further analyzed. Excision of the island by site-specific recombination probably requires an integrase and other co-factors as indicated. Abbreviations: see Fig. 9.1.

BIANCA HOCHHUT AND JÖRG HACKER

(Blum et al., 1994; Hacker et al., 1999). The molecular analysis revealed that PAI I$_{536}$ and PAI II$_{536}$ can excise from the chromosome by a site-specific process involving the DR sequences at the ends of the PAIs. After deletion, one copy of the DR is left in the chromosome, which is reminiscent of the excision mechanism of bacteriophages. Because PAIs are normally not able to replicate independently of their host chromosome, excision can result in loss of the element if the island fails to reintegrate into the chromosome. It is reasonable to speculate that the island-encoded integrases are involved in excision of the respective element. Indeed, the integrase gene associated with PAI II$_{536}$ seems to be required for deletion of this island (our unpublished results). The more recently described method of "island probing" (Rajakumar et al., 1997; Fig. 9.2) was applied to quantitate the frequency of deletion events of PAI II$_{536}$ under various growth conditions (Middendorf et al., 2004). An increase of the deletion frequency from 10^{-5} to 10^{-4} was observed when cells were grown at low temperatures (20°C), indicating that the deletion process could be influenced by specific environmental parameters. It is suggested that deletion of PAIs of UPEC strains may play a role in the adaptation of the pathogen from an acute state of infection to a chronic state where

Table 9.2. *Examples of unstable PAIs*

Organism	Pathogenicity island
Uropathogenic *E. coli*	PAI I$_{536}$
	PAI II$_{536}$
	PAI II$_{J96}$
REPEC 84/110–1	LEE
S. flexneri	SHI-1
	SHI-2
Y. pseudotuberculosis	HPI
Y. pestis	HPI
S. aureus	Sapi-1

the expression of highly antigenic factors can be disadvantageous. In this context, it is interesting that *leuX* is truncated when PAI II$_{536}$ excises from the chromosome, which affects expression of genes with a relatively high amount of *leuX* specific codons. For example, efficient expression of α-hemolysin requires an intact *leuX* allele (Dobrindt et al., 2002b). Also, translation of *fimB*, encoding a positive regulator of type 1 fimbriae in *E. coli*, is strongly dependent on the presence of *leuX*-specific tRNA$_5$^{Leu} (Ritter et al., 1997). This is an example of how a PAI may not only modulate virulence by the presence or absence of island-encoded factors, but also by affecting genes encoded elsewhere on the chromosome. Generally, pathogenic microbes that exhibit a higher genetic flexibility may have selective advantages over less flexible organisms and may be more competent in the colonization of new ecological niches.

Similarly to the PAIs of UPEC strain 536, other PAIs have been found to be unstable (Table 9.2). In fact, the SHI-1 PAI of *Shigella flexneri* was identified by its ability to delete from the chromosome with a relatively high frequency (Rajakumar et al., 1997). The LEE locus that mediates attaching and effacing lesions on epithelial cells has originally been found to be associated with *selC*, but has also been mapped in some bovine- or rabbit-specific isolates to be located next to either one of the two Phe-tRNA encoding genes. In contrast to the LEE islands integrated into *selC*, which are not flanked by direct repeats and which are very stable, the LEE locus of the rabbit EPEC strain 84/110–1 was found to be able to excise from the chromosome (Tauschek et al., 2002).

The HPI in *Yersinia* spp. shows a very diverse deletion behavior, depending on the respective strain background. In *Y. pseudotuberculosis*, the HPI deletes with a frequency of about 10^{-4} due to site-specific recombination between the flanking 17-bp DRs (Buchrieser et al., 1998). In contrast, the excision of the *Y. pestis* HPI is normally not precise but occurs as part of a bigger deletion of 102 kpb, including the pigmentation segment. This region deletes spontaneously at very high frequencies (2×10^{-3}), which is probably mediated by recombination between two flanking IS100 copies (Carniel, 2001). The *Y. pestis* HPI also possesses the capacity to excise alone because the integrase gene and the DRs are intact, but the high frequency of the bigger deletion may mask a less frequently occurring site-specific excision of the *Y. pestis* HPI. Finally, the *Y. enterocolitica* HPI is the most stable. This island lacks the potential of precise deletion because it carries a mutation within the integrase gene. In addition, there is no conserved 17-bp repeat on one site of the HPI.

The modification or deletion of the flanking DR sequences or the respective integrase genes is one way of stabilization (also termed *homing*) that has probably occurred in many other islands. For example, most islands of *S. enterica* do not possess integrase genes and are not flanked by direct repeats. In the case of the *vap* island of *D. nodosus*, an island-encoded maintenance system is believed to mediate the stable inheritance of the island (Bloomfield et al., 1997). The *toxA-vapA* operon is similar to the *higA-higB* operon of the *E. coli* plasmid Rts1 that encodes a stable toxin molecule and an unstable antidote molecule. Cells that lose the element will get killed by the toxin if it is no longer inactivated by the antidote protein. The careful analysis of other PAIs will show whether some of them may encode corresponding systems for their stable maintenance.

MOBILIZATION OF PATHOGENICITY ISLANDS

PAIs are believed to have been acquired by horizontal gene transfer, but little is known about the mechanisms that have been involved in the mobilization of PAIs. Besides a few exceptions, intra- or intercellular transfer of PAIs has not been observed in natural settings or under laboratory conditions. However, there are some examples of mobilization of PAIs. In the previous paragraph, the role of island-associated integrases in deletion processes has already been discussed. It is also assumed that integrases have played a critical role in the introduction of the islands into the recipient's chromosome. Rakin et al. (2001) were able to reconstitute the putative attachment site (*attP*) of the HPI and to show recombination with the chromosomal *attB* site. Integration occurred site specifically into any of the *asn*-tRNA genes without preference for a certain allele. Furthermore the recombination was dependent on an intact integrase gene. It is suggested that there may be a preference for integration due to recombination between *attB* and *attP* over excision of the island because no extrachromosomal form of the HPI has been detected (Rakin et al., 2001). Another indication of a functional site-specific recombination system was the observation that in particular strains of *Y. pseudotuberculosis* the HPI is able to translocate over short time periods from one target site to another. The island could be located in any of the asparagine-specific tRNA genes, *asnT, asnW,* or *asnV* (Buchrieser et al., 1998). The wide dissemination of HPI-like elements among enterobacteria implies lateral transfer by an as yet unknown mechanism (Rakin et al., 2001).

In at least two cases, intercellular transfer of a PAI has been reported. When cell-free phage preparations from a *V. cholerae* strain that was labeled with a kanamycin resistance gene inserted into the *Vibrio* pathogenicity island (VPI) were mixed with a VPI-negative strain, transfer of the VPI into the recipient strain could be observed (Karaolis et al., 1999). This experiment suggested that the VPI can be mobilized by transduction, and it was proposed that VPI may be a functional bacteriophage. However, additional evidence for the existence of a VPI-specific phage is lacking. An alternative mechanism of transfer of VPI among *V. cholerae* isolates by the generalized transducing phage CP-T1 has more recently been discovered (O'Shea and Boyd, 2002). Similarly, the SaPI1 family of elements from *S. aureus* is not capable of self-transfer; however, it can be mobilized by propagation of the staphylococcal generalized transducing phages $\phi80\alpha$ and $\phi13$ (Lindsay et al., 1998). During phage growth, SaPI1 is excised from its chromosomal *att* site and circularized; SaPI1 can replicate autonomously in the presence of $\phi80\alpha$ and is efficiently encapsidated into special small phage heads corresponding to its size (Ruzin et al., 2001). After transduction with high frequencies, SaPI1 reintegrates into the recipient's chromosome. This step is mediated by the island-encoded integrase. Taken together, SaPI1 and $\phi80\alpha$ have an intimate relationship similar to that between coliphages P4 and P2, the classical example of mobilization of a defective prophage (P4) by a helper phage (P2; Lindqvist et al., 1993). To complete a lytic cycle and to produce phage particles, P4 requires P2 to supply capsid, tail, and lysis genes. To achieve this, P4 can activate expression of the P2-specific late genes.

In this context, it is interesting that enterobacterial PAIs are frequently associated with P4-related integrases, which had led to the hypothesis that P4- or P2-related phages might have contributed to the acquisition of these PAIs. However, at least under laboratory conditions, transfer of PAI-specific DNA by P2 bacteriophages has not been shown.

SaPI1 and VPI can be considered as examples for PAIs that seem to have been derived from bacteriophages and that are mobilizable by transduction, but this may not be the only way PAIs have been acquired. In the last few years, a number of elements that have been described as "conjugative transposons," "constins" or "conjugative genomic islands" have been reported in various species of gram positives and gram negatives. They are normally integrated into the chromosome, but can excise and subsequently be transferred by conjugation (Hochhut and Waldor, 1999; Salyers et al., 1995; Scott and Churchward, 1995). Like many phages and PAIs, they encode a site-specific recombinase and carry the genes required for mating pair formation and conjugative DNA metabolism reminiscent of conjugative plasmids. Thus, it is likely that some PAIs may have been derived from integrating plasmids that have lost the genes required for replication and mobilization in expense of a more stable inheritance with the host bacterium. This is supported by the recent identification of an *E. coli* isolate that carries an HPI-derivative with an additional 35-kb region with similarity to conjugative plasmids (Schubert et al., 2004).

PATHOGENICITY ISLANDS ARE A SUBSET OF GENOMIC ISLANDS

The analysis of the growing number of completely determined bacterial DNA sequences has revealed that the concept of PAIs is not limited to the subgroup of microbes that has the ability to cause diseases. The comparison of DNA sequences has allowed identifying similar genetic structures, even in nonpathogenic bacteria. Here they may encode factors that are involved in symbiosis, resistance, or metabolism, and contribute to the adaptation to a specific ecological niche of their host organism (Table 9.3). Therefore, PAIs and related elements are of more general biological relevance than initially anticipated and have been summarized as "genomic islands" that represent blocks of DNA with signatures of mobile genetic elements (Hacker and Carniel, 2001). Depending on the functions they encode, they are termed *PAIs*, symbiosis islands, or *metabolic or resistance islands* (Hacker and Kaper, 2000). Generally, genomic islands often mediate a selective advantage to the island-carrying organism within a population under specific environmental conditions.

GENOMIC ISLANDS AND THEIR ROLE IN MICROBIAL EVOLUTION

The genome size of different variants of the same species or closely related species can vary by more than one megabase and can be accounted

Table 9.3. *Examples of genomic islands*

Organism	Description	Functions	Size [kbp]	Localization
Enterobacteriaceae	CTnscr94	Sucrose uptake and metabolism	100	*pheV* or *pheU*
P. putida	*clc* element	Chlorocatechol degradation	105	*gly* tRNA
M. loti	Symbiosis island	Nitrogen fixation	500	*phe* tRNA
S. enterica DT104	STI1	Antibiotic resistance	43	*thdf*
V. cholerae	SXT	Antibiotic resistance	100	*prfC*
Providencia rettgeri	R391	Antibiotic resistance	89	*prfC*
S. aureus	*mec* locus	Antibiotic resistance	52	Intergenic

for by the acquisition of plasmids, bacteriophages, or genomic islands. Most bacterial pathogens show evidence of horizontal gene transfer that has led together with the introduction of point mutations and DNA rearrangement to the creation of new genetic variants (Dobrindt and Hacker, 2001). In this chapter, we described the evidence that suggests that PAIs and genomic islands in general have been introduced by lateral gene transfer processes. In contrast to other mobile genetic elements that can still be actively transferred among populations and may play an important role in the fast adaptation to changing environmental conditions, such as the presence of antibiotics, PAIs have been major contributors to the development of pathogenic microbes in the scale of macroevolution. The acquisition of PAIs results in the occupation of new niches due to the functions they encode (e.g., by breaking the host cellular and anatomic barriers and the avoidance of defense mechanisms). For example, *E. coli* is normally part of the flora of the large bowel, but the presence of certain PAIs resulted in the development of new pathotypes that were able to colonize the small intestine or the urinary tract. In addition, some PAIs still show a tendency to internal or total deletions and rearrangements that seem to play a role in the adaptation during an acute infection.

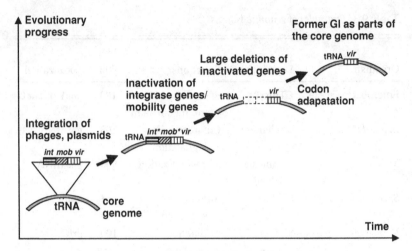

Figure 9.3. Stages of pathogenicity island (PAI) evolution. The evolutionary progress (*vertical line*) is pictured in a time scale (*horizontal line*). Grey lines represent the core genome, and PAI-specific DNA is represented by boxes. Mutations are indicated by asterisks. Abbreviations: tRNA, transfer tRNA gene; *int*, integrase gene; *mob*, mobility gene; *vir*, virulence-associated gene. (Adapted from Hacker and Kaper, 2000.)

Although the origin of genomic islands is mostly unknown, it has been speculated that integrating phages or plasmids may have represented PAI precursors or pre-PAIs (Hacker et al., 1997; Lee, 1996). DNA rearrangements, point mutations, and deletion events subsequently modified the elements' structures and mobilization capacities (Fig. 9.3). The various types of PAIs as we find them now in the bacterial chromosomes may represent different stages of these evolutionary processes. The SaPI1 island of *S. aureus* is an example of an early state as it is still mobilizable, whereas the PAIs of *S. enterica* are very ancient and are no longer mobile. If PAIs become part of the chromosome, the genotype of a bacterium is permanently altered indicating that genomic islands are important for the creation of new species.

REFERENCES

Barksdale, W. L., and Pappenheimer, A. M., Jr. (1954). Phage–host relationships in nontoxigenic and toxigenic diphtheria bacilli. *J. Bacteriol. 67*, 220–232.

Bloomfield, G. A., Whittle, G., McDonagh, M. B., Katz, M. E., and Cheetham, B. F. (1997). Analysis of sequences flanking the *vap* region of *Dichelobacter nodosus*: Evidence for multiple integration events, a killer system, and a new genetic element. Microbiology *143*, 553–562.

Blum, G., Ott, M., Lischewski, A., Ritter, A., Imrich, H., Tschäpe, H., and Hacker, J. (1994). Excision of large DNA regions termed pathogenicity islands from tRNA-specific loci in the chromosome of an *Escherichia coli* wild-type pathogen. Infect. Immun. *62*, 606–614.

Brown, J. S., Gilliland, S. M., and Holden, D. W. (2001). A *Streptococcus pneumoniae* pathogenicity island encoding an ABC transporter involved in iron uptake and virulence. Mol. Microbiol. *40*, 572–585.

Buchrieser, C., Brosch, R., Bach, S., Guiyoule, A., and Carniel, E. (1998). The high-pathogenicity island of *Yersinia pseudotuberculosis* can be inserted into any of the three chromosomal *asn* tRNA genes. Mol. *Microbiol. 30*, 965–978.

Burns, D. L. (1999). Biochemistry of type IV secretion. Curr. Opin. Microbiol. *2*, 25–29.

Carniel, E. (2001). The *Yersinia* high-pathogenicity island: An iron-uptake island. Microbes Infect. *3*, 561–569.

Censini, S., Lange, C., Xiang, Z., Crabtree, J. E., Ghiara, P., Borodovsky, M., et al. (1996). *cag*, A pathogenicity island of *Helicobacter pylori*, encodes type-I specific and disease-associated virulence factors. Proc. Natl. Acad. Sci. U S A *93*, 14648–14653.

Cheetham, B. F., and Katz, M. E. (1995). A role for bacteriophages in the evolution and transfer of bacterial virulence determinants. Mol. Microbiol. *18*, 201–208.

Christie, P. J. (2001). Type IV secretion: Intercellular transfer of macromolecules by systems ancestrally related to conjugation machines. Mol. Microbiol. *40*, 294–305.

Dillard, J. P., and Seifert, H. S. (2001). A variable genetic island specific for *Neisseria gonorrhoeae* is involved in providing DNA for natural transformation and is found more often in disseminated infection isolates. Mol. Microbiol. *41*, 263–277.

Dobrindt, U., Blum-Oehler, G., Nagy, G., Schneider, G., Johann, A., Gottschalk, G., and Hacker, J. (2002a). Genetic structure and distribution of four pathogenicity islands (PAI I_{536}-PAI IV_{536}) of uropathogenic *Escherichia coli* strain 536. Infect. Immun. *70*, 6365–6372.

Dobrindt, U., Emödy, L., Gentschev, I., Goebel, W., and Hacker, J. (2002b). Efficient expression of the α-haemolysin determinant in the uropathogenic *Escherichia coli* strain 536 requires the *leuX*-encoded $tRNA_5^{Leu}$. Mol. Genet. Genomics *267*, 370–379.

Dobrindt, U., and Hacker, J. (1999). Plasmids, phages and pathogenicity islands: lessons on the evolution of bacterial toxins. In J. Alouf and J. Freer, eds. *The comprehensive sourcebook of bacterial protein toxins* (New York: Academic Press), pp. 3–23.

Dobrindt, U., and Hacker, J. (2001). Whole genome plasticity in pathogenic bacteria. Curr. Opin. Microbiol. 4, 550–557.

Finn, C. W., Jr., Silver, R. P., Habig, W. H., Hardgree, M. C., Zon, G., and Garon, C. F. (1984). The structural gene for tetanus neurotoxin is on a plasmid. Science 224, 881–884.

Galan, J. E., and Collmer, A. (1999). Type III secretion machines: Bacterial devices for protein delivery into host cells. Science 284, 1322–1328.

Groisman, E. A., Blanc-Potard, A.-B., and Uchiya, K. (1999). Pathogenicity islands and the evolution of Salmonella virulence. In J. B. Kaper and J. Hacker, eds. Pathogenicity islands and other mobile virulence elements (Washington, DC: ASM Press), pp. 127–150.

Groisman, E. A., and Ochman, H. (1996). Pathogenicity islands: Bacterial evolution in quantum leaps. Cell 87, 791–794.

Groisman, E. A., and Ochman, H. (1997). How Salmonella became a pathogen. Trends Microbiol. 5, 343–349.

Hacker, J., Bender, L., Ott, M., Wingender, J., Lund, B., Marre, R., and Goebel, W. (1990). Deletions of chromosomal regions coding for fimbriae and hemolysins occur in vitro and in vivo in various extraintestinal Escherichia coli isolates. Microb. Pathog. 8, 213–225.

Hacker, J., Blum-Oehler, G., Janke, B., Nagy, G., and Goebel, W. (1999). Pathogenicity islands of extraintestinal Escherichia coli. In J. B. Kaper and J. Hacker, eds. Pathogenicity islands and other mobile virulence elements (Washington, DC: ASM Press), pp. 59–76.

Hacker, J., Blum-Oehler, G., Mühldorfer, I., and Tschäpe, H. (1997). Pathogenicity islands of virulent bacteria: Structure, function and impact on microbial evolution. Mol. Microbiol. 23, 1089–1097.

Hacker, J., and Carniel, E. (2001). Ecological fitness, genomic islands and bacterial pathogenicity. EMBO Rep. 2, 376–381.

Hacker, J., and Kaper, J. B. (1999). The concept of pathogenicity islands. In J. B. Kaper and J. Hacker, eds. Pathogenicity islands and other mobile virulence elements (Washington, DC: ASM Press), pp. 1–11.

Hacker, J., and Kaper, J. B. (2000). Pathogenicity islands and the evolution of microbes. Annu. Rev. Microbiol. 54, 641–679.

Hochhut, B., Jahreis, K., Lengeler, J. W., and Schmid, K. (1997). CTnscr94, a conjugative transposon found in enterobacteria. J. Bacteriol. 179, 2097–2102.

Hochhut, B., and Waldor, M. K. (1999). Site-specific integration of the conjugal Vibrio cholerae SXT element into prfC. Mol. Microbiol. 32, 99–110.

Iriarte, M., and Cornelis, G. R. (1999). The 70-kilobase virulence plasmid of yersiniae. In J. B. Kaper and J. Hacker, eds. Pathogenicity islands and other mobile virulence elements (Washington, DC: ASM Press), pp. 91–126.

Kaper, J. B., Mellies, J. L., and Nataro, J. P. (1999). Pathogenicity islands and other mobile genetic elements of diarrheagenic *Escherichia coli*. In J. B. Kaper and J. Hacker, eds. *Pathogenicity islands and other mobile virulence elements* (Washington, DC: ASM Press), pp. 33–58.

Karaolis, D. K. R., Johnson, J. A., Bailey, C. C., Boedeker, E. C., Kaper, J. B., and Reeves, P. R. (1998). A *Vibrio cholerae* pathogenicity island associated with epidemic and pandemic strains. Proc. Natl. Acad. Sci. U S A 95, 3134–3139.

Karaolis, D. K. R., Somara, S., Maneval, D. R. Jr., Johnson, J. A., and Kaper, J. B. (1999). A bacteriophage encoding a pathogenicity island, a type IV pilus and a phage receptor in cholera bacteria. Nature 399, 375–379.

Kim, J. F., and Alfano, J. R. (2002). Pathogenicity islands and virulence plasmids of bacterial plant pathogens. In J. Hacker and J. B. Kaper, eds. *Pathogenicity islands and the evolution of pathogenic microbes* (Berlin Heidelberg: Springer-Verlag), pp. 127–147.

Klose, K. E. (2001). Regulation of virulence in *Vibrio cholerae*. Int. J. Med. Microbiol. *291*, 81–88.

Lee, C. A. (1996). Pathogenicity islands and the evolution of bacterial pathogens. Infect. Agents Dis. *5*, 1–7.

Lindqvist, B. H., Deho, G., and Calendar, R. (1993). Mechanisms of genome propagation and helper exploitation by satellite phage P4. Microbiol. Rev. *57*, 683–702.

Lindsay, J. A., Ruzin, A., Ross, H. F., Kurepina, N., and Novick, R. P. (1998). The gene for toxic shock toxin is carried by a family of mobile pathogenicity islands in *Staphylococus aureus*. Mol. Microbiol. *29*, 527–543.

Lucas, R. L., and Lee, C. A. (2000). Unravelling the mysteries of virulence gene regulation in *Salmoella typhimurium*. Mol. Microbiol. *36*, 1024–1033.

Mellies, J. L., Elliott, S. J., Sperandio, V., Donnenberg, M. S., and Kaper, J. B. (1999). The Per regulon of enteropathogenic *Escherichia coli*: Identification of a regulatory cascade and a novel transcriptional activator, the locus of enterocyte effacement (LEE)-encoded regulator (Ler). Mol. Microbiol. *33*, 296–306.

Middendorf, B., Hochhut, B., Leipold, K., Dobrindt, U., Blum-Oehler, G., and Hacker, J. (2004). Instability of pathogenicity islands in uropathogenic *Escherichia coli* 536. J. Bacteriol. *186*, 3086–3096

Mikesell, P., Ivins, B. E., Ristroph, J. D., and Dreier, T. M. (1983). Evidence for plasmid-mediated toxin production in *Bacillus anthracis*. Infect. Immun. *39*, 371–376.

Morschhäuser, J., Vetter, V., Emödy, L., and Hacker, J. (1994). Adhesin regulatory genes within large, unstable DNA regions of pathogenic *Escherichia coli*:

Cross-talk between different adhesin gene clusters. Mol. Microbiol. *11*, 555–566.

Moss, J. E., Cardozo, T. J., Zychlinsky, A., and Groisman, E. A. (1999). The *selC*-associated SHI-2 pathogenicity island of *Shigella flexneri*. Mol. Microbiol. *33*, 74–83.

Odenbreit, S., and Haas, R. (2002). *Helicobacter pylori*: Impact of gene transfer and the role of the *cag* pathogenicity island for host adaptation and virulence. In J. Hacker and J. B. Kaper, eds. *Pathogenicity islands and the evolution of pathogenic microbes* (Berlin Heidelberg: Springer-Verlag), pp. 1–22.

O'Shea, Y. A., and Boyd, E. F. (2002). Mobilization of the *Vibrio* pathogenicity island between *Vibrio cholerae* isolates mediated by CP-T1 generalized transduction. FEMS Microbiol. Lett. *214*, 153–157.

Parsot, C., and Sansonetti, P. J. (1999). The virulence plasmid of shigellae: An archipelago of pathogenicity islands? In J. B. Kaper and J. Hacker, eds. *Pathogenicity islands and other mobile virulence elements* (Washington, DC: ASM Press), pp. 151–165.

Rajakumar, K., Sasakawa, C., and Adler, B. (1997). Use of a novel approach, termed island probing, identifies the *Shigella flexneri* she pathogenicity island which encodes a homolog of the immunoglobulin A protease-like family of proteins. Infect. Immun. *65*, 4606–4614.

Rakin, A., Noelting, C., Schropp, P., and Heesemann, J. (2001). Integrative module of the high-pathogenicity island of *Yersinia*. Mol. Microbiol. *39*, 407–415.

Reiter, W.-D., Palm, P., and Yeats, S. (1989). Transfer RNA genes frequently serve as integration sites for prokaryotic genetic elements. Nucleic Acids Res. *17*, 1907–1914.

Ritter, A., Gally, D. L., Olsen, P. B., Dobrindt, U., Friedrich, A., Klemm, P., and Hacker, J. (1997). The PAI-associated *leuX* specific tRNA$_5$^Leu affects type 1 fimbriation in pathogenic *E. coli* by control of FimB recombinase expression. Mol. Microbiol. *25*, 871–882.

Ruzin, A., Lindsay, J., and Novick, R. P. (2001). Molecular genetics of SaPI1—A mobile pathogenicity island in *Staphylococcus aureus*. Mol. Microbiol. *41*, 365–377.

Salyers, A. A., Shoemaker, N. B., Stevens, A. M., and Li, L.-Y. (1995). Conjugative transposons: An unusual and diverse set of integrated gene transfer elements. Microbiol. Rev. *59*, 579–590.

Schubert, S., Dufke, S., Sorsa, J., and Heesemann, J. (2004). A novel integrative and conjugative element (ICE) of *Escherichia coli*: the putative progenitor of the *Yersinia* high-pathogenicity island. Mol. Microbiol. *51*, 837–848.

Scott, J. R., and Churchward, G. G. (1995). Conjugative transposition. Ann. Rev. Microbiol. *49*, 367–397.

Segal, E. D., Cha, J., Lo, J., Falkow, S., and Tompkins, L. S. (1999). Altered states: Involvement of phosphorylated CagA in the induction of host cellular growth changes by *Helicobacter pylori*. Proc. Natl. Acad. Sci. U S A 96, 14559–14564.

Shankar, N., Baghdayan, A. S., and Gilmore, M. S. (2002). Modulation of virulence within a pathogenicity island in vancomycin-resistant *Enterococcus faecalis*. Nature 417, 746–750.

Sullivan, J. T., and Ronson, C. W. (1998). Evolution of rhizobia by acquisition of a 500 kb symbiosis island that integrates into a phe-tRNA gene. Proc. Natl. Acad. Sci. U S A 95, 5145–5149.

Tauschek, M., Strugnell, R. A., and Robins-Browne, R. M. (2002). Characterization and evidence of mobilization of the LEE pathogenicity island of rabbit-specific strains of enteropathogenic *Escherichia coli*. Mol. Microbiol. 44, 1533–1550.

Thorne, C. B. (1985). Genetics of *Bacillus anthracis*. In L. Lieve, P. F. Bonventre, J. A. Morello, S. Schlesinger, S. D. Silver, and H. C. Wu, eds. *Microbiology-85* (Washington, DC: ASM Press), pp. 56–62.

Vokes, S. A., Reeves, S. A., Torres, A. G., and Payne, S. M. (1999). The aerobactin iron transport system genes in *Shigella flexneri* are present within a pathogenictiy island. Mol. Microbiol. 33, 63–73.

Waldor, M. K., and Mekalanos, J. J. (1996). Lysogenic conversion by a filamentous phage encoding cholera toxin. Science 272, 1910–1914.

Williams, K. P. (2002). Integration sites for genetic elements in prokaryotic tRNA and tmRNA genes: Sublocation preference of integrase subfamilies. Nucleic Acids Res. 30, 866–875.

CHAPTER 10

Biological consequences for bacteria of homologous recombination

Diarmaid Hughes and Tobias Norström

(351)

The mechanisms of homologous recombination in bacteria (described in Chapter 1) are ancient and highly conserved. The basic requirement is two DNA sequences that share sequence identity over a minimum distance, typically at least 40 to 50 nucleotides (Shen and Huang, 1986; Watt et al., 1985). These identical sequences are brought together to create, and eventually resolve, a recombinant molecule, by the actions of enzymes such as RecA, RecBCD/RecF, and RuvABC, or their equivalents (Kowalczykowski et al., 1994). Homologous recombination serves several purposes in the cell. The most fundamental of these, according to current understanding, is to facilitate the completion of chromosome replication (Cox, 2001; Smith, 2001). Replication forks frequently stall or break, and homologous recombination can solve this problem using the sister homolog as a template (Kuzminov, 1995). A related problem is that after replication bacterial chromosomes are frequently entangled and require homologous recombination to disentangle them.

In addition to its housekeeping roles, homologous recombination can shuffle the order of genes in a genome by recombination between repetitive sequences. It can also facilitate the incorporation of foreign DNA, although this is also achieved by site-specific recombination (Ochman et al., 2000). Lateral DNA transfer in bacteria can alleviate the effects of Muller's ratchet (Andersson and Hughes, 1996; Muller, 1964), and provide bacteria with access to a very large gene pool potentially containing important innovative properties (Ochman et al., 2000). The focus of this chapter is not on the housekeeping functions of homologous recombination, but rather on homologous recombination as a process that shuffles genome organization and facilitates lateral DNA transfer. In particular, we are concerned with the physiological and evolutionary biological consequences for bacteria of altering genomes by homologous recombination.

RECOMBINATION WITHIN A LINKAGE GROUP

Recombination between two sequences sitting in direct orientation on the same chromosome can result in *duplication*, or *deletion*, of all the genetic material between the repeated sequences. The genetic material that is deleted may be broken down by nuclease activity, but can also recombine back into the chromosome by homologous recombination. If the repeated sequence involved is present in several copies (e.g., seven *rrn* operons in *Salmonella* spp.), then the reintegration event may occur at a novel position, resulting in a *translocation*. In contrast, recombination between repeated sequences sitting in inverse orientation on the same chromosome can result in *inversion* of the genetic material between the two repeated sequences.

RECOMBINATION BETWEEN LINKAGE GROUPS: LATERAL DNA TRANSFER

One possible consequence of homologous recombination involving laterally transferred DNA is the introduction into the genome of a completely novel gene, or set of genes, a *genetic island*. In most such cases, recombination after transfer is probably independent of homologous recombination (Ochman et al., 2000). However, homologous recombination can occur when the incoming DNA carries two separated regions of identity with the recipient chromosome, bounding a region of nonidentity – the genetic island. For example, the homology regions might be widespread and common sequences such as IS elements. This type of event could also simultaneously result in the loss of a region of the recipient chromosome, which will limit its frequency if essential genes are among those lost. In the same way, a plasmid (or other circular DNA molecule) could integrate into the chromosome by homologous recombination. More recently, several examples of recombination of foreign DNA facilitated by homology in one region and microhomology in another have been described. Interestingly, in these cases, recombination is frequently associated with both incorporation of foreign DNA and deletion of recipient DNA (de Vries and Wackernagel, 2002; Prudhomme et al., 2002). Another possible consequence of lateral DNA transfer is the creation of a *genetic mosaic* without the introduction of any entirely new genes. In this case, the two DNA molecules would share sequence similarity along their length, with the level of identity very high in some regions and low in others. Areas

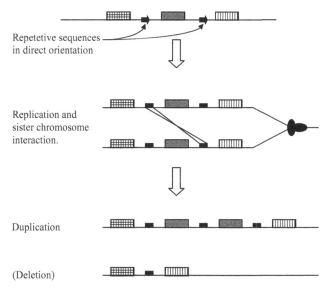

Repetitive sequences
in direct orientation

Replication and
sister chromosome
interaction.

Duplication

(Deletion)

Figure 10.1. *Inter*chromosomal recombination between repetitive sequences during replication, leading to gene duplication in one daughter cell and a deletion event in the other.

of very high identity would serve as sites for homologous recombination reactions, carrying with them an intervening region of lower identity. This type of lateral DNA transfer would typically involve closely related species.

INTRAGENIC HOMOLOGOUS RECOMBINATION AND ITS BIOLOGICAL CONSEQUENCES

Tandem duplication

Recombination between directly orientated repeat sequences can result in the duplication of the intervening genetic material, regardless of how large this is (Fig. 10.1). Extensive experimental investigation in *Salmonella enterica* serovar Typhimurium (common name, *Salmonella typhimurium*) has shown that all regions of the chromosome (with the exception of the terminus region) undergo frequent spontaneous duplication (Anderson et al., 1976; Anderson and Roth, 1981). The frequencies observed varied from $>10^{-2}$ to $\sim10^{-4}$. Frequently, the repeat sequences involved are pairs of the seven *rrn* operons scattered around the chromosome, but other repeat sequences are also involved (Haack and Roth, 1995; Jessop and Clugston, 1985; Lin et al., 1984; Shyamala et al., 1990). Duplications of up to one-third of the chromosome in size have been cataloged (Straus, 1975). The duplications are unstable because of the large region of tandemly duplicated sequence generated, and secondary recombination events can result in a deletion and return to the original chromosome sequence. However, tandem duplications can be maintained by selection. Thus, selecting a culture of *S. typhimurium* for cells with the ability to use malate, or other poor carbon sources, typically selects for bacteria carrying large chromosomal duplications (Sonti and Roth, 1989; Straus, 1975). Apparently, the bacteria can improve their growth on the selective nutrient by duplicating genes involved in uptake, transport, and metabolism of the limiting nutrient. Once the selection pressure is relieved, the large duplication will be lost spontaneously. This has advantages for the bacteria over the alternative of selecting a mutant. First, if the selection requires mutations in several genes its rate will be extremely low ($\sim10^{-10}$ per mutation required). Second, mutational changes are typically irreversible. Thus, mutations might trap the bacteria with a genotype that is suboptimal in a situation where the nutrient poor environment might only be transient. Responding by generating duplications avoids each of these problems: (1) the rate of duplication is extremely high ($\sim10^{-2}$–10^{-4}); and (2) the alteration is easily reversed once the selection pressure is relaxed. Thus, duplications provide bacterial genomes with the possibility of a flexible, high-frequency, and reversible way to respond to various selections and environments.

Similar results have been obtained when independent lines of *E. coli* were grown for 2000 generations at an elevated temperature, 41.5°C (Riehle et al., 2001). Several duplications and deletions were detected, including duplications selected at the same chromosomal region in independent lines. The rearrangements occurred through homologous recombination between multiple copies of insertion sequences and other repeat sequences. Four genes previously shown to play roles in stress responses were identified in this common region of duplication. The strains carrying duplications had elevated fitness relative to the wild-type at the high temperature, and the increase in fitness coincided with the occurrence of the duplication. Selecting for antibiotic resistance can also select for gene duplication or amplification. Thus, selection for increasing levels of ampicillin resistance in *E. coli* results in amplification of the β-lactamase-producing gene *ampC* on up to 10 tandemly repeated 10-kb regions (Edlund et al., 1979; Edlund and Normark, 1981). Similarly, selecting for mecillinam resistance in *E. coli* frequently selects mutants with an amplification of the *ftsQAZ* region, generated by homologous recombination between repeated sequences (Vinella et al., 2000).

Since the late-1980s, the phenomenon of "directed," or adaptive, mutations (Cairns et al., 1988) has generated controversy and stimulated research into the mechanisms by which bacteria respond to selection. The observation underlying the controversy is that if bacteria carrying a *lac⁻* frameshift mutation on an F′-factor are plated on minimal medium with lactose as the sole carbon source, *lac⁺* colonies arise on the selection medium at a surprisingly high frequency over a period of several days. After much research, a satisfying solution to the problem has emerged. The *lac⁻* mutation is translationally leaky, and it is located in a region of the F′-factor that can easily be amplified by homologous recombination between repeat sequences. The current model is that spontaneous amplification of the leaky *lac⁻* region results in some bacteria in the population with an improved growth rate on lactose (Andersson et al., 1998; Hendrickson et al., 2002). Selection for Lac⁺ colonies selects for cells within this subpopulation that further amplify the leaky *lac⁻* region, further enhancing growth. When the *lac* region is amplified, the probability of a spontaneous mutation to *lac⁺* is increased greatly. Once this mutation occurs, the *lac⁺* cell can grow rapidly and the selection to maintain the amplification is relaxed (the amplification is lost spontaneously). Thus, the phenomenon of apparently directed-*lac⁺* mutations is in essence a specific example of the genetic flexibility afforded bacteria by their ability to duplicate or amplify regions of their genomes. A phenomenon associated with this selection of Lac⁺ colonies is genomewide hypermutation (reviewed in Rosenberg, 2001) in a subset of the Lac⁺ colonies. The explanation for this association between *lac⁺* mutation and genomewide mutation illustrates both the pitfalls of unexpected coincidence, and the power of genetic experimentation, to create and solve difficult problems. It turns out that the chromosomal region on the F′-factor includes, in addition to the *lac* operon, the *dinB* gene, coding for the error-prone DNA polymerase IV. In the process of *lac* amplification, about 10% of the colonies also amplify the *dinB* gene on the F′-factor, and it is in these colonies that a high level of genomewide associated mutations occur (Slechta et al., 2003). Thus, in the case of *lac* adaptive mutation, there is a situation where homologous recombination facilitating gene amplification can initially solve a problem in a transient manner (an unstable gene array with a Lac⁺ phenotype), but then increases the probability of a mutation in the selected gene (*lac⁺*) and of unselected mutations around the genome.

DUPLICATIONS AND DIVERGENCE

Duplications, according to classic evolutionary theory should relieve selection pressure on a gene and potentially facilitate divergent evolution. The generation of a new function would be facilitated if the duplication could be stabilized to allow time for divergent evolution. This appears to be problematic with tandem duplications because of their inherent genetic instability. An alternative is that tandem duplication and subsequent deletion generates a fragment that reintegrates at a novel position on the genome (translocation). This would result in a nontandem duplication that might be more stable. Although the divergence of duplicate genes to a state where they perform different functions might be a very improbable event, on a long time scale it might nevertheless be very significant. This is especially so when one considers that the rate of substrate formation (generation of duplicated sequences) is extremely high. The rate-limiting factors will be in part the stabilization of the duplication and in part the probability that a new function can be derived from an existing gene (the alternative is that the duplicated genes degenerate). Evidence from genome sequence analysis strongly suggests that duplications are maintained frequently enough to act as an important source of new functions (Prince and Pickett, 2002; Yanai et al., 2000). In the cyanobacterium *Nostoc linckia*, a ~3-kb region carrying the circadian clock genes (*kaiABC*) has been subject to duplication and diversification in recent evolutionary history (Dvornyk et al., 2002). Several of these duplicated genes have degenerated into pseudogenes, but some appear to have acquired adaptive significance, increasing fitness. The duplication of the *kai* genes probably helps cyanobacteria to adapt to extreme and fluctuating environments.

DELETION

Deletions due to homologous recombination can arise by both intra- and interchromosomal recombination between directly repeated sequences (Figs. 10.1 and 10.2). The significance of deletions (involving unique genetic material) is that they create an irreversible alteration to the genotype (except in cases where they lead to a translocation). The rate of deletion formation will thus be limited by the physiological consequences of losing genetic material. This would appear to place a very severe limitation on deletions as a significant player in bacterial genetics. Indeed, in clinical isolates of *Mycobacterium tuberculosis*, there is a good correlation between the accumulation of deletions and reduced pathogenicity (Kato-Maeda et al., 2001). In *Ochrobacterium intermedium*, recombination between two 16S rDNA sequences deletes one, and the intervening 150 kb of DNA, without lethal effect, but the resulting strains suffer a 30% reduction in growth rate (Teyssier et al., 2003). The inability of many clinical isolates of *Streptococcus mutans* to use β-glucosides has recently been proposed to be due to homologous recombination resulting in a 4-kb chromosomal deletion (Robinson et al., 2003). The prevalence of such strains indicates the lack of a strong selection against strains carrying the deletion.

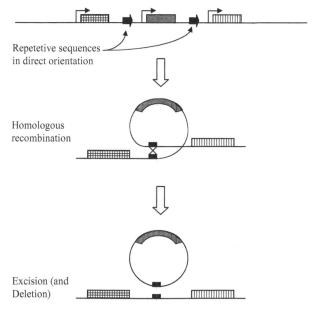

Repetetive sequences
in direct orientation

Homologous
recombination

Excision (and
Deletion)

Figure 10.2. Deletion of a fragment by intrachromosomal homologous recombination between direct repeat sequences.

However, there are interesting situations where deletions can play an important biological role. One is in the aftermath of the acquisition of foreign DNA into the genome (discussed in the section 'Deletions related to the acquisition of foreign DNA'), another is in situations where selection pressure on (parts of) the bacterial genome is relieved, leading to "reductive evolution."

Deletions and reductive evolution

During evolution free-living bacteria have occasionally been taken up by (or invaded) eukaryotic cells. Comparing free-living and endosymbiotic bacteria reveals a good correlation between genome size, repeat content, and bacterial lifestyle (Tamas et al., 2001). Thus, free-living bacteria tend to have large genomes, many repeated sequences, and few pseudogenes. In constrast, obligate intracellular bacteria tend to have small genomes, few repeat sequences, and many pseudogenes (Frank et al., 2002). The establishment of a symbiotic relationship between bacteria and eukaryotic cells probably relieves selection pressure on large parts of the bacterial genome. The evidence from current analysis of genome sequences is that symbiosis, where selection pressure is relieved on parts of the bacterial genome and the population may

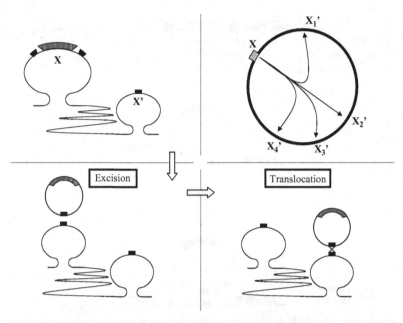

Figure 10.3. Any sequence found in more than two copies on the chromosome (*rrn* operons, IS elements, etc.) can, with the aid of homologous recombination, give rise to translocation events. The translocation is preceded by a homologous recombination-facilitated excision event. The destiny of the excised segment is then narrowed down to three alternatives: (1) destruction, (2) reintegration at the same location (at X in the figure), or (3) reintegration at another location, X′ in the figure, having sequence similarity to the excised fragment.

go through small bottlenecks, is closely associated with a reductive evolution in bacterial genome size. For example, in *Neisseria meningiditis* (genome size 2.27 Mb) repeat sequences longer than 200 bp make up 6.2% of the genome, whereas in *Buchnera ephidicola*, an aphid endosymbiont (genome size 0.65 Mb), no repeats of sequences longer than 200 bp are found (Frank et al., 2002). This process of genome reduction probably proceeded through homologous recombination at repeated sequences, removing blocks of genes, and through the accumulation of small deletions within genes (Frank et al., 2002). There are two lessons to be drawn from these situations and examples. One is that when selection pressure is relieved spontaneous mutation (and deletion by recombination) will inactivate and delete gene sequences. The consequence is that the genome will gradually reduce in size toward the set of genes that are subject to (at least intermittent) selection. The second is that deletion is an irreversible event and so the bacteria will be unable to return to its original genetic state, although lateral DNA transfer can potentially

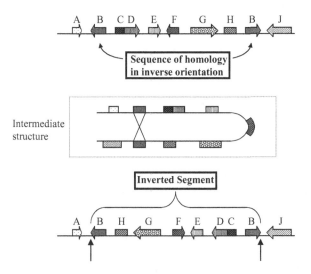

Figure 10.4. Homologous recombination involving homologous sequences situated in inverse orientation generating a chromosomal inversion.

ameliorate this effect. This means that bacteria that adapt to a new niche by a process involving deletion of unnecessary genes may become trapped in that niche (become specialists). The process of niche specialization is probably a significant evolutionary consequence of deletion events.

TRANSLOCATION AND INVERSION

In this section, we discuss both translocations and inversions within genomes. Although the events differ in how they arise, their biological consequences are often similar. Deletion, as a result of homologous recombination between two directly repeated sequences, generates a fragment of DNA that can potentially reintegrate at another location of the same repeat sequence, creating a translocation (Fig. 10.3). The translocated DNA can be in either orientation. Inversions occur through homologous recombination between inverted repeat sequences on the same chromosome (Fig. 10.4).

Rates of translocation and inversion are probably much lower than rates of duplication with equivalent repetitive sequences (Schmid and Roth, 1983). Experimentally, translocation has been shown to occur in *Salmonella typhimurium* using a 1-kbp sequence homologous to part of the *histidine* operon (Mahan and Roth, 1988). The rate of translocations related to *rrn* operon recombination in *S. typhimurium* is calculated to be 10^{-5} to 10^{-8} per cell/generation (Hughes, 1999). Inversion involving recombination between the ~1-kb *tuf* genes in *S. typhimurium* occurs at a rate of ~10^{-8} (Hughes, 2000a). The back rate for both translocations and inversions is expected to be the same as the forward rate – the repetitive recombining sequences are the same size and in the same locations. Thus, translocations and inversions should be relatively stable events, unlike tandem duplications.

Extensive studies of inversion in *S. typhimurium* and *E. coli* have revealed that all regions of the chromosome can be inverted without lethal effects (Rebello et al., 1988; Segall et al., 1988). However, inversions that disrupt the organization of the terminus region do not occur spontaneously, although they are not lethal (Meisel et al., 1994; Segall et al., 1988; Segall and Roth, 1989). The reason is apparently because their formation would require chromosome replication to move in the wrong direction through *Ter* sites (reviewed in Hughes, 1999).

POTENTIAL FITNESS COSTS OF TRANSLOCATION AND INVERSION EVENTS

Chromosome balance

Replication in *E. coli* and *S. typhimurium* begins at *oriC* and proceeds bidirectionally to TER. Consequently, the chromosome consists of two replichores of approximately equal size. This replichore balance may be a selected parameter designed to optimize the probability of each replication fork reaching TER at the same time. This could minimize replication time and facilitate the orderly separation and partitioning of sister chromosomes at the completion of replication. A translocation or inversion event could disrupt this balance and the efficiency of these processes.

Gene dosage

Translocation or inversion within a replichore will alter gene dosage. This is because genes situated closer to *oriC* are present in more copies during replication than those situated further away. Gene dosage is correlated with gene expression level (Schmid and Roth, 1987), and alterations in expression may cause reduced fitness.

Conflicts between transcription and replication

Many highly expressed genes are transcribed in the same direction as they are replicated. Translocations or inversions that reverse gene orientation might cause collisions between replication and transcription (Liu and Alberts, 1995).

Local superhelical context of the chromosome

Moving genes to a new position on the chromosome might alter gene expression independently of gene dosage and orientation (Charlebois and St. Jean, 1995; Scheirer and Higgins, 2001).

Novel joints

Translocations and inversions are conservative in the sense that they do not result in the gain or loss of any nucleotides. However, each creates two novel joints in the chromosome at the boundaries of the recombination event. Thus, functional units (e.g., an operon) can be broken and novel units created at the joints. This has obvious potential for altering gene expression levels within the novel units.

MEASURED FITNESS COSTS OF TRANSLOCATIONS AND INVERSIONS

There is a paucity of experimental measurements of fitness costs associated with genome rearrangements. Laboratory strains of *E. coli* were constructed so the 40-kb *rrnB-rrnE* segment was translocated to *rrnC*, *rrnD*, *rrnG*, and *rrnH* operons respectively. Each of these translocation strains was viable, but all of them had their fitness, measured as growth rate, reduced by up to 5% in rich medium (Hill and Harnish, 1982). During long-term evolution experiments (10,000 generations) with *E. coli*, inversions and deletions mediated by homologous recombination between copies of IS elements in the genome were common (Schneider et al., 2000). These latter data show that although translocations and inversions can have an immediate negative impact on bacterial fitness, rearranged chromosomes can go to fixation.

UNEVEN DISTRIBUTION OF TRANSLOCATIONS AND INVERSIONS AMONG BACTERIA

Genome rearrangements have been extensively studied in collections of natural isolates of closely related different *Salmonella* species (Liu and Sanderson, 1995a, 1995b, 1995c, 1996, 1998). The analysis reveals, not surprisingly, that recombination between *rrn* operons is a major cause of large rearrangements, but some involve IS*200* sequences (Alokam et al., 2002). However, a major surprise is that it also reveals a striking difference in the frequency of translocations and inversions among the different *Salmonella* lineages. Thus, rearranged chromosomes (taking *S. typhimurium* LT2 as the standard) are very frequent in *S. typhi*, *S. paratyphi*, *S. gallinarum*, and *S. pullorum*, but very infrequent in *S. typhimurium* and other *Salmonella* lineages. The rearrangements include many different translocations and inversions. A factor common to the lineages with frequent rearrangements is that they are host specialists. For example, *S. typhi* infects only humans, whereas *S. typhimurium* infects humans and many other warm-blooded vertebrates. One model put forward to explain the difference in rearrangement frequency is that the genomes of different *Salmonella* are under different intensities of selection pressure. According to the model, *S. typhimurium* must maintain an optimum genome arrangement because it is exposed to a great variety of selection pressures going between different hosts, including humans. This purifying selection maintains the ancestral genome organization. *S. typhi*, however, although it is exposed to selection by the human immune system, is never exposed to other selection regimes, and so in relative terms is evolving with less selection. In the case of *S. typhi*, many translocations and inversions would only have a trivial effect on its overall fitness. Given a high rate of generation of spontaneous rearrangement and a low level of purifying selection, the net result is a polymorphic population containing multiple genome arrangements.

An alternative model is that the rearrangements are selected for in the host specialists like *S. typhi* because of the advantage they provide in escaping the immune system. There is abundant evidence that *Salmonella* have many different fimbrial operons and are under strong selective pressure to continually evolve alternate adhesins (Edwards et al., 2002; Townsend et al., 2001). According to this model, variants that had an advantage in one host might be strongly counterselected in another host, resulting in host generalists having less variety in genome arrangement than host specialists. However, there has been no systematic study to test whether there is any correlation between genome rearrangements in *Salmonella* and antigenic variability.

Comparisons between more distantly related species typically show frequent genome rearrangements (reviewed in Hughes, 2000b). Comparison of the sequenced genomes of *Yersinia pestis* KIM, biovar Mediaevalis (Deng et al., 2002) with that of *Yersinia pestis* CO92, biovar Orientalis (Parkhill et al., 2001) reveals that, although more than 95% of the sequence is shared, the gene order is extensively rearranged. Multiple inversions, mostly involving IS sequences as regions of homology, appear to have occurred. The inversions may have placed two *rrn* operons in tandem. This arrangement could have facilitated the loss of one *rrn* operon, in CO92, by deletion (Deng et al., 2002). Remarkably, a comparison of *Yersinia* with *E. coli* (separated by approximately 375 million years) reveals that, although relative gene order is totally scrambled, gene orientations and distance from the origin of replication are highly conserved (Deng et al., 2002). Whether this conservation is a result of fitness selection or a consequence of the mechanisms causing rearrangement remains an open question. A more recent genomic analysis shows that inverted repeats are underrepresented in more than 50 sequenced bacterial genomes and that the avoidance of these repeats correlates with reported chromosomal stability (Achaz et al., 2003).

ANTIGENIC VARIATION ASSOCIATED WITH GENOME REARRANGEMENTS

The generation of antigenic variation is an important mechanism used by pathogenic bacteria to adapt to eukaryotic hosts. The variation often involves the rearrangement of genes encoding major antigens. This can be achieved either by site-specific recombination (reviewed in Henderson et al., 1999) or by homologous recombination. Although site-specific recombination is the most prevalent mechanism, homologous recombination is used in some bacteria and examples are presented in the following sections.

Deletion

Encapsulated *Haemophilus influenzae* type b produce nonencapsulated variants at a high frequency (up to 0.3%). The genetic basis of this instability is the presence in the genome of an 18-kb tandem repeat involved in type b capsule expression. Loss of one copy of the tandem repeat by homologous recombination results in loss of capsule production (Hoiseth et al., 1986).

Gene conversion

Neisseria gonorrhoeae has also evolved systems to vary the antigenicity of different surface antigens (Seifert, 1996). Antigenic variation of the major subunit of the pilus, pilin, occurs by unidirectional, homologous recombination (gene conversion) between 1 of 19 silent loci and the expression locus *pilE* (Serkin and Seifert, 1998).

Inversion

Genomic analysis of the group A *Streptococci* (GAS) reveals that invasive "flesh-eating" strains carry a large genomic inversion relative to other GAS strains. Interestingly, the sites of recombination are within two prophage-encoding regions and result in the exchange of genes coding for superantigens and mitogenic factors, resulting in new phage derivatives (Nakagawa et al., 2003).

Homologous recombination is also important in generating antigenic variation in *Campylobacter fetus*. *C. fetus* cells are covered by surface layer proteins (SLPs) critical for virulence. Each cell possesses eight *sap* homologs, lacking promoters and tightly clustered in the genome, and each capable of expressing an SLP. Variation in SLP expression occurs by nested inversions fusing a region carrying a *sap* promoter to a cassette lacking its own promoter (Dworkin et al., 1997). The *sap* inversion events are RecA dependent (Tu et al., 2001).

An outbreak of whooping cough occurred in Canada in 1989 to 1991 caused by *Bordetella pertussis* (Stibitz and Yang, 1999). The outbreak was apparently clonal in origin, yet patient isolates taken over this short period revealed 14 different restriction fragment length polymorphism classes with 10 different genome organizations involving large chromosomal inversions. The *B. pertussis* genome contains about 100 copies of a 1-kb sequence, IS481, providing many targets for homologous recombination. Nine of the 10 inversions are about the same axis. Whether these inversions improve fitness or generate antigenic variation has not been tested.

There are other examples of inversions distinguishing closely related bacteria, including *Pseudomonas aeruginosa* from cystic fibrosis patients (Römling et al., 1997) and *Neisseria meningitidis* epidemic strains (Frøholm et al., 2000), but there is a paucity of data on the biological consequences of these differences. The variety of arrangements being observed in natural and clinical isolates could reflect the dynamics of chromosome rearrangement in these organisms without implying much about fitness. We do not yet know how transient these populations are. Thus, it is unclear whether we are observing a snapshot of rearrangement events that occur at a high frequency (and disappear rapidly), or whether we are observing a situation where many different forms co-exist in nature on a relatively long-term basis (a bacterial polymorphism). Experiments should be made to test this because

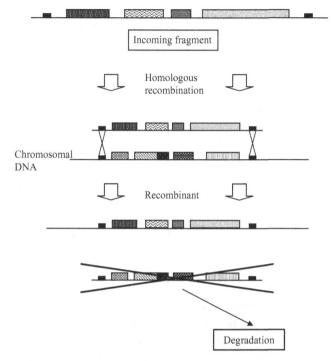

Figure 10.5. The incorporation of novel genes is facilitated by flanking sequences able to form Holliday junctions with homologous sequences on the chromosome. This might result in some loss of resident genes from the chromosome.

not only is it of intrinsic interest in understanding bacterial evolution, but it is also of interest in understanding the genetic dynamics of infectious disease organisms.

ACQUISITION OF FOREIGN DNA BY HOMOLOGOUS RECOMBINATION AND ITS BIOLOGICAL CONSEQUENCES

DNA fragments acquired by lateral transfer, lacking their own origin of replication, must be integrated into the genome by recombination to survive. These fragments may enter the recipient cell by transformation, conjugation, or bacteriophage-mediated transduction. Integration into the chromosome by homologous recombination implies regions of sequence identity between the two molecules. This requirement would favor integration of DNA from closely related sources, although some sequences, such as transposons, might be present among distantly related organisms and act as regions for homologous recombination (Fig. 10.5). Given local sequence homology, the overall genetic relatedness need not be high as shown by the experimental evidence that, at least in the laboratory, six different plant species could transfer DNA into the soil bacterium *Acitinobacter spp.* BD413, where it was recombined into the chromosome (Tepfer et al., 2003).

ACQUISITION OF GENETIC ISLANDS

The evidence for biologically significant lateral transfers from distant sources is apparent from a comparison of closely related bacteria occupying different biological niches. Among the best examples are the *Salmonella* and *E. coli* group of bacteria where various pathogenicity islands have entered the original common genome and contributed enormously to the current separation of these different groups (Edwards et al., 2002; Ochman et al., 2000). The evolutionary impact of lateral DNA transfer is that it can result in the acquisition of a complete biological system, opening up a novel biological niche for the recipient bacteria. In this context, it is interesting that the very existence of operons in bacteria is also evidence for the importance of lateral transfer during evolution (Lawrence and Roth, 1996; Ochman et al., 2000). A recipient bacterium acquiring a complete operon gets a biologically active unit. A bacterium that in contrast acquires one gene from an operon probably gets a useless gene that will degrade through lack of selection. Clearly, bacteria have for a very long time been acquiring successful inventions from each other. Interestingly, bacterial mismatch repair genes that are responsible for inhibiting recombination between imperfectly matched sequences in *E. coli* seem themselves to be in a state of flux due to lateral DNA transfer (Denamur et al., 2000). This suggests that during evolution bacteria frequently have periods with reduced levels of mismatch detection, and a coincidentally elevated potential for recombination of imperfectly homologous foreign DNA. Some of the most topical examples of lateral gene transfer with biological consequences concern the acquisition of antibiotic resistance genes by pathogenic bacteria. Although it is clear that lateral gene transfer plays a major role, for example, in the acquisition of vancomycin resistance by *Enterococci*, it is not obvious that this is related to homologous recombination. In many cases, the lateral transfer involves the activity of conjugative plasmids and transposons.

In addition to contributing new genetic information to a bacterium, acquisition of new DNA may have other consequences. One is that the recombination might result in the deletion of genes in the recipient. This consequence is likely to be relatively trivial because by definition nothing essential can be deleted. A more significant long-term consequence is that the newly acquired ability to occupy a new niche (as a result of the genetic acquisition) might render some parts of the original genome redundant. The relief of selection on these parts of the genome is expected to result in their degeneration and ultimate loss but might also facilitate the evolution of novel functions. Another consequence of acquisition of a significant amount of foreign DNA could be to disturb the organization of the chromosome [e.g., the length of each chromosome arm (distance from origin to terminus), or the packing of the chromosome into the nucleoid]. Imbalances of this type will act as a selection pressure favoring a series of additional rearrangements that might include inversions, deletions, and translocations, until a stable situation is achieved. On another level, the acquisition of foreign DNA might introduce a large region of sequence with a very different nucleotide composition to the rest of the genome. This is useful to scientists in that it assists in identifying laterally acquired DNA. For the host bacteria, however, it may initially cause problems related to increased demands for particular tRNA isoaccepters, possibly resulting in increased levels of translational errors throughout the proteome. For the acquired sequence, the likelihood is that it will gradually evolve toward the nucleotide composition of the host. Genomic nucleotide composition is in large part determined by the balance between the spectrum of mutations and the particular repair capacities of the bacterium. A complication to this equation is that the acquired DNA might itself code for novel DNA repair enzymes or polymerases or contribute to occupation of a new niche associated with a different spectrum of mutagenesis, and so contribute to an alteration in the nucleotide composition of the host genome. In conclusion, there is a general expectation that the acquisition of foreign DNA might be associated with a subsequent period of genetic rearrangement and sequence evolution. However, the requirement for sequence similarity means that homologous recombination will serve primarily to reassort variation among closely related taxa and is much less likely than site-specific recombination to facilitate the introduction of novel traits (reviewed in Hacker and Kaper, 2000, and Ochman et al., 2000).

Deletions related to the acquisition of foreign DNA

The divergence of *E. coli* from *S. typhimurium* over a period in excess of 100 Myrs has been associated with the acquisition of approximately 755 genes through at least 234 horizontal transfer events (Lawrence and Ochman, 1988; Ochman et al., 2000). However, bacterial genomes do not appear to continually increase in size. There is a flux in the gain and loss of genetic information (Ochman et al., 2000). Loss implies deletion, which in turn implies that some sequences are not essential for growth and survival. Incoming foreign DNA is by definition nonessential and can be deleted without loss of viability. If the foreign DNA provides a selective advantage on the recipient bacteria (e.g., ability to exploit a new niche), it will not be deleted without loss of this desirable phenotype. In contrast, the positive effect of the foreign DNA might be associated with only a part of its length, meaning that the remainder will not be under selection and can thus be deleted by spontaneous recombination events without a loss of fitness. A second possibility is that the foreign DNA includes genetic information that is already present on the recipient genome. This will relieve selection pressure on these sections, and deletions could occur spontaneously to remove either the original or new copies of these genes. A third possibility is that the occupation of a novel niche (facilitated by the presence of the newly acquired foreign DNA) might in itself relieve selection pressure on parts of the genome that are no longer important for survival in this new environment. One general biological consequence of these acquisitions and deletions is that diverging bacteria will tend to become isolated in separate ecological niches. The divergence of *E. coli* and *Salmonella* is probably an example of the type of separation.

HOMOLOGY-DIRECTED ILLEGITIMATE RECOMBINATION

A more recent experiment demonstrates the difficulty of predicting the consequences of acquiring DNA by horizontal transfer. Experimental transformation of *Acetinobacter sp.* (de Vries and Wackernagel, 2002) or *Streptococcus pneumoniae* (Claverys et al., 1980; Prudhomme et al., 2002) with foreign DNA carrying a short region homologous to the recipient chromosome, frequently resulted in insertion–deletion events (Fig. 10.6). It was observed that segments of foreign DNA would replace chromosomal sequences adjacent to the region of homology with the resident chromosome. What appears to happen is that a single region of homology (as short as ~150 nts) serves as a homologous recombination anchor between resident and foreign DNA, and this anchoring facilitates subsequent illegitimate recombination at regions of microhomology between the resident chromosome and foreign DNA. The amount of DNA inserted and deleted in these recombination events is of the order of a few kbs. These experiments demonstrate that transformation by foreign DNA can insert foreign genes in a genome without the necessity for multiple or extensive regions of homology. Experiments supporting similar conclusions have also been reported in *E. coli* (Kusano et al., 1997). The general significance of this phenomenon in nature will depend on the probability of a laterally transferred piece of DNA carrying at least one region of homology with the recipient chromosome. Insertion sequences are an obvious example of a widespread and common sequence class that could serve as recombination anchors to promote the capture of foreign DNA and the associated deletion of resident sequences.

DIARMAID HUGHES AND TOBIAS NORSTRÖM

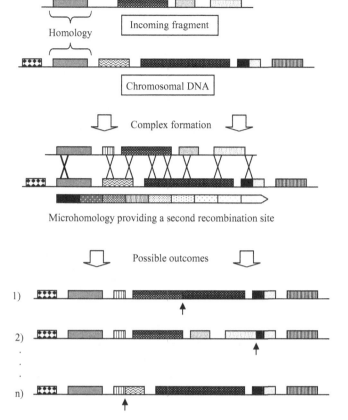

Figure 10.6. Genetic material with only partial sequence similarity can be incorporated into a recipient chromosome with the aid of homologous recombination. A homologous "anchor sequence" creates the initial recombination complex. This facilitates further recombination between foreign and resident DNA involving sequences with microhomology.

MOSAIC GENES

When lateral gene transfers occur between closely related bacteria, there is expected to be reasonable sequence similarity along the length of the incoming fragment and an equivalent region of the recipient chromosome. The expectation is that homologous recombination will occur between sequences that are identical, but that intervening sequences might differ significantly. The result will be the creation of a mosaic gene or genes (Fig. 10.7). There are very good examples of this type of bacterial evolution with great clinical significance. Transformable bacteria have acquired target-mediated

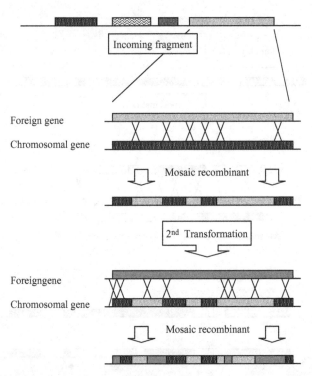

Figure 10.7. DNA taken up from the environment recombines with a homeologous sequence on the chromosome, forming a mosaic gene sequence. The resulting mosaic gene structure may be subjected to further rounds of recombination with foreign DNA resulting in the creation of complex sequences phylogenies.

antibiotic resistance by lateral transfer of fragments of chromosomal genes. Examples include penicillin-resistant binding proteins (PBPs) in *Streptococcus pneumoniae* and pathogenic *Neisseria* species, and sulfonamide-resistant dihydropteroate synthase in *Neisseria meningiditis* (reviewed in Maiden, 1998).

Penicillin resistance

In most bacteria, resistance to β-lactam antibiotics is caused by the acquisition of plasmids expressing β-lactamase. However, in a number of bacterial genera, including *Streptococcus*, *Neisseria*, and *Haemophilus*, resistance is most often target mediated, due to the presence of altered PBPs with low affinity for β-lactams. In penicillin-resistant *N. meningitidis*, the *penA* gene (encoding PBP2) is typically highly diverse and different from the gene in susceptible strains. The diversity is the result of blocks of nucleotide sequence within *penA* that are up to 22% divergent from the equivalent sequences in susceptible isolates (Campos et al., 1992; Zhang et al., 1990). Sequence comparisons and laboratory experiments suggest that one of the sources of the foreign DNA in these *penA* mosaics is the *penA* gene of *Neisseria flavescens* (Bowler et al., 1994; Spratt et al., 1989).

Penicillin-resistant strains of *Neisseria gonorrhoeae*, the gonococcus, sometimes carry a plasmid expressing β-lactamase. Resistance in strains with no β-lactamase activity is a result of acquisition of foreign DNA at four different loci, each one of which individually provides only low-level resistance (Faruki and Sparling, 1986). Two of these genes are PBP genes (PBP1 and PBP2), whereas the other two affect the permeability of the outer membrane. Mosaic *penA* alleles (PBP2) have been found in most (but not all) of these penicillin-resistant isolates (Spratt, 1988; Spratt et al., 1992). Sequence comparisons suggest that the source of foreign DNA is likely to be the related species *N. flavescens* and *N. cinera* (Spratt et al., 1992).

Strains of *Streptococcus pneumoniae* with high-level resistance to penicillin are common and are globally distributed (Dowson et al., 1994). Frequently, these strains are also resistant to cephalosporin, creating a serious clinical treatment problem (Coffey et al., 1995a). All the high-level β-lactam-resistant clinical isolates of *S. pneumoniae* identified have several low-affinity PBPs, typically PBP2x, PBP2b, and PBP1a (Maiden, 1998). *S. pneumoniae* has five high-molecular-weight PBPs susceptible to the action of β-lactam antibiotics. The genetic basis of resistance has been determined experimentally by transforming susceptible strains with cloned PBP genes from resistant isolates. Experiments of this type have shown that resistant strains must accumulate low-affinity variants of different PBPs (Barcus et al., 1995; Hakenbeck et al., 1998). The resistance-associated PBP genes are genetic mosaics differing in parts of their sequence by up to 23% from the sequence present in susceptible strains (Coffey et al., 1995b; Hakenbeck, 1995). Possible sources of the foreign DNA include the commensal strains *S. mitis* and *S. oralis*, which themselves are normally susceptible (Dowson et al., 1993; Sibold et al., 1994). There is evidence that resistance can evolve in these commensal streptococcal species prior to transfer to *S. pneumoniae* (reviewed in Hakenbeck, 2000). Homologous recombination is also implicated in the spread of resistance to other antibiotics in *S. pneumoniae*. Sequence analysis of the plasmid-borne tetracycline resistance determinant TetM shows that it has a mosaic structure, suggesting that homologous recombination has contributed to its evolution and heterogeneity (Oggioni et al., 1996). It has also been shown experimentally that fluoroquinolone resistance (due to altered *gyrA* and *parC*) can be transformed into *S. pneumoniae* from ciprofloxacin-resistant *S. oralis* (González et al., 1998).

SULFONAMIDE RESISTANCE IN *NEISSERIA MENINGITIDIS*

As with the β-lactam resistance in pneumococci, both experiments and sequence comparisons between susceptible and sulfonamide-resistant *Neisseria* strongly suggest that lateral DNA transfer of foreign *dhps* genes followed by homologous recombination is responsible for creating antibiotic-resistant mosaic genes (Fermer et al., 1995; Rådström et al., 1992).

COSTS OF RESISTANCE

One of the advantages, for bacteria, of becoming antibiotic resistant by acquisition of mosaic genes, may be that it imposes little or no cost on the resistant strain (Hakenbeck et al., 1998). Resistance is not associated with additional synthesis costs, and there should be little or no disruption of normal cell physiology. Furthermore, if the resistance phenotype is selectively neutral, then it will not decrease in frequency even if the selective antibiotic pressure is removed.

SEROTYPE SWITCHES AND TRANSFORMATION COMPETENCE

Streptococcal strains frequently switch serotype as a result of homologous recombination (Coffey et al., 1998). The polysaccharide capsule protects pneumococcal cells from engulfment by polymorphonuclear leucocytes and is their main virulence factor (reviewed in García et al., 2000). At least 90 different capsular serotypes have been documented (reviewed in García et al., 2000). Among the well-documented multiply antibiotic-resistant clones of *S. pneumoniae*, it appears that serotype 19F has arisen on several occasions as a variant of the major Spanish multiresistant serotype 23F clone by recombination at the capsular locus (Coffey et al., 1998). Frequent switching of serotype by recombination may thus have implications for the long-term efficacy of pneumococcal vaccines that protect against only a few serotypes (Coffey et al., 1998). Interestingly, the genes encoding capsular polysaccharide biosynthesis are located just downstream of the *pbp1a* locus, suggesting that by transformation a novel PBP variant and a novel capsular type might be inherited at the same time.

The importance of mosaic genes in the evolution of *S. pneumoniae* is of course related to its transformation competence. Thus, it comes perhaps as no surprise that the *comCD* genes, responsible for the transformation competence of *S. pneumoniae*, are themselves genetic mosaics (Håverstein et al., 1997). Thus, the incorporation of foreign DNA can alter the phenotype of *Streptococcus*, and in that way alter the range of the related species with which it will in the future exchange DNA. The possibility to make whole genome comparisons using DNA chip technology has facilitated a broad genomic comparison between individuals of the species *S. pneumoniae*, *S. mitis*, and *S. oralis* (Hakenbeck et al., 2001). The conclusion from that comparison is that the degree of interspecies DNA transfer and recombination is so great that there is a smooth transition between members of these species.

Recombinational exchanges between *E. coli* and *S. typhimurium* are severely inhibited by the action of the MutSHL mismatch repair system (Matic et al., 1996). It seems logical that the similar Hex system in *S. pneumoniae* would also inhibit homeologous recombination between different pneumo-cocci. The surprise is that, although single mismatches reduce recombination by up to 20-fold, interspecies transformation and recombination is not inhibited (Humbert et al., 1995). The Hex system is instead saturated by the excessive number of mismatches typical of an interspecies recombination, leading to a situation where transformation efficiency increases with divergence up to ~10%. It appears that *S. pneumoniae* has evolved to favor recombination of foreign DNA, whereas *E. coli* and *S. typhimurium* have evolved to abort interspecies recombination (Claverys et al., 2000). Interestingly, in *E. coli*, sequences comparisons between natural isolates show that the mismatch repair genes are genetic mosaics (Denamur et al., 2000). This strongly suggests that the increase in interspecific recombination rate associated with inactivation of mismatch repair is associated with horizontal gene transfer in *E. coli*.

VIRULENCE FACTORS AND ANTIGENIC VARIATION

Bacteria minimize the number of epitopes on their surface accessible to the host immune system. As a consequence, they maximize the probability that altering a single epitope will result in evasion of the immune system. There are several good examples where lateral DNA transfer and homologous recombination to create mosaic genes is associated with the generation of antigenic variation. Group B *Streptococci* are a cause of neonatal sepsis and meningitis, as well as serious invasive infections in adults. The virulence of group B *Streptococci* is strongly associated with a family of repeat-containing surface-associated antigens, exemplified by alpha C and Rib (Li et al., 1997). Alpha C protein contains a series of identical, 246-bp tandem repeat units, in addition to other partially repeated and unique regions. Antigen diversity is

373

generated by horizontal DNA transfer and subsequent homologous recombination (Lachenauer et al., 2000). This results in proteins of varying sizes according to the number of tandem repeats within the corresponding gene.

Antigenic diversity has also been generated by recombination in the group A *Streptococci* (reviewed in Kehoe et al., 1996). The M protein genes (*emm*) encode fibrillar cell-surface proteins with highly variable amino-terminal regions protruding from the cell. Sequence analysis suggests that much of this variation has occurred in response to selection, probably from host antibodies. The variation is due both to mutational alterations and to intergenomic recombination involving the horizontal transfer of *emm*-like genes among strains, resulting in a mosaic structure in *emm*-like genes (Bessen and Hollingshead, 1994; Podbielski et al., 1994; Whatmore and Kehoe, 1994).

The porin protein (Por) is the major outer membrane protein of *N. gonorrhoeae*. It is encoded by a single copy gene that exhibits great allelic variability. This variability is believed to have been selected by the immunological responses of the host populations. Genetic analysis of *por* genes reveals a mosaic organization with high levels of similarity in some regions and high levels of diversity (~18% nt differences) in other regions of the gene (Fudyk et al., 1999). Furthermore, large regions of complete identity were found in *por* genes from otherwise divergent strains. This suggests that homologous recombination of different *por* sequences to create novel mosaics may be an important strategy by which *N. gonorrhoeae* generates *por* gene diversity. Antigenic variation as a result of horizontal DNA transfer and recombination has also been reported for the *pil* (Kriz et al., 1999) and *opa* genes (Hobbs et al., 1994, 1998) of *N. meningitidis*. The evolution of the *opa* genes included gene duplication by translocation (Hobbs et al., 1998).

SUMMARY

Depending on the bacterial species and the time scale of interest, both site-specific and homologous recombination events make important contributions to the generation of bacterial diversity and variation. Some of the consequences of homologous recombination for bacterial physiology and evolution are illustrated in Fig. 10.8. On a short time scale, lateral DNA transfer and its incorporation by homologous recombination is clearly very important for bacteria that are naturally competent. The consequences of these events are well documented in the creation of genetic mosaics resistant to antibiotics and the generation of antigenic variation. The effect of frequent lateral transfer between closely related bacterial species is to effectively enlarge their

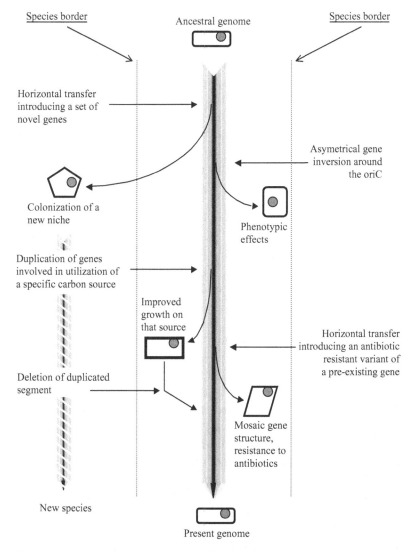

Figure 10.8. The picture depicts the progression of a bacterial species evolving through time. New variants diverge from the main ancestral line as genetic changes are introduced. Subspecies and variants are created, some of which soon become extinct, whereas others prove successful and persist. Some changes are so great that their effect is to push the novel bacteria over the "species border." This change in lifestyle will, in turn, lead to further genetic change as a consequence of the alterations in selection pressure.

gene pools and thus their genetic flexibility. The acquisition of completely novel properties may be an event dominated by site-specific recombination mechanisms, but the phenomenon of homology-directed illegitimate recombination suggests that homologous recombination can also play a role. In bacteria that are not naturally competent, the role of homologous recombination in generating variation will be restricted mostly to rearranging the genome organization. This can have significant short-term physiological effects, allowing bacteria to create flexible and reversible responses to various selections, including adaptation to poor nutrition, the generation of antigenic variation, and survival in the presence of antibiotics.

DIARMAID HUGHES AND TOBIAS NORSTRÖM

REFERENCES

Achaz, G., Coissac, E., Netter, P., and Rocha, E.P.C. (2003) Associations between inverted repeats and the structural evolution of bacterial genomes. *Genetics* **164**, 1279–1289.

Alokam, S., Liu, S.L., Said, K., and Sanderson, K.E. (2002) Inversions over the terminus region in *Salmonella* and *Escherichia coli*: IS*200*s as the sites of homologous recombination inverting the chromosome of *Salmonella enterica* serovar Typhi. *J. Bacteriol.* **184**, 6190–6197.

Anderson, R.P., and Roth, J R. (1981) Spontaneous tandem genetic duplications in *Salmonella typhimurium* arise by unequal recombination between rRNA (*rrn*) cistrons. *Proc. Natl. Acad. Sci. U S A* **78**, 3113–3117.

Anderson, R.P., Miller, C.G., and Roth, J.R. (1976) Tandem duplications of the histidine operon observed following generalized transduction in *Salmonella typhimurium*. *J. Mol. Biol.* **105**, 201–218.

Andersson, D.I., and Hughes, D. (1996) Muller's ratchet decreases fitness of a DNA-based microbe. *Proc. Natl. Acad. Sci. U S A* **93**, 906–907.

Andersson, D.I., Slechta, E.S., and Roth, J.R. (1998) Evidence that gene amplification underlies adaptive mutability of the bacterial *lac* operon. *Science* **282**, 1133–1135.

Barcus, V.A., Ghanekar, K., Yeo, M., Coffey, T.J., and Dowson, C.G. (1995) Genetics of high level penicillin resistance in clinical isolates of *Streptococcus pneumoniae*. *FEMS Microbiol. Lett.* **126**, 299–303.

Bessen, D.E., and Hollingshead, S.K. (1994) Allelic polymorphism of *emm* loci provides evidence for horizontal gene spread in group A streptococci. *Proc. Natl. Acad. Sci. U S A* **91**, 3280–3284.

Bowler, L.D., Zhang, Q.-Y., Riou, J.Y., and Spratt, B.G. (1994) Interspecies recombination between the *penA* genes of *Neisseria meningitidis* and commensal *Neisseria* species during the emergence of penicillin resistance in

N. meningitidis: Natural events and laboratory simulation. *J. Bacteriol.* **176**, 333–337.

Cairns, J., Overbaugh, J., and Miller, S. (1988) The origin of mutants. *Nature* **335**, 142–145.

Campos, J., Fuste, M.C., Trujillo, G., Saez-Nieto, J., Vazquez, J., Loren, J.G., Vinas, M., and Spratt, B.G. (1992) Genetic diversity of penicillin-resistant *Neisseria meningitidis. J. Clin. Infect.* **166**, 173–177.

Charlebois, R.L., and St. Jean, A. (1995) Supercoliing and map stability in the bacterial chromosome. *J. Mol. Evol.* **41**, 15–23.

Claverys, J.P., Lefevre, J.C., and Sicard, A.M. (1980) Transformation of *Streptococcus pneumoniae* with *S. pneumoniae*-lambda phage hybrid DNA: Induction of deletions. *Proc. Natl. Acad. Sci. U S A* **77**, 3534–3538.

Claverys, J.-P., Prudhomme, M., Mortier-Barrière, I., and Martin, B. (2000) Adaptation to the environment: *Streptococcus pneumoniae*, a paradigm for recombination-mediated genetic plasticity. *Mol. Microbiol.* **35**, 251–259.

Coffey, T.J., Daniels, M., McDougal, L.K., Dowson, C.G., Tenover, F.C., and Spratt, B.G. (1995a) Genetic analysis of clinical isolates of *streptococcus pneumoniae* with high-level resistance to expanded-spectrum cephalosporins. *Antimicrob. Agents Chemother.* **39**, 1306–1313.

Coffey, T.J., Dowson, C.G., Daniels, M., and Spratt, B.G. (1995b) Genetics and molecular biology of β-lactam-resistant pneumococci. *Microb. Drug Resist.* **1**, 29–34.

Coffey, T.J., Enright, M.C., Daniels, M., Morana, J.K., Morana, R., Hryniewicz, W., et al. (1998) Recombinational exchanges at the capsular polysaccharide locus lead to frequent serotype changes among natural isolates of *Streptococcus pneumoniae. Mol. Microbiol.* **27**, 73–83.

Cox, M.M. (2001) Recombinational DNA repair of damaged replication forks in *Escherichia coli. Annu. Rev. Genet.* **35**, 53–82.

Denamur, E., Lecointre, G., Darlu, P., Tenaillon, O., Acquaviva, C., Sayada, C., et al. (2000) Evolutionary implications of the frequent horizontal transfer of mismatch repair genes. *Cell* **103**, 711–721.

Deng, W., Burland, V., Plunkett, G., III, Boutin, A., Mayhew, G.F., Liss, P., et al. (2002) Genome sequence of *Yersinia pestis* KIM. *J. Bacteriol.* **184**, 4601–4611.

de Vries, J., and Wackernagel, W. (2002) Integration of foreign DNA during natural transformation of *Acinetobacter sp.* by homology-facilitated illegitimate recombination. *Proc. Natl. Acad. Sci. U S A* **99**, 2094–2099.

Dowson, C.G., Coffey, T.J., Kell, C., and Whiley, R.A. (1993) Evolution of penicillin resistance in *Streptococcus pneumoniae*: The role of *Streptococcus mitis* in the formation of a low affinity PBOP2B in *S. pneumoniae. Mol. Microbiol.* **9**, 635–643.

Dowson, C.G., Coffey, T.J., and Spratt, B.G. (1994) Origin and molecular epidemiology of penicillin-binding-protein-mediated resistance to β-lactam antibiotics. *Trends Microbiol.* **2**, 361–366.

Dvornyk, V., Vinogradova, O., and Nevo, E. (2002) Long-term microclimatic stress causes rapid adaptive radiation of *kaiABC* clock gene family in a cyanobacterium, *Nostoc linckia*, from 'Evolution Canyons' I and II, Israel. *Proc. Natl. Acad. Sci. U S A* **99**, 2082–2087.

Dworkin, J., Shedd, O.L., and Blaser, M.J. (1997) Nested DNA inversion of *Campylobacter fetus* S-layer genes is *recA* dependent. *J. Bacteriol.* **179**, 7523–7529.

Edlund, T., and Normark, S. (1981) Recombination between short DNA homologies causes tandem duplication. *Nature* **292**, 269–271.

Edlund, T., Grundström, T., and Normark, S. (1979) Isolation and characterization of DNA repetitions carrying the chromosomal beta-lactamase genes of *Escherichia coli* K-12. *Mol. Gen. Genet.* **173**, 115–125.

Edwards, R.A., Olsen, G.J., and Maloy, S.R. (2002) Comparative genomics of closely related salmonellae. *Trends Microbiol.* **10**, 94–99.

Faruki, H., and Sparling, P.F. (1986) Genetics of resistance in a non-β-lactamase-producing gonococcus with relatively high-level penicillin resistance. *Antimicrob. Agents Chemother.* **30**, 856–860.

Fermer, C., Kristiansen, B.E., Sköld, O., and Swedberg, G. (1995) Sulphonamide resistance in *Neisseria meningitidis* as defined by site-directed mutagenesis could have its origin in other species. *J. Bacteriol.* **177**, 4669–4675.

Frank, A.C., Amiri, H., and Andersson, S.G. (2002) Genome deterioration: Loss of repeated sequences and accumulation of junk DNA. *Genetica* **115**, 1–12.

Frøholm, L.O., Kolstø, A.-B., Berner, J.-M., and Caugant, D.A. (2000) Genomic rearrangements in *Neisseria meningitidis* strains of the ET-5 complex. *Curr. Microbiol.* **40**, 372–379.

Fudyk, T.C., Maclean, I.W., Simonsen, J.N., Njagi, E.N., Kimani, J., Brunham, R.C., and Plummer, F.A. (1999) Genetic diversity and mosaicism at the *por* locus of *Neisseria gonorrhoeae. J. Bacteriol.* **181**, 5591–5599.

García, E., Llull, D., Munoz, R., Mollerach, M., and López, R. (2000) Current trends in capsular polysaccharide biosynthesis of *Streptococcus pneumoniae. Res. Microbiol.* **151**, 429–435.

González, I., Georgiou, M., Alcaide, F., Balas, D., Liñares, J., and de la Campa, A.G. (1998) Fluoroquinolone resistance mutations in the *parC, parE*, and *gyrA* genes of clinical isolates of viridans group streptococci. *Antimicrob. Agents Chemother.* **42**, 2792–2798.

Haack, K.R., and Roth, J.R. (1995) Recombination between chromosomal IS *200* elements supports frequent duplication formation in *Salmonella typhimurium. Genetics* **141**, 1245–1252.

Hacker, J., and Kaper, J.B. (2000) Pathogenicity islands and the evolution of microbes. *Annu. Rev. Microbiol.* **54**, 641–679.

Hakenbeck, R. (1995) Target-mediated resistance to β-lactam antibiotics. *Biochem. Pharmacol.* **50**, 1121–1127.

Hakenbeck, R. (2000) Transformation in *Streptococcus pneumoniae*: Mosaic genes and the regulation of competence. *Res. Microbiol.* **151**, 453–456.

Hakenbeck, R., König, A., Kern, I., van der Linden, M., Keck, W., Billot-Klein, D., et al. (1998) Acquisition of five high-M_r penicillin-binding protein variants during transfer of high-level β-lactam resistance from *Streptococcus mitis* to *Streptococcus pneumoniae. J. Bacteriol.* **180**, 1831–1840.

Hakenbeck, R., Balmelle, N., Weber, B., Gardes, C., Keck, W., and de Saizieu, A. (2001) Mosaic genes and mosaic chromosomes: Intra- and interspecies genomic variation of *streptococcus pneumoniae. Infect. Immun.* **69**, 2477–2486.

Håverstein, L.S., Hakenbeck, R., and Gaustad, P. (1997) Natural competence in the genus *Streptococcus*: Evidence that streptococci can change pherotype by interspecies recombinational exchanges. *J. Bacteriol.* **179**, 6589–6594.

Henderson, I.R., Owen, P., and Nataro, J.P. (1999) Molecular switches – The ON and OFF of bacterial phase variation. *Mol. Microbiol.* **33**, 919–932.

Hendrickson, H., Slechta, E.S., Bergthorsson, U., Andersson, D.I., and Roth, J.R. (2002) Evidence that 'directed' adaptive mutation and general hypermutability result from growth with a selected gene amplification. *Proc. Natl. Acad. Sci. U S A* **99**, 2164–2169.

Hill, C.W., and Harnish, B.W. (1982) Transposition of a chromosomal segment bounded by redundant rRNA genes into other rRNA genes in *Escherichia coli. J. Bacteriol.* **149**, 449–457.

Hobbs, M.M., Seiler, A., Achtman, M., and Cannon, J.G. (1994) Microevolution within a clonal population of pathogenic bacteria: Recombination, gene duplication and horizontal genetic exchange in the *opa* gene family of *Neisseria meningitidis. Mol. Microbiol.* **12**, 171–180.

Hobbs, M.M, Malorny, B., Prasad, P., Morelli, G., Kusecek, B., Heckels, J.E., et al. (1998) Recombinational reassortment among *opa* genes from ET-37 complex *Neisseria meningitidis* isolates of diverse geographical origins. *Microbiology* **144**, 157–166.

Hoiseth, S.K., Moxon, E.R., and Silver, R.P. (1986) Genes involved in *Haemophilus influenzae* type b capsule expression are part of an 18-kilobase tandem duplication. *Proc. Natl. Acad. Sci. U S A* **83**, 1106–1110.

Hughes, D. (1999) Impact of homologous recombination on genome organization and stability. In R.L. Charlebois, ed. Organization of the prokaryotic genome, pp. 109–128. Washington, DC: American Society for Microbiology.

Hughes, D. (2000a). Co-evolution of the *tuf* genes links gene conversion with the generation of chromosomal inversions. *J. Mol. Biol.* **297**, 355–364.

Hughes, D. (2000b) Evaluating genome dynamics: The constraints on rearrangements within bacterial genomes. *Genomebiology* **1**, 0006.1–0006.8.

Humbert, O., Prudhomme, M., Hakenbeck, R., Dowson, C.G., and Claverys, J.P. (1995) Homeologous recombination and mismatch repair during transformation in *Streptococcus pneumoniae*: Saturation of the Hex mismatch repair system. *Proc. Natl. Acad. Sci. U S A* **92**, 9052–9056.

Jessop, A.P., and Clugston, C. (1985) Amplification of the ArgF region in strain HfrP4X of *E. coli* K-12. *Mol. Gen. Genet.* **201**, 347–350.

Kato-Maeda, M., Rhee, J.T., Gingeras, T.R., Salmon, H., Drenkow, J., Smittipat, N., and Small, P.M. (2001) Comparing genomes within the species *Mycobacterium tuberculosis*. *Genome Res.* **11**, 547–554.

Kehoe, M.A., Kapur, V., Whatmore, A.M., and Musser, J.M. (1996) Horizontal gene transfer among group A streptococci: Implications for pathogenesis and epidemiology. *Trends Microbiol.* **4**, 436–443.

Kowalczykowski, S.C., Dixon, D.A., Eggleston, A.K., Lauder, S.D., and Rehauer, W.M. (1994) Biochemistry of homologous recombination in *Escherichia coli*. *Microbiol. Rev.* **58**, 401–465.

Kriz, P., Giorgini, D., Musilek, M., Larribe, M., and Taha, M.-K. (1999) Microevolution through DNA exchange among strains of *Neisseria meningitidis* isolated during an outbreak in the Czech Republic. *Res. Microbiol.* **150**, 273–280.

Kusano, K., Sakagami, K., Yokochi, T., Naito, T., Tokinaga, Y., Ueda, E., and Kobayashi, I. (1997) A new type if illegitimate recombination is dependent on restriction and homologous interaction. *J. Bacteriol.* **179**, 5380–5390.

Kuzminov, A. (1995) Collapse and repair of replication forks. *Mol. Microbiol.* **16**, 373–384.

Lachenauer, C.S., Creti, R., Michel, J.L., and Madoff, L.C. (2000) Mosaicism in the alpha-like protein genes of group B streptococci. *Proc. Natl. Acad. Sci. U S A* **97**, 9630–9635.

Lawrence, J.G., and Ochman, H. (1998) Molecular archaeology of the *Escherichia coli* genome. *Proc. Natl. Acad. Sci. U S A* **95**, 9413–9417.

Lawrence, J.G., and Roth, J.R. (1996) Selfish operons: Horizontal transfer may drive the evolution of gene clusters. *Genetics* **143**, 1843–1860.

Li, J., Kasper, D.L., Ausubel, F.M., Rosner, B., and Michel, J.L. (1997) Inactivation of the α C protein antigen gene, *bca*, by a novel shuttle/suicide vector results in attenuation of virulence and immunity in group B *Streptococcus. Proc. Natl. Acad. Sci. U S A* **94**, 13251–13256.

Lin, R.J., Capage, M., and Hill, C.W. (1984) A repetitive DNA sequence, *rhs*, responsible for duplications within the *Escherichia coli* K-12 chromosome. *J. Mol. Biol.* **177**, 1–18.

Liu, B., and Alberts, B.M. (1995) Head-on collision between a DNA replication apparatus and RNA polymerase transcription complex. *Science* **267**, 1131–1136.

Liu, S.-L., and Sanderson, K.E. (1995a) Rearrangements in the genome of the bacterium *Salmonella typhi. Proc. Natl. Acad. Sci. U S A* **92**, 1018–1022.

Liu, S.-L., and Sanderson, K.E. (1995b) I-*Ceu*I reveals conservation of the genome of independent strains of *Salmonella typhimurium. J. Bacteriol.* **177**, 3355–3357.

Liu, S.-L., and Sanderson, K.E. (1995c) The chromosome of *Salmonella paratyphi* A is inverted by recombination between *rrn*H and *rrn*G. *J. Bacteriol.* **177**, 6585–6592.

Liu, S.-L., and Sanderson, K.E. (1996) Highly plastic chromosomal organization in *Salmonella typhi. Proc. Natl. Acad. Sci. U S A* **93**, 10303–10308.

Liu, S.-L., and Sanderson, K.E. (1998) Homologous recombination between *rrn* operons rearranges the chromosome in host-specialized species of *Salmonella. FEMS Microbiol. Lett.* **164**, 275–281.s

Mahan, M.J., and Roth, J.R. (1988) Reciprocality of recombination events that rearrange the chromosome. *Genetics* **120**, 23–35.

Maiden, M.C.J. (1998) Horizontal genetic exchange, evolution, and spread of antibiotic resistance in bacteria. *Clin. Infect. Dis.* **27**(Suppl 1), S12–S20.

Matic, I., Taddei, F., and Radman, M. (1996) Genetic barriers among bacteria. *Trends Microbiol.* **4**, 69–72.

Meisel, L., Segall, A., and Roth, J.R. (1994) Construction of chromosomal rearrangements in *Salmonella* by transduction: Inversions of nonpermissive segments are not lethal. *Genetics* **137**, 919–932.

Muller, H.J. (1964) The relation of recombination to mutational advance. *Mutat. Res.* **1**, 2–9.

Nakagawa, I., Kurokawa, K., Yamashita, A., Nakata, M., Tomiyasu, Y., Okahashi, N., Kawabata, S., et al. (2003) Genome sequence of an M3 strain on *Streptococcus pyogenes* reveals a large-scale genomic rearrangement in invasive strains and new insights into phage evolution. *Genome Res.* **13**, 1042–1055.

Ochman, H., Lawrence, J.G., and Groisman, E.A. (2000) Lateral gene transfer and the nature of bacterial innovation. *Nature* **405**, 299–304.

Oggioni, M.R., Dowson, C.G., Maynard Smith, J., Provvedi, R., and Pozzi, G. (1996) The tetracycline resistance gene *tet*(M) exhibits mosaic structure. *Plasmid* **35**, 156–163.

Parkhill, J., Wren, B.W., Thomson, N.R., Titball, R.W., Holden, M.T., Prentice, M.B., et al. (2001) Genome sequence of *Yersinia pestis*, the causative agent of plague. *Nature* **413**, 523–527.

Podbielski, A., Krebs, B., and Kaufhold, A. (1994) Genetic variability of the *emm* locus of the large vir regulon of group A streptococci: Potential intra- and intergenomic recombination events. *Mol. Gen. Genet.* **243**, 691–698.

Prince, V.E., and Pickett, F.B. (2002) Splitting pairs: The diverging fates of duplicating genes. *Nature Rev. Genet.* **3**, 827–837.

Prudhomme, M., Libante, V., and Claverys, J.-P. (2002) Homologous recombination at the border: Insertion-deletions and the trapping of foreign DNA in *Streptococcus pneumoniae*. *Proc. Natl. Acad. Sci. U S A* **99**, 2100–2105.

Rådström, P., Fermer, C., Kristiansen, B.-E., Jenkins, A., and Sköld, O. (1992) Transformational exchanges in the dihydropterate synthase gene of *Nesiieria meningitidis*: A novel mechanism for the acquisition of sulfonamide resistance. *J. Bacteriol.* **174**, 6386–6393.

Rebello, J.-E., François, V., and Louarn, J.-M. (1988) Detection and possible role of two large nondivisible zones on the *Escherichia coli* chromosome. *Proc. Natl. Acad. Sci. U S A* **85**, 9391–9395.

Riehle, M.M., Bennett, A.F., and Long, A.D. (2001) Genetic architecture of thermal adaptation in *Escherichia coli*. *Proc. Natl. Acad. Sci. U S A* **98**, 525–530.

Robinson, W.G., Old, L.A., Shah, D.S., and Russell, R.R. (2003) Chromosomal insertions and deletions in *Streptococcus mutans*. *Caries Res.* **37**, 148–156.

Römling, U., Schmidt, K.D., and Tümmler, B. (1997) Large genome rearrangements discovered by the detailed analysis of 21 *Pseudomonas aeruginosa* clone C isolates found in environment and disease habitats. *J. Mol. Biol.* **271**, 386–404.

Rosenberg, S.M. (2001) Evolving responsively: Adaptive mutation. *Nat. Rev. Genet.* **2**, 504–515.

Scheirer, K.E., and Higgins, N.P. (2001) Transcription induces a supercoil domain barrier in bacteriophage Mu. *Biochimie* **83**, 155–159.

Schmid, M.B., and Roth, J.R. (1983) Selection and endpoint distribution of bacterial inversion mutations. *Genetics* **105**, 539–557.

Schmid, M.B., and Roth, J.R. (1987) Gene location affects expression level in *Salmonella typhimurium*. *J. Bacteriol.* **169**, 2872–2875.

Schneider, D., Duperchy, E., Coursange, E., Lenski, R.E., and Blot, M. (2000) Long-term experimental evolution in *Escherichia coli*. IX. Characterization of insertion sequence-mediated mutations and rearrangements. *Genetics* **156**, 477–488.

Segall, A.M., and Roth, J.R. (1989) Recombination between homologies in direct and inverse orientation in the chromosome of *Salmonella typhimurium*: Intervals which are nonpermissive for inversion formation. *Genetics* **122**, 737–747.

Segall, A.M., Mahan, M.J., and Roth, J.R. (1988) Rearrangement of the bacterial chromosome: Forbidden inversions. *Science* **241**, 1314–1318.

Seifert, H.S. (1996) Questions about gonococcal pilus and phase-antigenic variation. *Mol. Microbiol.* **21**, 433–440.

Serkin, C.D., and Seifert, H.S. (1998) Frequency of pilin antigenic variation in *Neisseria gonorrhoeae. J. Bacteriol.* **180**, 1955–1958.

Shen, P., and Huang, H.V. (1986) Homologous recombination in *Escherichia coli*: Dependence on substrate length and homology. *Genetics* **112**, 441–457.

Shyamala, V., Schneider, E., and Ames, G.F. (1990) Tandem chromosomal duplications: Role of REP sequences in the recombination event at the jointpoint. *EMBO J.* **9**, 939–946.

Sibold, C., Henrichsen, J., König, A., Martin, C., Chalkley, L., and Hakenbeck, R. (1994) Mosaic *pbpX* genes of major clones of penicillin-resistant *streptococcus pneumoniae* have evolved from *pbpX* genes of a penicillin-sensitive *Streptococcus oralis. Mol. Microbiol.* **12**, 1013–1023.

Slechta, E.S., Bunny, K.L., Kugelberg, E., Kofoid, E., Andersson, D.I., and Roth, J.R. (2003) Adaptive mutation: General mutagenesis is not a programmed response to stress but results from rare amplification of *dinB* with *lac. Proc. Natl. Acad. Sci. U S A* **100**, 12847–12852.

Smith, G.R. (2001) Homologous recombination near and far from DNA breaks: Alternative roles and contrasting views. *Annu. Rev. Genet.* **35**, 243–274.

Sonti, R.V., and Roth, J.R. (1989) Role of gene duplications in the adaptation of *Salmonella typhimurium* to growth on limiting carbon sources. *Genetics* **123**, 19–28.

Spratt, B.G. (1988) Hybrid penicillin-binding proteins in penicillin-resistant strains of *Neisseria gonorrhoeae. Nature* **332**, 173–176.

Spratt, B.G., Zhang, Q.-Y., Jones, D.M., Hutchison, A., Brannigan, J.A., and Dowson, C.G. (1989) Recruitment of a penicillin-binding protein gene from *Neisseria flavescens* during the emergence of penicillin-resistance in *Neisseria meningitidis. Proc. Natl. Acad. Sci. U S A* **86**, 8988–8992.

Spratt, B.G., Bowler, L.D., Zhang, Q.-Y., Zhou, J., and Smith, J.M. (1992) Role of interspecies transfer of chromosomal genes in the evolution of penicillin resistance in pathogenic and commensal *Neisseria* species. *J. Mol. Evol.* **34**, 115–125.

Stibitz, S., and Yang, M.-S. (1999) Genomic plasticity in natural populations of *Bordetella pertussis. J. Bacteriol.* 181, 5512–5515.

Straus, D.S. (1975) Selection for a large genetic duplication in *Salmonella typhimurium. Genetics* 80, 227–237.

Tamas, I., Klasson, L., Canbäck, B., Näslund, A.K., Eriksson, A-S., Wernegreen, J.J., et al. (2001) 50 Million years of genomic stasis in endosymbiotic bacteria. *Science* 296, 2376–2379.

Tepfer, D., Garcia-Gonzales, R., Mansouri, H., Seruga, M., Message, B., Leach, F., and Perica, M.C. (2003) Homology-dependent DNA transfer from plants to a soil bacterium under laboratory conditions: Implications in evolution and horizontal gene transfer. *Transgenic Res.* 12, 425–437.

Teyssier, C., Marchandin, H., Siméon De Bouchberg, M., Ramuz, M., and Jumas-Bilak, E. (2003) Atypical 16S rRNA gene copies in *Ochrobactrum intermedium* strains reveal a large genomic rearrangement by recombination between *rrn* copies. *J. Bacteriol.* 185, 2901–2909.

Townsend, S.M., Kramer, N.E., Edwards, R., Baker, S., Hamlin, N., Simmonds, M., et al. (2001) *Salmonella enterica* serovar Typhi possesses a unique repertoire of fimbrial gene sequences. *Infect. Immun.* 69, 2894–2901.

Tu, Z.-C., Ray, K.C., Thompson, S.A., and Blaser, M.J. (2001) *Campylobacter fetus* uses multiple loci for DNA inversion within the 5' conserved regions of *sap* homologues. *J. Bacteriol.* 183, 6654–6661.

Vinella, D., Cashel, M., and D'ari, R. (2000) Selected amplification of the cell division genes *ftsQ-ftsA-ftsZ* in *Escherichia coli. Genetics* 156, 1483–1492.

Watt, V.M., Ingles, C.J., Urdea, M.S., and Rutter, W.J. (1985) Homology requirements for recombination in *Escherichia coli. Proc. Natl. Acad. Sci. U S A* 82, 4768–4772.

Whatmore, A.M., and Kehoe, M.A. (1994) Horizontal gene transfer in the evolution of group A streptococcal *emm*-like genes: Gene mosaics and variation in Vir regulons. *Mol. Microbiol.* 11, 363–374.

Yanai, I., Camacho, C.J., and DeLisi, C. (2000) Predictions of gene family distributions in microbial genomes: Evolution by gene duplication and modification. *Phys. Rev. Lett.* 85, 2641–2644.

Zhang, Q.-Y., Jones, D.M., Saez-Nieto, J.A., Perez-Trallero, E., and Spratt, B.G. (1990) Genetic diversity of penicillin-binding protein 2 genes of penicillin-resistant strains of *Neisseria meningitidis* revealed by fingerprinting of amplified DNA. *Antimicrob. Agents Chemother.* 34, 1523–1528.

DIARMAID HUGHES AND TOBIAS NORSTRÖM

CHAPTER 11

Horizontal gene transfer and bacterial genomic legacies

James R. Brown

385

Molecular biologists have long used viruses, plasmids, transposons, and other "vectors" as tools to directly manipulate the genetic makeup of experimental organisms. In nature, these tool vectors originated in species, usually bacteria, as facilitators of horizontal (also known as lateral) gene transfer (HGT). In contrast to vertical inheritance, where the transmission of genetic material occurs vertically from parent to offspring, HGT refers to the horizontal exchange of genes between distantly related strains and species. As described in this volume, there are many examples of HGT between species of bacteria, such as that mediated by plasmids and phages, which bear genes responsible for pathogenicity and antibiotic resistance. HGT is also known to occur in eukaryotes; for example, DNA transposons have been suggested as being horizontally transferred between different species of the fruitfly *Drosophila* (Bushman, 2002). These are examples of HGT on a relatively recent evolutionary timescale. However, HGT might have had a pivotal evolutionary role in more ancient times. Comparative analyses of molecular data that are exploding from genome sequencing projects indicates that HGT might have been the main driving force behind the evolution of cellular life (Brown, 2003).

The reason for believing the occurrence of ancient HGT is relatively simple. In an evolutionary context, genes are not found where they are expected to be. The most fundamental subdivisions of living organisms are the three urkingdoms or domains of life: the Archea (traditionally called "archaebacteria"), Bacteria (traditionally called "eubacteria"), and Eucarya (interchangeable here and elsewhere with the term "eukaryote"; Woese, Kandler, and Wheelis, 1990). Some genes from selected species of the Bacteria are more closely related to versions in the Archea. Conversely, some eukaryotic genes appear to have originated in the Bacteria rather than the Archea, the supposed sister domain to eukaryotes (Brown and Doolittle, 1997; Doolittle et al., 1990; Golding and Gupta, 1995; Smith, Feng, and Doolittle, 1992). These are not rare exceptions; rather, ancient HGT might have moved about a very significant portion of all known genes (Doolittle, 1999b).

The importance of understanding ancient HGT extends beyond the realm of academic interest. Hospital clinics around the world increasingly report bacterial infections that are resistant to many commercially available antibiotics (Felmingham and Washington, 1999). This phenomenon is partially linked to the exchange among bacteria of resistance-determining genes, some of which might have originated long ago from species of Archea or eukaryotes (Brown, Zhang, and Hodgson, 1998). HGT among different species of soil-dwelling bacteria, as well as between plants and their bacterial symbionts, raises concerns about bioengineered plants or bacteria-exchanging genes with natural species (Dröge, Pühler, and Selbitschka, 1998).

In the decades to come, critical advances in biomedical research will be linked to the sequence of the human genome; thus, a thorough understanding of the evolutionary relationships of human genes is crucial. A high-profile case of the wrongful evolutionary relationships was reported in the first publication of the draft human genome sequence by the International Human Genome Sequencing Consortium, where it was concluded that more than 100 human genes were directly transferred from bacteria (International Human Genome Sequencing Consortium, 2001). Subsequent studies that employed more rigorous database searches and evolutionary analyses showed the claims of bacteria-to-vertebrate HGT to be unfounded (Roelofs and Van Haastert, 2001; Salzberg et al., 2001; Stanhope et al., 2001).

This chapter considers the evidence for ancient HGT and its impact in reconstructing the universal tree of life. Ancient HGT is defined here as the exchange of genes between species of different domains. The availability of complete genome sequences from species belonging to each of the three domains now permits large-scale comparative genomic analyses to address fundamental questions in cellular evolution. However, the answers have not been definitive. Although some studies suggest that ancient HGT has been extensive, others advise caution when interpreting evolutionary patterns of gene distributions and gene phylogenies. Given that the prokaryotes were likely the first cellular life on earth, it is appropriate to review advances in understanding ancient HGT from the perspective of the contributions of bacteria, their legacy, to the genomes of eukaryotes and the Archea.

DETECTION OF HORIZONTAL GENE TRANSFER

HGT can be detected using several different computational methods, each with notable advantages and caveats (Eisen 2000; Garcia-Vallvé, Romeu, and Palau, 2000; Grauer and Li, 2000; Lawrence and Ochman, 1998, 2002). First, phylogenetic trees of proteins or DNA sequences can show incongruent relationships among taxa, which might be the result of HGT. However, lineage-specific gene losses, convergence, and unequal mutation rates in different taxa can cause similar effects. Because protein sequences change more slowly than their DNA coding regions, protein phylogenetic trees are best for determining ancient HGT events, whereas nucleotide trees are better suited for studying closely related species.

Second, shifts in nucleotide composition between neighboring sequences, such as a marked increase in the frequencies of G/C bases over A/T bases, could indicate that a gene or noncoding region from another species has been inserted into the genome. However, intragenome regional biases in nucleotide composition can introduce a false signal. Further, any signal of HGT will be degraded as the base composition of a foreign piece of DNA grows more similar to that of the host genome or "ameliorates" over time, eventually becoming unrecognizable as an exogenous segment. Therefore, analyses of nucleotide base frequency likely yield better estimates of relatively recent rather than ancient HGT events (Ochman, Lawrence, and Groisman, 2000; Ragan, 2001, 2002).

Third, the presence of genes in a particular species that are also found in distant relatives but not closely related species suggests the occurrence of HGT. Multiple HGT genes might be physically linked (such as an operon) and/or encode components of a common pathway or multiprotein complex, which would suggest selection for a new physiological function. However, gene order on the chromosome is often poorly conserved even among closely related species and pathways could evolve through the acquisition of genes from different sources. The alternative theory of extensive, multiple occurrences of gene loss in the intervening lineages must also be considered.

RECONSTRUCTING THE UNIVERSAL TREE OF LIFE: A BRIEF HISTORY

At this point, it is worthwhile to recapitulate, at least briefly, the development of the universal tree of life concept. In the late 1970s, Woese, Fox, and co-workers (Fox et al., 1977; Woese and Fox, 1977) constructed dendograms

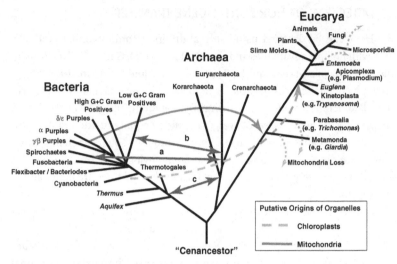

Figure 11.1. Schematic phylogeny of the universal tree of life. In addition to putative HGT events associated with the origin of mitochondria and chloroplasts *(green and purple arrows)*, HGT involving other groups have also been observed, such as between (a) Spirochetes and Archea, (b) between low G+C gram-positive bacteria and Archea, and (c) between thermophilic Bacteria and Archea (see text). The diagram shows the relative evolutionary positions of various groups of organism on the tree, but branch lengths are not to scale. Among eukaryotes there are protist groups, once called the Archezoa, that are believed to have secondarily "lost" or modified their mitochondria-like organelles *(yellow dashed arrows)*. (Adapted from Brown 2003.)

based on 16S small subunit (SSU) ribosomal RNA molecules, which showed, among other things, the existence of a group of highly divergent methanogenic "bacteria" that they subsequently christened the "archaebacteria." In 1990, Woese, Kandler, and Wheelis (1990) proposed replacing the bipartite division of life forms of prokaryotes and eukaryotes with a tripartite scheme of three "urkingdoms" or domains, the Archea, Bacteria, and Eucarya (Fig. 11.1). Archea were clearly neither entirely bacterial nor eukaryotic in their cellular or genomic features (reviewed in Brown and Doolittle, 1997; Olsen and Woese, 1997). More recently, Graham et al. (2000) found 351 archeal-specific "phylogenetic footprints," or combinations of genes, that are uniquely shared by two or more archeal species but are not found in either bacteria or eukaryotes.

Although controversial, the universal tree of life provides a basis on which hypotheses about the evolution of cellular life can be built. It is important to differentiate between the phylogenetic trees based solely on SSU rRNA molecules, and the universal tree of life, which in this review refers to the proposed evolutionary relations between cellular life forms. The universal tree of life is a concept based on multiple pieces of evidence that include SSU rRNA phylogenetic trees, phylogenies of ancient duplicated protein genes (some that provide a computational means for rooting the tree), and the distributions of genes and biochemical pathways among the different domains.

There are three salient features of the universal tree. First, there is the monophyly of the three domains. Each domain forms a separate clade, and none of the three major lineages evolved from a specific species or group in another. All three domains are joined together at the "trunk" or base of the universal tree. Second, Archea and eukaryotes are more closely related to each other than either group is to the Bacteria. Archea and eukaryotes share many important genes involved in the DNA replication and repair, transcription, and translation, which are absent in the Bacteria (Graham et al., 2000; Olsen and Woese, 1997). The closer relation between Archea and eukaryotes is also supported by rootings of the universal tree based on paralogous gene families, which originated before the three domains of life (Brown and Doolittle, 1995; Gogarten et al., 1989; Iwabe et al., 1989).

The third fundamental characteristic of the tree is the early evolution of thermophilic Bacteria. Thermophilic species of bacteria, such as *Aquifex aoelicus* and *Thermotoga maritima*, are the earliest branching lineages in phylogenies based on rRNA molecules and many proteins (Olsen, Woese, and Overbeek, 1994). The proposal that thermophilic bacteria are the most ancient organisms has lent support to theories that cellular life itself evolved in hot environments.

However, subsequent phylogenetic analyses of many proteins, including several of those originally used to root the universal tree, revealed marked exceptions to the universal tree paradigm (Hilario and Gogarten, 1993; Woese et al., 2000). Phylogenetic analyses using alternative methods and expanded protein data sets have raised questions about the monophyly of the Archea and the rooting of the universal tree in the Bacteria (Baldauf, Palmer, and Doolittle, 1996; Lopez, Forterre, and Philippe, 1999). HGT between many different groups has been proposed (Fig. 11.1), including Spirochetes and Archea (Brown, 2001; Teichmann and Mitchison, 1999; Wolf et al., 1999), low G+C gram-positive bacteria and Archea (Benachenhou-Lahfa, Forterre, and Labedan, 1993; Boucher and Doolittle, 2000; Brown et al., 1994), and thermophilic Bacteria and Archea (Aravind et al., 1998).

HOW EXTENSIVE IS ANCIENT HGT?

Prior to the publication in 1995 of the first complete bacterial genome sequence, that of *Haemophilus influenzae* (Fleischmann et al., 1995), and the launch of the "genome revolution." a diversity of genes from species of Archea, Bacteria, and eukaryotes had been analyzed using phylogenetic methods. Many such phylogenetic trees of single proteins had irreconcilable topologies with the universal tree (Smith, Feng, and Doolittle, 1992). In fact, different studies showed that only between 20% to 50% of protein trees supported the monophyly of domains and the sisterhood of Archea and eukaryotes (Brown and Doolittle, 1997; Feng, Cho, and Doolittle, 1997; Gupta and Golding, 1996; Roger and Brown, 1996). The majority of protein trees depicted the paraphyly or polyphyly of one or more of the domains. These incongruencies, which were violations of the universal tree of life topology, have been taken as evidence of complex and extensive interdomain HGT. Coupled with HGT is the loss of genes from specific lineages whose functions may or may not be replaced by an acquired gene. Extensive debates have developed over whether universal tree reconstruction is a practical and relevant exercise if HGT had widely occurred in the early stages of cellular evolution (Doolittle, 1999b; Martin, 1996; Pennisi, 1998, 1999).

Estimates of the number of genes acquired by HGT vary considerably with the detection methodology. A sense of the possible scope of HGT in the Bacteria comes from comparisons of closely related species. Lawrence and Ochman (1998) compared the G/C composition and codon bias in the genomes of two closely related γ-proteobacteria, *Escherichia coli* and *Salmonella typhimurium*. They estimated that *E. coli* acquired 18% of its genes over 234 HGT events since it diverged from the *Salmonella* lineage, nearly 100 million years ago, which would mean a rate of genetic exchange around 16 kilobases per million years. However, Gracia-Vallvé et al. (2000) used similar methodologies and a wider variety of species – 17 from the Bacteria and 7 from the Archea – and estimated much lower, but still considerable, percentages of transferred genes, ranging from 1.5% to 14.5%, depending on the species.

Comparative analyses of whole genome sequences using DNA or protein sequence homology search algorithms such as BLAST (Altschul et al., 1997) also indicate that HGT was extensive between domains. Hyperthermophilic archeal and bacterial species appear to share many genes; for example, nearly 24% of the genes in *Thermotoga maritima*, a bacterium that thrives at 80°C, are most similar to their archaeal counterparts (Nelson et al., 1999). Bacterial pathogens of plants and animals appear to have acquired a much smaller but still significant number of genes from their respective hosts (Koonin, Makarova, and Aravind, 2001). However, a BLAST score is not always a reliable indicator of evolutionary relatedness (Koski and Golding, 2001). Several studies have suggested that HGT among hyperthermophilic bacterial and archaeal species has been falsely assigned because both groups are the earliest evolved lineages in their respective domains, and thus, might share genes by common ancestry (Kyrpides and Olsen, 1999). Genes from deep branching species are susceptible to arbitrary clustering with genes of outgroup species. For example, BLAST results might show that genes from *T. maritima* are most similar to genes in archaeal species (the outgroup) simply because there are few orthologs available in the database from other hyperthermophilic bacterial species (Logsdon and Faguy, 1999).

Although phylogenetic trees are time consuming to construct and care must be taken to reduce various methodological errors (Grauer and Li, 2000), the anomalous placement of a particular species or several species in a phylogenetic tree remains the best indicator of ancient HGT. This frequently occurs, and in fact, very few proteins converge on exactly the same tree topology. Does this mean that we can not use phylogenies of proteins to reconstruct the universal tree of life?

SUPPORT FOR THE UNIVERSAL TREE DESPITE HGT

In a group of species, various genes might have been lost, duplicated, or horizontally transferred, resulting in different genes having unique phylogenetic tree topologies. Thus, it is very difficult to select a single gene as the best molecular marker for recreating the evolutionary history of those species. However, more recent studies using multiple genes from entire genomes suggest the existence of an underlying phylogenetic signal that is generally congruent with the universal tree. Basically, two approaches have been taken to build universal trees from whole genomes. The first, called phylogenetic profiling, determines the presence or absence of genes across the genomes of many species. Bork and colleagues (Huynen, Snel, and Bork, 1999; Snel, Bork, and Huynen, 1999) and Fitz-Gibbon and House (1999) constructed data matrices and trees based on the fraction of orthologous proteins shared in the genomes of species from the three domains of life. Both studies showed the three domains of life are monophyletic, and confirm the early origin of thermophiles by placing the hyperthermophilic bacterium, *Aquifex aoelicus*, on one of the deepest evolutionary branches in the Bacteria. The universal tree therefore seems to prevail, despite reports of extensive interdomain HGT.

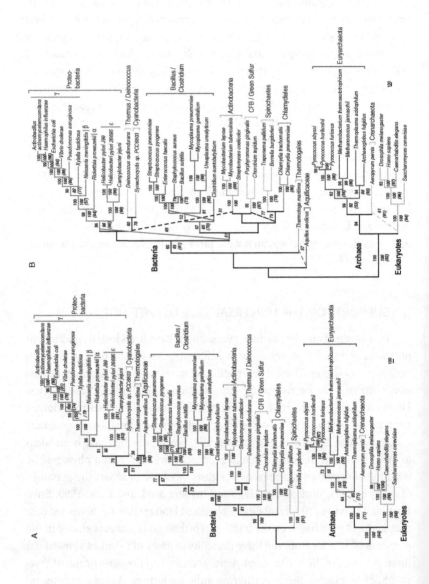

However, Doolittle (1999a) suggested that such whole genome trees based on the presence or absence of genes are biased toward data on genes that are fluid during evolution (i.e., genes need to be either lost or acquired in two or more genomes to be phylogenetically informative). Thus, information contained in the amino acid sequences of highly conserved genes found in all species are not used by this approach. Conversely, because so few proteins are universally conserved orthologs, most known genes are being considered in the phylogenetic profiling method.

A second approach to universal tree reconstruction relies on the combination, or concatenation, of many conserved proteins into a single data set for phylogenetic analysis. Phylogenies based on such large, multiple-sequence alignments are potentially more robust and representative of the evolutionary relationships among species because the number of informative sites and sampled gene loci are increased. Such an approach has been used to determine the phylogenetic relationships among photosynthetic bacteria (Xiong, Inoue, and Bauer, 1998), protists (Baldauf et al., 2000), and yeast (Rokas et al., 2003) but have met with less success in resolving the evolutionary relationships among disparate groups of bacteria (Hansmann and Martin, 2000). We found strong support for domain monophyly in universal trees based on a combined multiple-sequence alignment of 23 proteins conserved across 45 species of Bacteria, Archea, and eukaryotes (Brown et al., 2001). However, an inconsistency was that Spirochetes (represented by two pathogenic species), rather than thermophiles, were found to be the earliest branching group of Bacteria (Fig. 11.2A). The subsequent removal from the data set of nine proteins for which the Bacteria were not monophyletic in single protein trees rectified the placement of thermophiles as the deepest branching

Figure 11.2. Universal trees based on combined protein data sets. **(A)** Minimal-length maximum parsimony universal trees based on 23 combined protein data sets is shown. Spirochetes are placed as the lowest branching Bacteria. Numbers along the branches show the percent occurrence of nodes in 1000 bootstrap replicates of maximum parsimony (MP; plain text; Swofford, 1999) and neighbor joining (NJ; italicized text; Felsenstein, 1993) analyses or greater than 50% of 1000 quartet puzzling steps of maximum likelihood analysis (ML; in parentheses; Strimmer and von Hasseler, 1996). Dashed lines show occasional differences in branching orders in neighbor-joining trees. Scale bar represents 100 amino acid residue substitutions. CFB stands for the Cytophaga-Flexibacter-Bacteroides group of bacteria. **(B)** Universal tree based on 14 combined protein data sets. Minimal-length MP universal tree based on 14 proteins, with 9 horizontal gene transfer proteins removed, is shown. The tree shows Thermophiles as the basal group in Bacteria. All trees drawn using the software TREEVIEW (Page, 1996). (Adapted from Brown et al., 2001.)

species in the Bacteria, which illustrates the effect of HGT on tree topologies (Fig. 11.2B).

Another novel analysis using whole genome data suggests that the evolution relationships of cellular life are better represented as a "ring" rather than a "tree" (Rivera and Lake 2004). The ring phylogeny actually recapitulates previous hypotheses that it was the fusion or union of two prokaryotic organisms that led to the evolution of the first eukaryote.

One of the problems of the combined protein sequence method is the lack of highly conserved proteins found across all species. Koonin, Aravind, and Kondrashov (2000) estimated that among 21 completely sequenced genomes there are more than 2100 ancient protein families, but fewer than 80 of them are universally found in all species. As more genomes are sequenced, the number of conserved proteins inevitably decreases. For example, the 70-kDa heat shock protein (HSP70), once believed to occur universally, is now known to be absent from several species of Archea (Gribaldo et al., 1999).

HOW ARE GENES TRANSFERRED ACROSS DOMAINS?

Studies of genes in contemporary species or in laboratory cell lines hold few clues to understanding the molecular basis of interdomain HGT. There are no known naturally occurring vectors (plasmids, phages, or transposable elements) responsible for interdomain gene transfer, and there is little evidence that eukaryotic cells are naturally competent although this is a well-known characteristic of many bacterial species. The closest example of such a promiscuous vector is the tumor-inducing or Ti plasmid of *Agrobacterium*, which facilitates the transfer of bacterial DNA to plant somatic cells and its expression (Zambryski, Tempe, and Schell, 1989). These bacterial infections create large woody protrusions, called crown galls, which can be seen on the trunks of fruit trees. The frequency of naturally occurring, stable germ-line gene transfers by Ti plasmids is unknown and might be limited to half a dozen genes implicated in hairy root disease, which is the mutation of root cells caused by *A. rhizogenes* (Bushman, 2002). *Wolbachia*, an α-proteobacterial endosymbiont, has reportedly transferred at least 11 kb of DNA to the X chromosome of its insect host (Kondo et al., 2002). However, the coding regions of the inserted DNA are riddled with stop codons and frameshift mutations, so these transferred genes are likely nonfunctional. The human genome is strewn with fragments of human mitochondrial DNA, which also appear to be pseudogenes (Tourmen et al., 2002). In the lab, mammalian cells have been transformed by bacterial cells (Courvalin, Goussard, and Grillot-Courvalin, 1995). For example, genes have been successfully transferred by

Agrobacterium to human Hela cells (Kunik et al., 2001). However, there is no evidence yet for natural occurrences of HGT from bacteria to eukaryotic germ cells – an essential condition if ancient HGT between these two domains is to be detected (Kurland, 2000). Gene transfers in the opposite direction, from eukaryotes to bacteria, occur at extremely low frequencies, based on the observed transfer rates of recombinant genes from plants to soil bacteria (Dröge, Pühler, and Selbitschka, 1998). The means by which ancient HGT might have occurred has been the subject of intense speculation. Most proposals suggest that HGT was extensive in the early stages of cellular evolution, when the barriers to genetic exchange between Archea, Bacteria, and eukaryotes were weak. Eukaryotes themselves might have emerged from some fusion of the genomes of an archaeal and bacterial species (reviewed in Brown and Doolittle, 1997). In their hydrogen hypothesis, Martin and Müller (1998) postulated that the first eukaryote evolved with the engulfment of a hydrogen-producing α-proteobacterial symbiont by a hydrogen-dependent archaeal host. The syntrophic hypothesis of Moreira and López-García (1998) is somewhat similar, although they propose a δ-proteobacterium as the hydrogen-producing symbiont. Horiike et al. (2002) reversed this scenario by proposing a bacterial host and an archaeal endosymbiont, although under their theory the origin of the nuclear envelope (Poole and Penny, 2002), as well as trying to distinguish from other bacterial endosymbiotic theories (Rotte and Martin, 2001), are problematic. Regardless, symbioses between proteobacteria and methanogenic archea have been observed in nature (Martin and Müller, 1998; Moreira and López-García, 1998). Other unusual symbioses are also known to occur; for example, intracellular β-proteobacteria endosymbionts found in mealybug insects have, in turn, their own γ-proteobacterial symbionts (von Dohlen et al., 2001). Also, there is a more recent report of two archaeal species potentially dependent on mutual endosymbiosis (Huber et al., 2002).

It is worth noting that not all evidence points to the eukaryotes evolving from the fusion of two prokaryotic cells (i.e., an archaeon and a bacterium). Hartman and Fedorov (2002) identified 347 proteins specific to eukaryotes, which they suggest supports the existence of a third type of protoeukaryote, the chronocyte. The chronocyte would have had a cytoskeleton, a feature not found in contemporary archaeal or bacterial species, which would have allowed it to engulf prokaryotic organisms, some of which were perhaps destined to be organelle-forming endosymbionts. The chronocyte theory is not new (for a review, see Brown and Doolittle, 1997) and it suffers, as other theories, from the lack of extant examples. Doolittle (1998) suggested that, as a consequence of endosymbiosis, the bacterial gene content of the host eukaryotic genome would logically "ratchet up" if an introduced gene assumed a new useful and indispensable function.

Putative ancient HGT occurrences might be artifacts of very early precellular evolutionary processes. Woese (2002) suggested that HGT was the primary force in early cellular evolution, predominating over the vertical inheritance of genes, because early cells were extremely simple, loosely organized, and essentially communal. These early "cells" had highly error-prone DNA replication and expression systems, which facilitated the integration of DNA from other cells in the ecosystem. As cells became more structured, a "Darwinian threshold" was reached where vertical inheritance predominated as the main path of gene flow, relegating HGT to being a rarer event. Others have suggested that the observed incidents of ancient HGT might be explained if the universal tree was not rooted in the Bacteria but in eukaryotes (Forterre and Philippe, 1999) or in some subdivision of the Bacteria, such as the gram positives (Gupta, 1998). However, the theory of eukaryotes evolving first has several difficulties, including explanations about the reduction in genome sizes to prokaryote levels, the loss of introns, and physiological adaptations (i.e., all living eukaryotes presently have or once had bacterial endosymbionts). Rooting the tree of life in specific bacterial species also requires the postulating of overly complex evolutionary scenarios.

EVIDENCE FOR HGT BETWEEN DOMAINS

Archea and Eucarya

At this point, it seems worthwhile to review current knowledge on the relationships between the domains of life and how it relates to the status of ancient HGT. The close evolutionary relationship between Archea and Eucarya is evident from the phylogenetic analyses of many "informational" proteins – those involved in DNA replication, transcription, and translation (reviewed in Olsen and Woese, 1997). Lake and colleagues (Rivera et al., 1998) suggested that informational genes are less likely to be transferred between genomes than operational genes, that is, those involved in cell metabolism. Informational gene products, at least qualitatively, have additional interactions, which might limit genetic exchange and fixation (Jain, Rivera, and Lake, 1999). Additional support for this view is the observation that, although operons and regulatory elements might be rearranged in different bacterial genomes, translation and ribosomal-related genes are generally restricted to a particular neighborhood in the chromosome (Lathe, Snel, and Bork, 2000).

Neither ribosomal RNAs nor proteins have been candidates for interdomain transfer, although ribosomal protein S14 and rRNAs might have been exchanged among distantly related groups of bacteria (Brochier, Philippe, and Moreira, 2000; Ueda et al., 1999; Yap, Zhang, and Wang, 1999). Trees based on ancient gene paralogs also position Archea and eukaryotes as sister groups, although it has been suggested that such results are idiosyncratic due to more rapid rates of evolutionary change in Bacteria (Brinkman and Philippe, 1999).

Among the three possible universal tree scenarios, only trees depicting the Archea and eukaryotes as sister groups also show, albeit rarely, all three domains as simultaneously monophyletic (Brown and Doolittle, 1997). If extensive polyphyly or paraphyly is evidence for HGT then, by default, monophyly indicates evolution in the absence of HGT. Although domain monophyly appears to be rare in single gene trees, it is supported by whole genome analyses. However, genome data from new species of Bacteria and Archea will continue to revise lists of gene-species distributions and phylogenetic trees. Similar situations are emerging for eukaryotes where, until more recently, sequence data were mainly available from the "crown" species – fungi, plants, and animals. Certain species of anaerobic protists have pyruvate:ferredoxin oxidoreductase genes that are also found in species of Bacteria and Archea but are absent from aerobic protists and metazoans (Horner, Hirt, and Embley, 1999; Horner et al., 2002). Some highly conserved genes show different prokaryotic origins in protists and eukaryotes (Chihade et al., 2000). As the diversity of eukaryotes becomes appreciated, there will no doubt emerge stronger evidence for HGT with particular groups of Archea and Bacteria.

Archea and Bacteria

Unlike the trees showing Archea and Eucarya as sister groups, some protein trees, such as that of glutamine synthetase (Fig. 11.3), cluster Archea and Bacteria together with one or both domains portrayed as a paraphyletic or polyphyletic group (Brown et al., 1994). This pattern indicates the probable occurrence of HGT. The genes and species implicated in HGT between Archea and Bacteria are highly varied. The evolutionary patterns of some genes suggest that HGT has occurred between Archea and gram-positive Bacteria (Benachenhou-Lahfa, Forterre, and Labedan, 1993; Gupta and Golding, 1996). For example, hyperthermophilic archaeal and bacterial species share reverse gyrase, possibly an adaptation to living at extremely high temperatures (Forterre et al., 2000). Catalase-peroxidase might have been exchanged between Archea and pathogenic proteobacteria (Faguy and Doolittle, 2000).

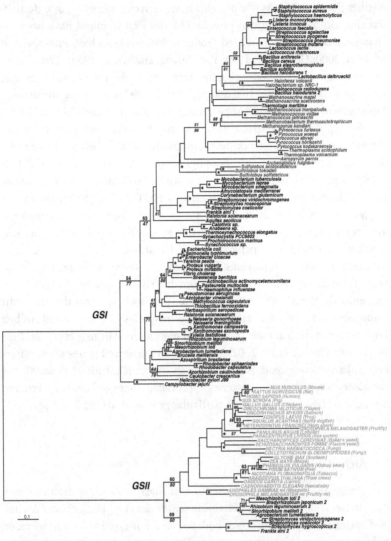

Figure 11.3. Phylogenetic tree for glutamine synthetase suggests HGT between the three domains. Although it is unrooted, the tree based on the glutamine synthetase protein sequence is incongruent with the canonical universal tree. Species of Archea (red) and Bacteria (black), mainly gram-positive groups, are mixed and therefore are not monophyletic. In contrast, eukaryotes (green) are monophyletic, in keeping with the universal tree; however, Bacteria, rather than Archea, are its closest outgroup. In addition to HGT between Archea and Bacteria, shown how here by the fact that bacteria and archea are polyphyletic, several bacterial species have both a bacterial *GSI* isoform and an eukaryotic *GSII* isoform. This protein tree was constructed using the NJ method

Two component signal transduction systems in the Archea, as well as fungi and slime molds, were also probably acquired from Bacteria (Koretke et al., 2000).

What are the alternative scenarios to HGT in these instances? Perhaps eukaryotes evolved first, with Archea and Bacteria having undergone extreme genome simplification or "streamlining" (Forterre and Philippe, 1999). However, a number of complex streamlining events would have had to occur, namely, the requisite reduction of genome size, the loss of introns, the reorganization of genes into operons, and the adaptation of a single origin of replication. Furthermore, such models need to account for the retention of eukaryotic DNA replication, transcription, and translation proteins by the Archea, on the one hand, and the commonality of bacterial and eukaryotic membrane structures, on the other. Possibly, genes found only in Bacteria and Archea are not the result of HGT between these two domains at all; rather these represent genes that were lost early in the evolution of eukaryotes.

Bacteria and Eucarya

Clues to the directionality of HGT can sometimes be inferred from the species distribution of the gene in question. If the gene occurs in many species of one domain but only a few species of another, then it is likely that a species from the more populated group was the donor. An example of eukaryote-to-bacteria transfer is glutaminyl-tRNA synthetase, which appears throughout eukaryotes but only in a limited number of bacterial species, mainly in Proteobacteria (Brown and Doolittle, 1999; Lamour et al., 1994).

The origin of mitochondria and plastids from different bacterial endosymbionts has been a widely accepted hypothesis for several decades (Margulis, 1970). However, the extent of additional gene transfer from bacteria to eukaryotic genomes is still being revealed. Phylogenies for many universal proteins, such as metabolic enzymes, show Archea, rather than Bacteria,

Figure 11.3 (*cont.*). (Felsenstein, 1993) based on pairwise distance estimates of the expected number of amino acid replacements per site (0.1 in the scale bars). Numbers above and below (in italics) branches show the percent occurrence of nodes in 1000 bootstrap replications of MP (Swofford, 1999) and NJ analyses (Felsenstein, 1993), respectively. Only values greater than 50% are shown. The symbol * indicates support greater than 70% for that node by both methods. For clarity, some plant, fungal, and animal GSII sequences that clustered with like groups were excluded from the final tree. (Adapted from Brown, 2003.)

as the outgroup, which suggests a close relationship between Bacteria and eukaryotes (reviewed in Kurland and Andersson, 2000). Further evidence for the early integration of bacterial genes into the eukaryotic genome comes from studies of proteins from simple protists such as *Giardia intestinalis* (previously *G. lamblia*; Diplomonadida) and *Trichomonas vaginalis* (Parabaslia), which commonly lack mitochondria (amitochondria); rRNA phylogenies indicate that these are the earliest evolved eukaryotic lineages. Early evolutionary views on these protists, once collectively called the "Archezoa," were that they evolved before mitochondrial biogenesis – the endosymbiosis driven fixation of an α-proteobacterium in an eukaryotic cell (Cavalier-Smith, 1993).

However, molecular studies have shown that these amitochondrial protists have several proteins targeted to the mitochondria in higher eukaryotes with putative bacterial origins (Clark and Roger, 1995; Germot, Philippe, and Le Guyader, 1996; Henze et al., 1995; Keeling and Doolittle, 1997; Keeling and McFadden, 1998; Roger, Clark, and Doolittle, 1996). Parabaslids, as well as some ciliates and fungi that lack mitochondria, have specialized organelles called hydrogenosomes to ferment pyruvate and produce hydrogen (Horner et al., 2002). The hydrogenosome lacks a genome, which might have helped trace its origins. However, phylogenetic analysis of hydrogenosome targeted proteins suggest a shared evolutionary path with mitochondria; thus, the hydrogenosome itself might be a highly derived mitochondria (Tielens et al., 2002; van der Giezen et al., 2002). More recently, a mitochondrial remnant organelle called the mitosome has been found in *G. intestinalis* where it seems to function in the maturation of iron-sulphur proteins (Tovar et al., 2003). Therefore, amitochondrial protists either secondarily lost or highly modified their mitochondria into other organelle-like structures.

More recent publications by Kurland, Andersson, and co-workers and by W. Martin and colleagues call for further consideration about the origins of the mitochondrial and chloroplast proteomes, respectively. The most striking aspect of mitochondrial evolution is the extent of the partnership between the host and the endosymbiont (Andersson et al., 1998; Kurland and Andersson, 2000). In the yeast *Saccharomyces cerevisiae*, there are about 400 nuclear proteins involved in mitochondrial interactions, but half of them have no bacterial counterparts (Karlberg et al., 2000). Thus, the evolution of endosymbiosis might have been a two-way street. The bacterial endosymbiont directly contributed some of its genes to the genome of the host (either an archaeon or early eukaryote). Experiments suggest that gene transfer, at least in yeast, is biased in the direction from the mitochondria to the nuclear genome (1 transfer event per 10^5 generations versus near negligible transfers from the nucleus to the mitochondria; Thorsness and Fox, 1990, 1993).

However, the host must have also targeted or invented nuclear genes for communication and support to the new organelle (Kurland and Andersson, 2000).

Martin and co-workers asked how many genes from cyanobacteria, the progenitor of the chloroplast, found their way into the plant nuclear genome (Martin et al., 2002; Rujan and Martin, 2001). They performed phylogenetic analysis on 9368 proteins (from a total of 24,990 proteins) encoded by the genome of the thale cress, *Arabidopsis thaliana*, and concluded that nearly 1700 genes in the nucleus originated from cyanobacteria (Martin et al., 2002). Extrapolating from this result, they estimated that 4500 or 18% (1700/9368) of the genes in the *Arabidopsis* genome originated in the cyanobacteria. Many of these proteins had cellular targets other than the chloroplast, which suggests that plastid endosymbiosis made a broader gene contribution to the plant genome beyond chloroplast biogenesis (Palenik, 2002).

Many eukaryotic genes appear to be evolutionary related to bacterial species other than cyanobacteria or α-proteobacteria. Rather than being other instances of bacteria to eukaryote HGT, these gene affinities might be due to undetermined relationships among bacterial species. For example, the bacterial genus *Chlamydia* has genes homologous to those of plants because it is evolutionary related to cyanobacteria and chloroplasts (Brinkman et al., 2002).

Still, there remain a large number of conserved eukaryotic genes that are closely related to orthologs in species of bacteria other than plastid or mitochondrial endosymbionts. How did these genes become integrated into the eukaryotic genome? Perhaps, as suggested by Doolittle (1998), endosymbiosis was a series of process of failed relationships, where the different bacterial house guests left behind remnant genes in the nucleus. Martin (1999) suggested that the fluidity of the genomes of all bacteria, including putative endosymbionts, could have been a factor. HGT between bacterial species, combined with gene mutation or deletion, results in a high turnover of genes in a bacterial genome over time. If endosymbiosis is transient and frequent, there could be genes from different sources introduced into the eukaryotic genome. In both models of Doolittle (1998) and Martin (1999), the accumulation of genes in the nucleus occurs at a much higher rate than either gene loss or gene transfer in the opposite direction, from the nucleus to the organelle. Empirical support for unidirectional HGT is the fact that organelle genomes are much smaller than those of putative bacterial endosymbionts, an outcome that might be driven by the conventional evolutionary processes, natural selection and genetic drift (Berg and Kurland, 2000; Blanchard and Lynch, 2000).

There is evidence from combined protein trees that Proteobacteria evolved early in bacterial evolution, soon after the emergence of thermophiles (Brown et al., 2001). If endosymbiosis experimentation occurred simultaneously, perhaps genes from different bacterial groups passaged through proteobacterial and cyanobacterial endosymbionts, and ultimately, some of their genetic diversity were deposited in the eukaryotic host genome. Canback, Andersson, and Kurland (2002) performed phylogenetic analysis of eight glycolytic enzymes, all of them functional in the mitochondrial compartment, and showed that no single bacterial group could be linked to the origin of eukaryotic enzymes. Although the authors interpreted this finding as disparaging to HGT theories, it could also provide support for the theory that the evolution of proteobacteria and the occurrence of endosymbiosis were near coincidental events.

Another example of controversies regarding endosymbiosis and ancient HGT concerns is valyl-tRNA synthetase, where eukaryotic versions, including that of the amitochondrial protist *T. vaginalis*, appeared to have evolved from bacteria (Brown and Doolittle, 1995; Hashimoto et al., 1998); this would suggest endosymbiotic gene transfer and replacement. However, the valyl-tRNA synthetase sequenced from the genome of *Rickettsia prowazekii*, an α-Proteobacterium, is clearly of archaeal origin (Andersson et al., 1998; Kurland and Andersson, 2000). Perhaps α-Proteobacteria exchanged valyl-tRNA synthetase with an archaeon after mitochondrial biogenesis; alternatively, eukaryotes might have received their valyl-tRNA synthetase from a bacterial endosymbiont other than an α-Proteobacterium. Phylogenetic analyses of some tRNA synthetases (i.e., alanlyl-tRNA synthetase in Chihade et al., 2000; methionyl-tRNA synthetase in Brown et al., 2003) show very different origins of these enzymes in protists relative to multicellular organisms, which suggests that the eukaryotic lineage occasionally experienced either interdomain HGT or substantial gene loss events throughout its evolutionary history.

CONCLUDING REMARKS

Controversies about ancient HGT will either be resolved or amplified as genomes from more taxa are sequenced. Certainly, any resolution of these debates awaits an understanding of the genome compositions of a greater number of species of lower eukaryotes and diverse groups of Bacteria and Archea. Methodologically, the detection of HGT is still evolving. Although phylogenetic analysis is the best indicator of HGT, there are still questions about the reliability of such predictions. Comparisons of BLAST homology scores, without supporting phylogenetic analyses, can lead to inflated reports of HGT occurrences. More thoughtful analysis of cellular pathways could better illuminate the extent and significant of putative HGT or gene loss events. The question that needs to be more frequently asked is: Given that no protein or RNA functions alone in the cell, what are the origins of its interacting partners or co-components in the same pathway?

The practical importance of understanding HGT should not be underestimated. The archaeon *Methanosacrina mazei* is an important generator of the greenhouse gas, methane, and appears to own much of its ecological success to ancient HGT (nearly 30% of its genes are from bacteria; Deppenmeier et al., 2002). HGT is a frequent means by which microbial pathogens acquire virulence (Wren, 2000). Antibiotics can become less effective in certain clinical isolates due to the HGT of its gene target from another species. For example, mupirocin is a topical agent used against *Staphylococcus aureus* infections and is derived from pseudomonic acid, a natural inhibitor of isoleucyl-tRNA synthetases. Resistance to mupirocin in some clinical isolates of *S. aureus* has been linked to the ancient HGT of a highly divergent isoleucyl-tRNA synthetase that has putative eukaryotic origins (Fig. 11.4; Brown, Zhang, and Hodgson, 1998). Antimicrobial compounds acting upon *Streptococcus pneumoniae* methionyl-tRNA synthetase also encountered natural resistance in isolates due to strain-specific HGT of eukaryote-type methionyl-tRNA synthetase (Gentry et al., 2003). Both pathogens appear to have more recently acquired their respective drug-resistant isoleucyl- and methionyl-tRNA synthetases through another bacteria donor, *Bacillus anthracis*, a soil-dwelling species that is the causative agent of anthrax (Brown et al., 2003).

New therapies against chronic disease will also benefit from a better understanding of the origin of human genes. For example, many components of the programmed cell death pathway, which is a potential target for cancer therapy (Huang and Oliff, 2001), have bacterial origins, perhaps via the mitochondrial endosymbiont (Koonin and Arvind, 2002). As discussed earlier, it is unlikely that ancient HGT resulted in the direct transfer of genes from bacteria to vertebrates, mainly humans. However, a more recent report of bacterial and fungal genes responsible for cellulose metabolism in the genome of the urochordate, *Ciona intestinalis*, warrants further investigation (Dehal et al., 2002).

Has HGT rendered the universal tree of life obsolete as a paradigm for understanding cellular evolution? Phylogenetic analyses of individual proteins, as well as large-scale comparisons of whole genomes, have shown that many genes, perhaps the vast majority, have been transferred between lineages at some time. However, the universal tree still seems to hold clues about the underlying framework of cellular evolution. The challenge is to try

Figure 11.4. Phylogeny of isoleucyl-tRNA synthetase protein sequences. Those species (*Staphylococcus aureus, Pseudomonas fluorescens, Bacillus anthracis,* and *B. cereus*) with both genes for IleRS1 and IleRS2 are indicated (in light and dark blue large font, respectively). Eukaryotic (uppercase, green text), archaeal (red text), and bacterial (black text) species are also indicated. Trees were constructed using the neighbor-joining (NJ) method (Felsenstein, 1993). Numbers along the branches show the percent occurrence of nodes in 1000 bootstrap replicates of NJ (plain text) and MP (italicized text) analyses (Swofford, 1999) or greater than 50% of 1000 ML (Strimmer and von Hasseler, 1996) quartet puzzling steps (in parentheses). An asterisk represents support for that node in more than 70% of NJ and MP bootstrap replications, as well as 60% of ML puzzle steps. Scale bar represents the estimated number of amino acid substitutions per site. (Adapted from Brown et al., 2003.)

to reconcile HGT with the underlying commonality of life in a new synthesis of early cellular evolution.

ACKNOWLEDGMENT

The support of Bioinformatics Division, Genetics Research, Glaxo-SmithKline R & D is gratefully acknowledged.

REFERENCES

Altschul, S. F., et al. (1997) Gapped BLAST and PSI-BLAST: A new generation of protein database search programs. *Nucleic Acids Res.* **25**, 3389–3402.

Andersson, S. G., et al. (1998) The genome sequence of *Rickettsia prowzekii* and the origin of mitochondria. *Nature* **396**, 133–140.

Aravind, L., Tatusov, R. L., Wolf, Y. I., Walker, D. R., and Koonin, E. V. (1998) Evidence for massive gene exchange between archaeal and bacterial hyperthermophiles. *Trends Genetics* **14**, 442–444.

Baldauf, S. L., Palmer, J. D., and Doolittle, W. F. (1996) The root of the universal tree and the origin of eukaryotes based on elongation factor phylogeny. *Proc. Natl. Acad. Sci. U S A* **93**, 7749–7754.

Baldauf, S. L., Roger, A. J., Wenk-Siefert, I., and Doolittle, W. F. (2001) A kingdom-level phylogeny of eukaryotes based on combined protein data. *Science* **290**, 972–977.

Benachenhou-Lahfa, N., Forterre, P., and Labedan, B. (1993) Evolution of glutamate dehydrogenase genes: Evidence for paralogous protein families and unusual branching patterns of the archaebacteria in the universal tree of life. *J. Mol. Evol.* **36**, 335–346.

Berg, O. G., and Kurland, C. G. (2000) Why mitochondrial genes are most often found in nuclei. *Mol. Biol. Evol.* **17**, 951–961.

Blanchard, J. L., and Lynch, M. (2000) Organellar genes: Why do they end up in the nucleus? *Trends Genetics* **16**, 315–320.

Boucher, Y., and Doolittle, W. F. (2000) The role of lateral gene transfer in the evolution of isoprenoid biosynthesis pathways. *Mol. Microbiol.* **37**, 703–716.

Brinkman, F. S. L., et al. (2002) Evidence that plant-like genes in *Chlamydia* species reflect an ancestral relationship between Chlamydiaceae, Cyanobacteria, and the chloroplast. *Genome Res.* **12**, 1159–1167.

Brinkman, H., and Philippe, H. (1999) Archaea sister group of bacteria? Indications from tree reconstruction artifacts in ancient phylogenies. *Mol. Biol. Evol.* **16**, 817–825.

Brochier, C., Philippe, H., and Moreira, D. (2000) The evolutionary history of ribosomal protein RpS14: Horizontal gene transfer at the heart of the ribosome. *Trends Genetics* **16**, 529–533.

Brown, J. R. (2001) Genomic and phylogenetic perspectives on the evolution of prokaryotes. *Syst. Biol.* **50**, 497–512.

Brown, J. R. (2003) Ancient horizontal gene transfer. *Nat. Rev. Genetics* **4**, 121–132.

Brown, J. R., and Doolittle, W. F. (1995) Root of the universal tree of life based on ancient aminoacyl-tRNA synthetase gene duplications. *Proc. Natl. Acad. Sci. U S A* **92**, 2441–2445.

Brown, J. R., and Doolittle, W. F. (1997) Archaea and the prokaryote to eukaryote transition. *Microbiol. Mol. Biol. Rev.* **61**, 456–502.

Brown, J. R., and Doolittle, W. F. (1999) Gene descent, duplication, and horizontal transfer in the evolution of glutamyl- and glutaminyl-tRNA synthetases. *J. Mol. Evol.* **49**, 485–495.

Brown, J. R., Douady, C. J., Italia, M. J., Marshall, W. E., and Stanhope, M. J. (2001) Universal trees based on large combined protein sequence datasets. *Nature Genetics* **28**, 281–285.

Brown, J. R., Gentry, D. R., Becker, J. A., Ingraham, K., Holmes, D. J., and Stanhope, M. J. (2003) Horizontal transfer of drug resistant aminoacyl-tRNA synthetases of anthrax and Gram-positive pathogens. *EMBO Rep.* **4**, 692–698.

Brown, J. R., Masuchi, Y., Robb, F. T., and Doolittle, W. F. (1994) Evolutionary relationships of bacterial and archaeal glutamine synthetase genes. *J. Mol. Evol.* **38**, 566–576.

Brown, J. R., Zhang, J., and Hodgson, J. E. (1998) A bacterial antibiotic resistance gene with eukaryotic origins. *Curr. Biol.* **8**, R365–R367.

Bushman, F. (2002) *Lateral DNA transfer.* Cold Spring Harbor, NY: Cold Spring Harbor Laboratory Press.

Canback, B., Andersson, S. G. E., and Kurland, C. G. (2002) The global phylogeny of glycolytic enzymes. *Proc. Natl. Acad. Sci. U S A* **99**, 6097–6102.

Cavalier-Smith, T. (1993) Kingdom protozoa and its 18 phyla. *Microbiol. Rev.* **57**, 953–994.

Chihade, J., Brown, J. R., Schimmel, P., and Ribas de Pouplana, L. (2000) Origin of mitochondria in relation to evolutionary history of eukaryotic alanyl-tRNA synthetase. *Proc. Natl. Acad. Sci. U S A* **97**, 12153–12157.

Clark, C. G., and Roger, A. J. (1995) Direct evidence for secondary loss of mitochondria in *Entamoeba histolytica. Proc. Natl. Acad. Sci. U S A* **92**, 6518–6521.

Courvalin, P., Goussard, S., and Grillot-Courvalin, C. (1995) Gene transfer from bacteria to mammalian cells. *C. R. Acad. Sci. Paris, Life Sciences* **318**, 1207–1212.

Dehal, P., et al. (2002) The draft genome of *Ciona intestinalis*: Insights into chordate and vertebrate origins. *Science* **298**, 2157–2167.

Deppenmeier, U., et al. (2002) The genome of *Methanosacrina mazei*: Evidence for lateral gene transfer between bacteria and archaea. *J. Mol. Microbiol. Biotechnol.* **4**, 453–461.

Doolittle, R. F., Feng, D. F., Anderson, K. L., and Alberro, M. R. (1990) A naturally occurring horizontal gene transfer from a eukaryote to prokaryote. *J. Mol. Evol.* **31**, 383–388.

Doolittle, W. F. (1998) You are what you eat: A gene transfer ratchet could account for bacterial genes in eukaryotic nuclear genomes. *Trends Genetics* **14**, 307–311.

Doolittle, W. F. (1999a) Lateral gene transfer, genome surveys and the phylogeny of prokaryotes. Technical Comments. *Science* **286**, 1443a.

Doolittle, W. F. (1999b) Phylogenetic classification and the universal tree. *Science* **284**, 2124–2128.

Dröge, M., Pühler, A., and Selbitschka, W. (1998) Horizontal gene transfer as a biosafety issue: A natural phenomenon of public concern. *J. Biotechnol.* **64**, 75–90.

Eisen, J. A. (2000) Horizontal gene transfer among microbial genomes: New insights from complete genome analysis. *Curr. Opin. Genetics Dev.* **10**, 606–611.

Faguy, D. M., and Doolittle, W. F. (2000) Horizontal transfer of catalase-peroxidase genes between Archaea and pathogenic bacteria. *Trends Genetics* **16**, 196–197.

Felmingham, D., and Washington, J. (1999) Trends in the antimicrobial suspectibility of bacterial respiratory tract pathogens – findings of the Alexander Project 1992–1996. *J. Chemother.* **11**, 5–21.

Felsenstein, J. (1993) PHYLIP (Phylogeny Inference Package) version 3.57c. Department of Genetics, University of Washington, Seattle. Available at: http://evolution.genetics.washington.edu/phylip.html

Feng, D.-F., Cho, G., and Doolittle, W. F. (1997) Determining divergence times with a protein clock: update and reevaluation. *Proc. Natl. Acad. Sci. U S A* **94**, 13028–13033.

Fitz-Gibbon, S. T., and House, C. H. (1999) Whole genome-based phylogenetic analysis of free-living organisms. *Nucleic Acids Res.* **27**, 4218–4222.

Fleischmann, R. D., et al. (1995) Whole-genome random sequencing and assembly of *Haemophilus influenzae* Rd. *Science* **269**, 496–512.

Forterre, P., Bouthier de la Tour, C., Philippe, H., and Duguet, M. (2000) Reverse gyrase from thermophiles: Probable transfer of a thermoadaptation trait from Archaea to Bacteria. *Trends Genetics* **16**, 152–154.

Forterre, P., and Philippe, H. (1999) Where is the root of the universal tree of life? *BioEssays* **21**, 871–879.

Fox, G. E., et al. (1977) Classification of methanogenic bacteria by 16S ribosomal RNA characterization. *Proc. Natl. Acad. Sci. U S A* **74**, 4537–4541.

Garcia-Vallvé, S., Romeu, A., and Palau, J. (2000) Horizontal gene transfer in bacterial and archaea complete genomes. *Genome Res.* **10**, 1719–1725.

Gentry, D. R., Ingraham, K. A., Stanhope, M. J., Rittenhouse, S., Jarvest, R. L., O'Hanlon, P. J., Brown, J. R., et al. (2003) Variable sensitivity to bacterial methionyl-tRNA synthetase inhibitors reveals sub populations of *Streptococcus pneumoniae* with two distinct methionyl tRNA synthetase genes. *Antimicrob. Agents Chemother.* **47**, 1784–1789.

Germot, A., Philippe, H., and Le Guyader, H. (1996) Presence of a mitochondrial-type 70-kDa heat shock protein in *Trichomonas vaginalis* suggests a very early mitochondrial endosymbiosis in eukaryotes. *Proc. Natl. Acad. Sci. U S A* **93**, 14614–14617.

Gogarten, J. P., et al. (1989) Evolution of the vacuolar H+-ATPase: Implications for the origin of eukaryotes. *Proc. Natl. Acad. Sci. U S A* **86**, 6661–6665.

Golding, G. B., and Gupta, R. S. (1995) Protein-based phylogenies support a chimeric origin for the eukaryotic genome. *Mol. Evol. Biol.* **12**, 1–6.

Graham, D. E., Overbeek, R., Olsen, G. J., and Woese, C. R. (2000) An archaeal genomic signature. *Proc. Natl. Acad. Sci. U S A* **97**, 3304–3308.

Grauer, D., and Li, W. H. (2000) *Fundamentals of molecular evolution*, 2nd ed. Sunderland, MA: Sinauer Associates.

Gribaldo, S., Lumia, V., Creti, R., Conway de Macario, E., Sanangelantoni, A., and Cammarano, P. (1999) Discontinuous occurrence of the hsp70 (dnaK) gene among Archaea and sequence features of HSP70 suggest a novel outlook on phylogenies inferred from this protein. *J. Bacteriol.* **181**, 434–443.

Gupta, R. H. (1998) Protein phylogenies and signature sequences: A reappraisal of evolutionary relationships among archaebacteria, eubacteria and eukaryotes. *Microbiol. Mol. Biol. Rev.* **62**, 1435–1491.

Gupta, R. S., and Golding, G. B. (1996) The origin of the eukaryotic cell. *Trends Biochem. Sci.* **21**, 166–171.

Hansmann, S., and Martin, W. (2000) Phylogeny of 33 ribosomal and six other proteins encoded in an ancient gene cluster that is conserved across prokaryotic genomes: Influence of excluding poorly alignable sites from analysis. *Int. J. Syst. Evol. Microbiol.* **50**, 1655–1663.

Hartman, H., and Fedorov, A. (2002) The origin of the eukaryotic cell: A genomic investigation. *Proc. Natl. Acad. Sci. U S A* **99**, 1420–1425.

Hashimoto, T., Sánchez, L. B., Shirakura, T., Müller, M., and Hasegawa, M. (1998) Secondary absence of mitochondria in *Giardia lamblia* and *Trichomonas vaginalis* revealed by valyl-tRNA synthetase phylogeny. *Proc. Natl. Acad. Sci. U S A* **95**, 6860–6865.

Henze, K. A., Badr, A., Wettern, M., Cerff, R., and Martin, W. (1995) A nuclear gene of eubacterial origin in *Euglena gracilis* reflects cryptic endosymbioses during protist evolution. *Proc. Natl. Acad. Sci. U S A* **92**, 9122–9126.

Hilario, E., and Gogarten, J. P. (1993) Horizontal transfer of ATPase genes – the tree of life becomes the net of life. *BioSystems* **31**, 111–119.

Horiike, T., Hamada, K., Kanaya, S., and Shinozawa, T. (2002) Origin of eukaryotic cell nuclei by symbiosis of Archaea in Bacteria is revealed by homology-hit analysis. *Nature Cell Biol.* **3**, 210–214.

Horner, D. S., Heil, B., Happe, T., and Embley, T. M. (2002) Iron hydrogenases – ancient enzymes in modern eukaryotes. *Trends Biochem. Sci.* **27**, 148–153.

Horner, D. S., Hirt, R. P., and Embley, T. M. (1999) A single eubacterial origin of eukaryotic pyruvate:ferredoxin oxidoreductase genes: Implications for the evolution of anaerobic eukaryotes. *Mol. Biol. Evol.* **16**, 1280–1291.

Huang, P., and Oliff, A. (2001) Signaling pathways in apoptosis as potential targets for cancer therapy. *Trend Cell Biol.* **11**, 343–348.

Huber, H., et al. (2002) A new phylum of Archaea represented by a nanosized hyperthermophilic symbiont. *Nature* **417**, 63–67.

Huynen, M., Snel, B., and Bork, P. (1999) Lateral gene transfer, genome surveys and the phylogeny of prokaryotes. Technical Comments. *Science* **286**, 1443a.

International Human Genome Sequencing Consortium. (2001) Initial sequencing and analysis of the human genome. *Nature* **409**, 860–892.

Iwabe, N., Kuma, K.-I., Hasegawa, M., Osawa, S., and Miyata, T. (1989) Evolutionary relationship of Archaea, Bacteria, and eukaryotes inferred from phylogenetic trees of duplicated genes. *Proc. Natl. Acad. Sci. U S A* **86**, 9355–9359.

Jain, R., Rivera, M. C., and Lake, J. A. (1999) Horizontal gene transfer among genomes: The complexity hypothesis. *Proc. Natl. Acad. Sci. U S A* **96**, 3801–3806.

Karlberg, O., Canbäck, B., Kurland, C. G., and Andersson, S. G. E. (2000) The dual origin of the yeast mitochondrial proteome. *Yeast Comp. Funct. Genomics* **17**, 170–187.

Keeling, P. J., and Doolittle, W. F. (1997) Evidence that eukaryotic triosephosphate isomerase is of alpha-proteobacterial origin. *Proc. Natl. Acad. Sci. U S A* **94**, 1270–1275.

Keeling, P. J., and McFadden, G. I. (1998) Origins of microsporidia. *Trends Microbiol.* **6**, 19–23.

Kondo, N., Nikoh, N., Ijichi, N., Shimada, M., and Fukatsu, T. (2002) Genome fragment of *Wolbachia* endosymbiont transferred to X chromosome of host insect. *Proc. Natl. Acad. Sci. U S A* **99**, 14280–14285.

Koonin, E. V., and Aravind, L. (2002) Origin and evolution of eukaryotic apoptosis: The bacterial connection. *Cell Death Differ.* **9**, 394–404.

Koonin, E. V., Aravind, L., and Kondrashov, A. S. (2000) The impact of comparative genomics on our understanding of evolution. *Cell* **101**, 573–576.

Koonin, E. V., Makarova, K. S., and Aravind, L. (2001) Horizontal gene transfer in prokaryotes: Quantification and classification. *Annu. Rev. Microbiol.* **55**, 709–742.

Koretke, K. K., Lupas, A. N., Warren, P. V., Rosenberg, M., and Brown, J. R. (2000) Evolution of two-component signal transduction. *Mol. Biol. Evol.* **17**, 1956–1970.

Koski, L. B., and Golding, B. (2001) The closest BLAST hit is often not the nearest neighbor. *J. Mol. Evol.* **52**, 540–542.

Kunik, T., Tzfira, T., Kapulnik, Y., Gafni, Y., Dingwall, C., and Citovsky, V. (2001) Genetic transformation of HeLa cells by *Agrobacterium*. *Proc. Natl. Acad. Sci. U S A* **98**, 1871–1876.

Kurland, C., and Andersson, S. G. E. (2000) Origin and evolution of the mitochondrial proteome. *Mol. Biol. Rev.* **64**, 786–820.

Kurland, C. G. (2000) Something for everyone: Horizontal gene transfer in evolution. *EMBO Rep.* **1**, 92–95.

Kyrpides, N. C., and Olsen, G. J. (1999) Archaeal and bacterial hyperthermophiles: Horizontal gene exchange or common ancestry? *Trends Genetics* **15**, 298–299.

Lamour, V., Quevillon, S., Diriong, S., N'Guyen, V. C., Lipinski, M., and Mirande, M. (1994) Evolution of the Glx-tRNA synthetase family: The glutaminyl enzyme as a case for horizontal gene transfer. *Proc. Natl. Acad. Sci. U S A* **91**, 8670–8674.

Lathe, W. C., Snel, B., and Bork, P. (2000) Gene context conservation of a higher order than operons. *Trends Biochem. Sci.* **25**, 474–479.

Lawrence, J. G., and Ochman, H. (1998) Molecular archaeology of the *Escherichia coli* "genome". *Proc. Natl. Acad. Sci. U S A* **95**, 9413–9417.

Lawrence, J. G., and Ochman, H. (2002) Reconciling the many faces of lateral gene transfer. *Trends Microbiol.* **10**, 1–3.

Logsdon, J. M., Jr., and Faguy, D. M. (1999) Evolutionary genomics: Thermotoga heats up lateral gene transfer. *Current Biol.* **9**, R747–R751.

Lopez, P., Forterre, P., and Philippe, H. (1999) The root of the tree of life in the light of the covarion model. *J. Mol. Evol.* **49**, 496–508.

Margulis, L. (1970) *Origin of eukaryotic cells.* New Haven CT: Yale University Press.

Martin, W. (1996) Is something wrong with the tree of life? *BioEssays* **18**, 523–527.

Martin, W. (1999) Mosaic bacterial chromosomes: A challenge en route to a tree of genomes. *BioEssays* **21**, 99–104.

Martin, W., and Müller, M. (1998) The hydrogen hypothesis for the first eukaryote. *Nature* **392**, 37–41.

Martin, W., et al. (2002) Evolutionary analysis of *Arabidopsis*, cyanobacterial, and chloroplast genomes reveals plastid phylogeny and thousands of cyanobacterial genes in the nucleus. *Proc. Natl. Acad. Sci. U S A* **99**, 12246–12251.

Moreira, D., and López-García, P. (1998) Symbiosis between methanogenic Archaea and δ-Proteobacteria as the origin of eukaryotes: The syntrophic hypothesis. *J. Mol. Evol.* **47**, 517–530.

Nelson, K. E., et al. (1999) Evidence for lateral gene transfer between Archaea and bacteria from genome sequence of *Thermotoga maritima*. *Nature* **399**, 323–329.

Ochman, H., Lawrence, J. G., and Groisman, E. A. (2000) Lateral gene transfer and the nature of bacterial innovation. *Nature* **405**, 299–304.

Olsen, G. J., and Woese, C. R. (1997) Archaeal genomics – an overview. *Cell* **89**, 991–994.

Olsen, G. J., Woese, C. R., and Overbeek, R. (1994) The winds of (evolutionary) change: Breathing new life into microbiology. *J. Bacteriol.* **176**, 1–6.

Page, R. D. M. (1996) TREEVIEW: An application to display phylogenetic trees on personal computers. *Comput. Appl. Biosci.* **12**, 357–358.

Palenik, B. (2002) The genomics of symbiosis: Host keep the baby and the bath water. *Proc. Natl. Acad. Sci. U S A* **99**, 11996–11997.

Pennisi, E. (1998) Genome data shake tree of life. *Science* **280**, 672–674.

Pennisi, E. (1999) Is it time to uproot the tree of life? *Science* **284**, 1305–1307.

Poole, A., and Penny, D. (2001) Does endosymbiosis explain the origin of the nucleus? *Nature Cell Biol.* **3**, E173.

Ragan, M. (2001) On surrogate methods for detecting lateral gene transfer. *FEMS Microbiol. Lett.* **201**, 187–191.

Ragan, M. (2002) Reconciling the many faces of lateral gene transfer: Response. *Trends Microbiol.* **10**, 3.

Rivera, M. C., Jain, R., Moore, J. E., and Lake, J. A. (1998) Genomic evidence for two functionally distinct gene classes. *Proc. Natl. Acad. Sci. U S A* **95**, 6239–6244.

Rivera, M. C., and Lake, J. A. (2004) The ring of life provides evidence for a genome fusion origin of eukaryotes. *Nature* **431**, 152–155.

Roelofs, J., and Van Haastert, P. J. (2001) Genes lost during evolution. *Nature* **411**, 1013–1014.

Roger, A. J., and Brown, J. R. (1996) A chimeric origin for eukaryotes re-examined. *Trends Biochem. Sci.* **21**, 370–371.

Roger, A. J., Clark, C. G., and Doolittle, W. F. (1996) A possible mitochondrial gene in the early-branching amitochondriate protist *Trichomonas vaginalis*. *Proc. Natl. Acad. Sci. U S A* **93**, 14618–14622.

Rokas, A., Williams, B. L., King, N., and Carroll, S. B. (2003) Genome-scale approaches to resolving incongruence in molecular phylogenies. *Nature* **425**, 798–804.

Rotte, C., and Martin, W. (2001) Does endosymbiosis explain the origin of the nucleus? *Nature Cell Biol.* **3**, E173.

Rujan, T., and Martin, W. (2001) How many genes in *Arabidopsis* come from cyanobacteria? An estimate from 386 protein phylogenies. *Trends Genetics* **17**, 113–119.

Salzberg, S. L., White, O., Peterson, J., and Eisen, J. A. (2001) Microbial genes in the human genome: Lateral transfer or gene loss? *Science* **292**, 1903–1906.

Smith, M. W., Feng, D.-F., and Doolittle, R. F. (1992) Evolution by acquisition: The case for horizontal gene transfers. *Trends Biochem. Sci.* **17**, 489–493.

Snel, B., Bork, P., and Huynen, M. A. (1999) Genome phylogeny based on gene content. *Nature Genetics* **21**, 108–110.

Stanhope, M. J., et al. (2001) Phylogenetic analyses of genomic and EST sequences do not support horizontal gene transfers between bacteria and vertebrates. *Nature* **411**, 940–944.

Strimmer, K., and von Haeseler, A. (1996) Quartet puzzling: A quartet maximum likelihood method for reconstructing tree topologies. *Mol. Biol. Evol.* **13**, 964–969.

Swofford, D. L. (1999) PAUP*. Phylogenetic Analysis Using Parsimony (*and Other Methods). Version 4. Sunderland, MA: Sinauer Associates.

Teichmann, S. A., and Mitchison, G. (1999) Is there a phylogenetic signal in prokaryote proteins? *J. Mol. Evol.* **49**, 98–107.

Thorsness, P. E., and Fox, T. D. (1990) Escape of DNA from mitochondria to nucleus in *Saccharomyces cerevisiae*. *Nature* **346**, 376–379.

Thorsness, P. E., and Fox, T. D. (1993) Nuclear mutations in *Saccharomyces cerevisiae* that affect the escape of DNA from mitochondria to the nucleus. *Genetics* **134**, 21–28.

Tielens, A. G. M., Rotte, C., van Hellemond, J. J., and Martin, W. (2002) Mitochondria as we don't know them. *Trends Biochem. Sci.*, **27**, 564–572.

JAMES R. BROWN

Tourmen, Y., Baris, O., Dessen, P., Jacques, C., Malthièry, Y., and Reynier, P. (2002) Structure and chromosomal distribution of human mitochondrial pseudogenes. *Genomics* **80**, 71–77.

Tovar, J., León-Avila, G., Sánchez, L. B., Sutak, R., Tachezy, J., van der Giezen, M., et al. (2003) Mitochondrial remnant organelles of *Giardia* function in iron-sulphur protein maturation. *Nature* **426**, 172–176.

Ueda, K., Seki, T., Kudo, T., Yoshida, T., and Kataoka, M. (1999) Two distinct mechanisms cause heterogeneity of 16S rRNA. *J. Bacteriol.* **181**, 78–82.

van der Giezen, M., et al. (2002) Conserved properties of hydrogenosomal and mitochondrial ADP/ATP carriers: A common origin for both organelles. *EMBO J.* **21**, 572–579.

von Dohlen, C. D., Kohler, S., Alsop, S. T., and McManus, W. R. (2001) Mealybug β-proteobacterial endosymbionts contain γ-proteobacterial symbionts. *Nature* **412**, 433–436.

Woese, C. R. (2002) On the evolution of cells. *Proc. Natl. Acad. Sci. U S A* **99**, 8742–8747.

Woese, C. R., and Fox, G. E. (1977) Phylogenetic structure of the prokaryotic domain: The primary kingdoms. *Proc. Natl. Acad. Sci. U S A* **51**, 221–271.

Woese, C. R., Kandler, O., and Wheelis, M. L. (1990) Towards a natural system of organisms: Proposal for the domains Archaea, Bacteria and Eucarya. *Proc. Natl. Acad. Sci. U S A* **87**, 4576–4579.

Woese, C. R., Olsen, G. J., Ibba, M., and Söll, D. (2000) Aminoacyl-tRNA synthetases, the genetic code, and the evolutionary process. *Microbiol. Mol. Biol. Rev.* **64**, 202–236.

Wolf, Y. I., Aravind, L., Grishin, N. V., and Koonin, E. V. (1999) Evolution of aminoacyl-tRNA synthetases – analysis of unique domain architectures and phylogenetic trees reveals a complex history of horizontal gene transfer events. *Genome Res.* **9**, 689–710.

Wren, B. W. (2000) Microbial genome analysis: Insight into virulence, host adaptation and evolution. *Nature Rev. Genetics* **1**, 30–39.

Xiong, J., Inoue, K., and Bauer, C. E. (1998) Tracking molecular evolution of photosynthesis by characterization of a major photosynthesis gene cluster from *Heliobacillus mobilis*. *Proc. Natl. Acad. Sci. U S A* **95**, 14851–14856.

Yap, W. H., Zhang, Z., and Wang, Y. (1999) Distinct types of rRNA operon exist in the genome of the actinomycete *Thermomonospora chromogena* and evidence for horizontal transfer of an entire rRNA operon. *J. Bacteriol.* **181**, 5201–5209.

Zambryski, P., Tempe, J., and Schell, J. (1989) Transfer and function of T-DNA genes from *Agrobacterium* Ti and Ri plasmids. *Cell* **56**, 193–201.

Index